KLIMA

EDITED BY

James Rodger Fleming and
Vladimir Jankovic

O S I R I S | 26

A Research Journal Devoted to the
History of Science and Its Cultural Influences

Osiris

Series editor, 2002–2012

KATHRYN OLESKO, *Georgetown University*

Volumes 17 to 27 in this series are designed to dissolve boundaries between history and the history of science. They cast science in the framework of larger issues prominent in the historical discipline but infrequently treated in the history of science, such as the development of civil society, urbanization, and the evolution of international affairs. They aim to open up new categories of analysis, to stimulate fresh areas of investigation, and to explore novel ways of synthesizing major historical problems that demand consideration of the role science has played in them. They are written not only for historians of science, but also for historians and other scholars who wish to integrate issues concerning science into courses on broader themes, as well as for readers interested in viewing science from a general historical perspective. Special attention is paid to the international dimensions of each volume's topic.

Cover Illustration:

"I believe there is a train under here somewhere!" March 9, 1966, Jamestown, North Dakota. Photographer: Mr. Bill Koch, North Dakota State Highway Dept. Credit: Collection of Dr. Herbert Kroehl, NGDC. Courtesy of NOAA National Weather Service Collection, http://www.photolib.noaa.gov/htmls/wea00958.htm.

OSIRIS   2011   SECOND SERIES   VOLUME 26

Revisiting *Klima*

by James Rodger Fleming and Vladimir Jankovic*

We have named this volume *Klima*. ("κλίμα period"). This terse but dynamic moniker fits the tone of the contributions, which revive a multivocal and inclusive understanding of a venerable but elusive term. We seek to decouple *Klima* from its current exclusive association with atmospheric sciences and revisit the implications of an ancient vocabulary for medical, geographical, agricultural, economic, racial, and other "endemic" concerns. If climate is not just about the weather, what is it? What does it seek to explain? When is climate invoked? By whom? For what purposes? How are other "key words" linked with it, such as culture, society, civilization, time, and change? In what ways is climate a proxy for other concerns such as regulation, industry, and identity? When is it not an explanation at all? Where is climate incarnated? And how does it matter?

Modern scientific climatology cannot answer these questions. Yet it is burdened with the enormous challenge of delineating how climate relates to social and economic life. This is perhaps inevitable, given that the definition of climate abstracts it from the "lived" experience and constructs it as a derived entity, a statistical index of averaged parameters across space and time. In 1897, the preeminent Austrian meteorologist Julius Hann wrote that "by *climate* we mean the sum total of the meteorological phenomena that characterize the average condition of the atmosphere at any one place on the earth's surface."[1] According to Karl Schneider-Carius, the pioneering weather pilot and meteorologist, "Climatology is concerned with the average states of weather, and the frequency of the different individual types of weather in their geographical distribution."[2] More recently, the World Meteorological Organization defined climate "as the 'average weather,' or more rigorously, as the statistical description in terms of the mean and variability of relevant quantities over a period of time ranging from months to thousands or millions of years."[3] In a broader, scientific

* James Rodger Fleming, STS Program, Colby College, 5881 Mayflower Hill, Waterville, ME 04901; jfleming@colby.edu. Vladimir Jankovic, Centre for the History of Science, Technology and Medicine, Simon Building, 2nd floor, University of Manchester, Manchester M13 9PL, U.K.; vladimir.jankovic@manchester.ac.uk.
The articles in this volume originated in a conference at Colby College in April 2009 supported by the National Science Foundation under grant no. 0843162. Any opinions, findings, and conclusions or recommendations expressed in this material are those of the authors and do not necessarily reflect the views of the National Science Foundation.

[1] Hann, *Handbook of Climatology*, trans. R. deC. Ward (New York, 1903), 1.

[2] Schneider-Carius, *Weather Science, Weather Research: History of Their Problems and Findings from Documents during Three Thousand Years*, trans. National Oceanic and Atmospheric Administration and National Science Foundation (Washington, D.C., 1975), 386.

[3] "Frequently Asked Questions: What is Climate?" World Meteorological Organization Web site, http://www.wmo.int/pages/prog/wcp/ccl/faq/faq_doc_en.html (accessed 25 January 2011); see also Paul Edwards, *A Vast Machine: Computer Models, Climate Data, and the Politics of Global Warming* (Cambridge, Mass., 2010), xiv.

sense, climate is understood as the status of the "Earth system" comprising the atmosphere, the hydrosphere, the cryosphere, the upper lithosphere, and the biosphere. Humans too have been granted roles within this nonlinear interacting system, even as human behavior remains resistant to deterministic modeling. With such an expansive vision, it is no surprise that the climatologist C. E. P. Brooks, echoing Rudyard Kipling's "nine and sixty ways of making tribal lays," quipped that "there are at least nine and sixty ways of constructing a theory of climatic change, and there is probably some truth in quite a number of them."[4]

Taking a longer view, however, the definition of climate as a statistical index is an anomaly. The reason is that such a definition is possible only in connection to an instrumental, quantitative, and weather-biased understanding of the atmosphere. Outside this context one is more likely to encounter climate as an agency rather than an index. Climate has more often been defined as what it *does* rather than what it *is*. This means that climate has not usually been seen as an indicator of weather trends, but as a force—and a resource—informing social habits, economic welfare, health, diet, and even the total "energy of nations."[5] In these domains of social life, climate as agency has helped translate matters of concern into matters of fact.[6]

Early modern scholars, for example, considered climate as a biospatial frame of reference to categorize the relationship between life, on the one hand, and atmospheric, hydrological, seismic, and mineral features, on the other. These naturalists viewed climate descriptively as a human experience related to social and natural productions local to a latitude or a tract of land. But they also saw climate prescriptively as the norm that connected environmental features with social potentials. In this sense, climate literally produced seasons and endemic disease, vegetation and diet, soil and vernacular architecture, customs and political organization. Climate was considered an agency organizing social experience as a result of the material circumstances of life.[7] More recently, Andrew Ross underscored this circumstance in relation to climate change politics, writing that climatology, "hitherto considered a second-class adjunct to the more exciting field of meteorology, or at best a branch of physics that had more in common with geography, has seen its object—knowledge about a stable archive of climate statistics—transformed into a volatile, political commodity of the first importance."[8]

CLIMATE ANXIETY

The world's current, some would say unprecedented, state of environmental apprehension is a product of a long historical process that elevated climate science to its privileged position of legitimacy. The global hype about the climatological future

[4] Brooks, *Climate through the Ages: A Study of the Climatic Factors and Their Variations,* 2nd rev. ed. (London, 1949), 7.

[5] See, e.g., Hubert Lamb, "An Approach to the Study of the Development of Climate and Its Impact in Human Affairs," in *Climate and History: Studies in Past Climates and Their Impact on Man,* eds. T. M. L. Wigley, M. J. Ingram, and G. Farmer (Cambridge, 1981), 291–309.

[6] Vladimir Jankovic and Christina Barboza, eds., *Weather, Local Knowledge and Everyday Life* (Rio de Janeiro, 2009).

[7] James Oliver Thomson, *History of Ancient Geography* (Cheshire, Conn., 1948), 106; Karen Ordahl Kupperman, "The Puzzle of the American Climate in the Early Colonial Period," *Amer. Hist. Rev.* 87 (1982): 1262–89; Vladimir Jankovic, "Climates as Commodities: Jean Pierre Purry and the Modelling of the Best Climate on Earth," *Stud. Hist. Phil. Mod. Phys.* 41 (2010): 201–7.

[8] Ross, "Is Global Culture Warming Up?" *Social Text* 28 (1991): 3–30, on 7.

was formulated early by Lord Zuckerman, who, as chief scientific adviser to British prime ministers Edward Heath and Harold Wilson, thought that "man's present political problems are miniscule in relation to what could result from major changes in climate, and someone from outer space viewing our globe . . . could well suppose that nations of today behave like people who quarrel violently and murderously over immediate trivialities on the fiftieth floor of some huge modern Tower of Babel, oblivious of the fact that it is blazing away merrily beneath them."[9] Some claim that global warming was recently "discovered"; others that it was constructed in its current anthropogenic form by methods and agreements over the *longue dureé*. Its history was assembled from privileged positions deemed reliable and authoritative, based on scholarship, data, experiments, models, technologies, and accredited bodies reaching back several centuries. Yet acquiring such knowledge was also unnerving. Meteorologist Jerome Namias observed in 1989 that "the greenhouse effect is now firmly part of our collective angst, along with nuclear winter, asteroid collisions, and other widely bruited global nightmares"; a decade earlier the geographer Yi-Fu Tuan observed in *Landscapes of Fear* that "to apprehend is to risk apprehensiveness. If we did not know so much, we would have less to fear."[10] As descriptive climatology morphed in recent decades into prescriptive climate dynamics, it gained a normative edge that is at once urgent and judgmental. What is to be *done* about climate change? Can climate ever be fully apprehended—or just feared? If the models are accurate, will our choices now somehow determine the course of the next century or even the fate of humanity? A multiplicity of responses looms with dynamic physical, moral, and behavioral factors that baffle even the most sophisticated modelers.

The index- and agency-based readings of "climate" are by no means mutually exclusive, despite the fact that the former has ruled modern science and has become part of folklore. Recently, however, climate as agency has made a spectacular reappearance through the risks associated with ongoing climate change. Climate change is routinely portrayed as something that "impacts" the economy, "affects" countries, "harms" national security, "hurts" the world's poor, and potentially "leads" to global conflict. A recent United Nations Development Programme report calls for a "*fight* against climate change," while the BBC and the British Met Office say that "*tackling* climate change will be one of the most important things this generation does."[11] Climate is clearly not just in metrics—it has become a power to reckon with, generating fortunes or churning out destruction. Recently, it has been repeatedly emphasized that the fate of the planet lies in epochal climacteric (or even climactic) decisions and commitments regarding climate change. This revival of climate as agency makes it a propitious time to examine what this turn of events means in the longer history of constructions of Klima. One way to approach the issue is by scaling it down.

[9] Quoted in Hubert Lamb, *Weather, Climate and Human Affairs* (London, 1982), 4.

[10] Namias, "The Greenhouse Effect as a Symptom of Our Collective Angst," *Oceanus* 32 (1989): 65–7, on 66; Tuan, *Landscapes of Fear* (New York, 1979), 6.

[11] United Nations Development Programme, "Global Human Development Report 2007/2008 Launched in Khartoum: Sudan Reiterates Commitment to Global Fight against Climate Change," press release, January 22, 2008, http://www.sd.undp.org/press%20hdr.htm; Olivier Boucher, "What Is Geoengineering?" UK Met Office Web site, http://www.metoffice.gov.uk/climatechange/science/explained/geoengineering.html (both sites accessed 28 January 2011; emphasis added).

THE CLIMATE OF OPINION NEAR THE GROUND

There is a layer of air within two meters of the ground, the noosphere, that is argu-ably the most important of all in Earth's atmosphere. It is located in what meteorolo-gists have come to call the troposphere, nestled in the "boundary layer," a turbulent, well-mixed zone at the very base of the sublunar realm. This is a space in which the "natural" atmosphere gets entangled with human energy.

Earth is unique in that it is the only body in the known universe to have such a layer, and the future of the planet will be decided by what transpires in these first two meters. The course and fates of empires begin here, where small disturbances with little initial energy can grow into enormous movements. This is the anthropocentric layer in which we express our opinions, some of which are quickly damped out, while others are recorded for posterity. It is the interdisciplinary sphere of human af-fairs, the most influential layer of our planet's atmosphere.

This layer has not been fully or even adequately explored, which is unusual, since it is so accessible to us—as intimately close as our next breath. Indeed, it has been consciously excluded from environmental analysis. In the second half of the nine-teenth century meteorologists adopted a policy of locating their instruments in shel-ters two meters above the ground. In an attempt to standardize their measurements and compare measurements over widespread areas, they began to consider the layer of air adjacent to the ground "a zone of disturbance" to be excluded from scientific analysis. This two-meter layer was the location of the killing zone for poison gas attacks during World War I, which was undoubtedly a motivating factor for the Ger-man meteorologist Rudolf Geiger, who wrote in 1927 that this zone of disturbance that had been so meticulously avoided in official meteorological observations was important to humans and other living things. He considered it "no longer sufficient" and even "misleading" to focus only on the "large scale climates"— the so-called macroclimates—emphasized by the imperial national services.[12]

Ironically, and despite Geiger's wishes, the zone of disturbance will probably never be fully explored. From the point of view of "synoptic" gaze and global climate, this layer remains "contaminated" and unrepresentative of the processes in the free at-mosphere. It does not contain information that could be used for charting a map of continental scale. From the human perspective, however, the layer is inexhaustible in meaning, teeming with change and chance, maintaining its biocentric character over the course of history.

The place of this layer in history and science epitomizes the purpose of this vol-ume. Deeply significant for all human transactions, this layer remains out of sight, its very proximity rendering it invisible. And this invisibility means that the modern sense of "climate" has been eroded to an abstract three-dimensional geophysical *sys-tem*, rather than an intimate ground-level *experience*. As a result of this dichotomy, the geophysical reading of climate contrasts with that of climate as constitutive of human affairs—an understanding that has long informed the Western perception of social history. For example, ever since antiquity, climate has been thought of as an element in economic modes of subsistence that, before mass production and division of labor, amounted to optimization of agricultural yield and trade. Such climatol-

---

[12] Rudolph Geiger, *The Climate Near the Ground*, trans. M. H. Stewart (Cambridge, Mass., 1950), xvii.

ogy attempted to explain why certain species of grain grew in one region rather than another, why a textile was worn here rather than elsewhere, and why agricultural methods yielded better results in one place than in another. Climatic distribution of natural objects and living creatures embodied providential balance in a definitive way: "We learn from experience that no animal nor vegetable is fitted for every climate; and from experience we also learn that there is no animal or vegetable but what is fitted for some climate, where it grows to perfection."[13]

The historical ubiquity of the assumption of causal linkages between bodies and climates, cultures and climates, and natural productions and climates can be seen as an example of what Michel Foucault called a classical episteme. For Foucault, in a classical episteme, adjacency of objects was not a mere "exterior relation between things, but the sign of relationship."[14] Proximity of objects to each other becomes a similitude and similitude becomes a causal connection. Classical geography exemplified this in taking it as an axiom that in any given geographic area, storms, soils, plants, beasts, and humans stood in a causal rather than an accidental relation to each other. They were not simply contiguous to each other, but also co-constituted each other. In other words, when bodies and things were found in a particular place, their very placement became a warrant of ontological affinity between and among them.[15] It was in this context that the scholars of nature working before instrumental climatology argued for an ontological relationship of bodies, races, cultures, and climates.

By the time of far-reaching socioeconomic change in eighteenth-century Europe, early modern naturalists had developed a range of doctrines on how the atmosphere affects human physiology, health, and everyday life. Sociometeorological correspondence spawned an extensive literature on topics covering everything from national characterology, to ambiental medicine, to city planning. Naturalists linked the climate with social welfare and gave an impetus to medical topography, health travel, altitude physiology, eudiometry, and ventilation. Physicians recorded weather to understand epidemics; others asked about the relationship between health and social change in the wake of industrialization; and colonists reported on the physiological effects of the tropics or extreme cold and discussed the climatological reasons behind the moral and political differences of nations. Such richness preceded the modern sense of climatology.[16]

Biocentric and anthropocentric readings of Klima found a champion in the German polymath and naturalist Alexander von Humboldt, for whom the term "climate," taken in its most general sense,

> indicates all the changes in the atmosphere which sensibly affect our organs, as temperature, humidity, variations in the barometrical pressure, the calm state of the air or the action of opposite winds, the amount of electric tension, the purity of the atmosphere or its admixture with more or less noxious gaseous exhalations, and, finally, the degree of ordinary transparency and clearness of the sky, which is not only important with respect to the increased radiation from the Earth, the organic development of plants, and

---

[13] Lord Kames, *Six Sketches on the History of Man* (Philadelphia, 1776), 4.

[14] Michel Foucault, *The Order of Things* (London, 2004), 20.

[15] Jim Egan, *Authorizing Experience: Refigurations of the Body Politic in Seventeenth-Century New England Writing* (Princeton, N.J., 1999), 16.

[16] Vladimir Jankovic, *Confronting the Climate: British Airs and the Making of Environmental Medicine* (New York, 2010).

the ripening of fruits, but also with reference to its influence on the feelings and mental condition of men.[17]

This view of climate found support among generations of medical practitioners such as Robert Scoresby Jackson, who extended Humboldt's phenomenology into a demographic vision in which climate was "the sum of all those physical forces which by their operation upon the constitutions of organized beings prohibit their permanent migration from one region of the earth's surface to another."[18] Lions are not found on icebergs; people from the tropics cannot live in Iceland. A decade earlier physician Thomas Burgess asked, "Has not Nature adapted the constitution of man to his hereditary climate? Is it consistent with nature's laws that a person born in England and attacked by consumption can be cured in a foreign climate?"[19] Stable and native climates enabled civilizations; changing and foreign climates disabled them. The historical emergence of societies was synonymous with settlement, and settlement was possible in quasipermanent climates that enabled planning and development.[20] Change of air in general—and change of local climate more specifically—was a subject of great concern among doctors and colonial promoters. What would happen to Protestant bodies in the sultry climates of the Orient? Europeans perceived colonization—which usually required settlers to move to new climatic zones—as a great risk. According to the Abbé Jean-Baptiste Du Bos, French priest and polymath, air that is wholesome to the inhabitants of one country can be a slow poison to strangers. Blood formed by the air and nourishments of Europe was thought incapable of mixing with the air or with the chyle produced by the food of America. According to Montesquieu, countries are cultivated in proportion not to their fertility, but to their liberty.[21]

Such arguments assumed that stable climates guaranteed prosperity, but they also downplayed the importance of climate as a statistics independent of biological and social dimensions. Was there a point of doing science with no use to it? "There is nothing more jejune and uninteresting," argued the London practitioner John Hennen early in the nineteenth century, "than a protracted enumeration of the daily variations of [atmospheric parameters] if the person who describes such occurrences does not deduce from them some practical information."[22] With the rising concerns over industrial pollution, agricultural failures, fluctuations of trade, insurance costs, and energy physics as well as the ice age debates, climate as agency became increasingly pertinent as it seemed to control or at least strongly influence human welfare and

---

[17] Humboldt, *Cosmos: A Sketch of the Physical Description of the Universe,* vol. 1, trans. E. C. Otté (New York, 1877), 317–8.

[18] Scoresby Jackson, *Medical Climatology or, a Topographical and Meteorological Description of the Localities Resorted to in Winter and Summer by Invalids of Various Classes, Both at Home and Abroad* (London, 1862), 2.

[19] Burgess, "Inutility of Resorting to the Italian Climate for the Cure of Pulmonary Consumption," *Lancet* 55 (1850): 591–4, on 591.

[20] For discussions of climate and social life, see Robert I. Rotberg and Theodore K. Rabb, eds., *Climate and History: Studies in Interdisciplinary History* (Princeton, N.J., 1981); Wigley, Ingram, and Farmer, *Climate and History* (cit. n. 5); J. D. Post, *The Last Great Subsistence Crisis in the Western World* (Baltimore, 1977); Robert Claiborne, *Climate, Man and History* (New York, 1970); Franklin Thomas, *The Environmental Basis of Society* (New York, 1925).

[21] Montesquieu, *Spirit of Laws* (1748), book 18, C 3.

[22] Hennen, *Sketches of the Medical Topography of the Mediterranean Comprising an Account of Gibraltar, the Ionian Islands and Malta* (London, 1830), xvi.

economic growth—even evolution. In 1880, an American lawyer considered climatology a science of "ventilation and hygiene" and predicted that governments would
become obliged to protect the climatological rights of their subjects, using state institutions to procure "to every citizen the needful amount of pure air."[23] In England
Robert Angus Smith unwittingly worked on one such project when he used "chemical climatology" to collaborate with manufacturers in curbing emissions in a way
that would raise the productivity and profits of the emitters. His fellow Mancunian
William Stanley Jevons wrote in 1866 on the impact of outdoor and leisure markets
on the autumnal trends in the Bank of England's decisions to raise interest rates.[24]

Clearly, climate is not just an index of the average weather. Both climatologists
and the general public have come to think of it as a sort of "mechanism" with major
implications for life in modern times.[25] The characterization of climate as either enabling or disabling was relevant especially among the generations struck by the increasing complexity of social, political, and economic transactions, together with the
expanding dependence of everyday work on technological systems and natural resources. The infrastructural interdependencies required by the energy-based economies produced new vulnerabilities and new risks.

"Governments and universities should devote meteorology and economic statistics
funds for research on the same scale as those devoted to astronomy, geology, physics
and chemistry," agued H. Stanley Jevons, William's son. "Knowledge of the weather
cycles and their correlation with crop cycles in different countries would also be of
great value to economists, as the foundation of an intensive statistical investigation
of industrial fluctuations."[26] In the United States climatologist Helmut Landsberg,
whose broad-ranging interests linked atmospheric and social phenomena, argued in
1946 that American climates should be seen as a "friendly element" to be tapped for
national benefits: "They constitute, if properly exploited, a very important natural
resource."[27] The deliberate "uses" of climate would cut expenditures in housing,
heating, airports, "all-weather" highways, dam construction, flood control, and wind
power. In Landsberg's terms, these were all parts of "the exploitation of climatic income." Soon after becoming the chief of the U.S. Weather Bureau, F. W. Reichelderfer claimed that weather information accounted for savings and profits of more than
$3 billion annually. He added that "permanent changes in climate could bring ruin to
our entire business structure [in which] two million businessmen every morning turn
at once to the weather report, [and] more than a million listen to the weather forecast
by radio once or more each day."[28] Now, entrepreneurs are headed to the bank to cash
in profits from their environmental accounting schemes to avoid climate change.

Reichelderfer's concern continued to inform the perception of short-term climate
change as disabling especially after the 1970s, a decade marked by extraordinarily
adverse weather events linked to economic downturn. Hubert Lamb's book *Climate,*

---

[23] Britton Armstrong Hill, *Liberty and Law,* 2nd ed. (St. Louis, 1880), 67.

[24] R. Angus Smith, *Air and Rain: The Beginnings of a Chemical Climatology* (London, 1872);
Jevons, "On the Frequent Autumnal Pressure in the Money Market and the Action of the Bank of England," *Journal of the Statistical Society of London* 29 (1866): 235–53.

[25] E.g., Joseph Fletcher, "Polar Ice and the Global Climate Machine," *Bull. Atom. Sci.*, December
1970, 40–7; and four decades later, Edwards, *Vast Machine* (cit. n. 3).

[26] H. Stanley Jevons, *The British Coal Trade* (London, 1915), 581.

[27] Landsberg, "Climate as a Natural Resource," *Scientific Monthly* 63 (1946): 293–8, on 293.

[28] F. W. Reichelderfer, "The How and Why of Weather Knowledge," in *Climate and Man: Yearbook
of Agriculture 1941*, U.S. Department of Agriculture (Washington, D.C., 1941), 128–53, on 128.

History and the Modern World (1982) summarized this in a discussion of the "expe-rience of 1972," during which the extraordinary heat and drought in Russia, China, India, and Australia caused grain shortages leading to massive death and migra-tion southward. Coffee harvests dropped in Ethiopia, Kenya, and Ivory Coast, and El Niño ruined anchovy fisheries in Peru and Ecuador. The net effect was the first drop in the world's total food production since 1945. Lamb thought that among "the leading scientific, technical and administrative institutions in the advanced countries, there was some confusion about how to interpret the climatic event and revise atti-tude to climate, even before the anxieties aroused by the unprecedented international economic crisis, which began to develop with the first (fourfold) oil price increase in 1973-4."[29] Which aspects of the crises of 1972 were short-lived, and which were the signals of longer-term trends? Was the crisis caused by or only precipitated by seasonal anomalies? An international workshop on climate issues reported that it was "exceedingly difficult to extract the climatic 'signal' from the 'noise' induced by other factors. . . . In 1972, for example, the effects of a series of climatic anomalies were greatly magnified by other factors to produce among other things unusually large changes in the world food prices."[30]

By the early twentieth century, recognition of emergent vulnerabilities informed bioclimatology and urban climatology. Bioclimatology grew out of medical, agri-cultural, and geographic attempts to understand the relationship between life and the "geographical envelope" on a comprehensive level. Working within this tradi-tion, German physiologist Adolf Loewy in 1924 defined climate as "the sum of all the atmospheric and terrestrial conditions, typical of a place, by which our state is directly influenced."[31] The centrality of human experience of climate and the treat-ment of climate as agency evolved also with the growth of urban climatology and the research in *Kleinklima* as a response to concerns over industrial hygiene and resi-dential quality of life. Early studies by Luke Howard, Emilien Renou, and August Schmaus reported the existence of urban heat islands, followed by the microclimatol-ogy of built spaces.[32] On a hot July day in 1934, U.S. Weather Bureau chief Willis R. Gregg, presiding over the dedication ceremonies at the air-conditioned house at the Century of Progress world's fair in Chicago, announced that there was "no longer any need for suffering from weather discomforts."[33] Yet now is the season of our climate discontent.

In a recent interview, the French art critic Jean Christophe Royoux and the German philosopher and media theorist Peter Sloterdijk exchanged ideas about, among other things, Sloterdijk's recent thinking about (atmo)spheres as a metaphysical concept of existence. Sloterdijk at one point digresses to say that "air has always been a medium that allowed humans to realize the fact that they're always already immersed in some-thing almost imperceptible and yet very real, and that this space of immersion domi-nates the changing sites of the soul down to its most intimate modifications. Ventila-

[29] Hubert Lamb, *Climate, History and the Modern World* (New York, 1982), 307.

[30] International Workshop on Climate Issues, *International Perspectives on the Study of Climate and Society* (Washington, D.C., 1978), 68.

[31] Quoted in Schneider-Carius, *Weather Science* (cit. n. 2), 389.

[32] Helmut Landsberg, *Physical Climatology* (State College, Pa., 1941); Landsberg, *The Urban Climate* (New York, 1981); Michael Hebbert and Vladimir Jankovic, "Hidden Climate Change: Urban Meteorology and the Scales of Real Weather," *Climatic Change* (forthcoming, 2011).

[33] Quoted in James Rodger Fleming, *Fixing the Sky: The Checkered History of Weather and Climate Control* (New York, 2010), 133.

tion is the profound secret of existence." Sloterdijk considers early twentieth-century gas warfare as the perversion of military art, in which for the first time on such a scale, political leaders condoned a lethal manipulation of the atmosphere and the type of warfare that "no longer kills by direct fire but by destroying the environment. The art of killing the environment is one of the big ideas of modern civilization. It contains the nucleus of contemporary terror: to attack not the isolated body of the adversary, but the body in its Umwelt."[34]

World War I was dominated by the crushing realities of trench warfare and the controlling influence of Generals "Mud" and "Winter." This, however, did not dampen the enthusiasm of promoters such as Alexander McAdie, head of the new Aviation Weather Advisory Service of the U.S. Navy, who crowed, "Who commands the air, commands all! . . . Henceforth the ships of the sky shall play the leading role, and the nation holding the mastery of the air will have in its palm the power to make or mar. . . . The strategies of warfare will be entirely different. Individuals will count for less; machines and *weather* will determine the victory."[35] Saturation bombing of civilian targets and the nuclear annihilation of Hiroshima and Nagasaki were one generation away. When control of weather and even climate emerged as a distinct possibility following World War II, General George C. Kenney, commander of the Strategic Air Command, announced, "The nation which first learns to plot the paths of air masses accurately and learns to control the time and place of precipitation will dominate the globe."[36] Such attitudes led to an all-out military effort in the atmospheric sciences and to secret cloud-seeding efforts in the jungles over North and South Vietnam, Laos, and Cambodia. In the twenty-first century, who is to say that the military will not be centrally engaged in climate change issues and climate control efforts, especially if environmental degradation triggers concerns about national sovereignty and security?[37]

All these approaches are important, but unprecedented, aspects of climate history. The themes are many, but the workers are, as yet, few. We argue that, historically speaking, climate discourse cannot be understood without paying tribute to a more inclusive—and a less reductionist—perception of geophysical reality, which we have tried to capture in the classical (later traditional) concept of Klima. The articles in this collection demonstrate the importance of this wider perception in describing the complex and elusive character of environmental thinking from the early modern era until the most recent efforts to model climate change. If there is a single common theme underlying our collection, it is a view of climate as a framing device in which the verities of life such as food, health, wars, housing, economy, social movement, or local identity change synchronically with Klima. Regardless of whether such domains can be shown to depend on atmospheric events, geophysical processes, human perceptions, or something yet more elusive, they have routinely been framed as if they did. Racial and mental differences, whether or not they may be shown to derive

[34] Peter Sloterdijk, "Foreword to the Theory of Spheres," available on the Web site of the Manchester Architecture Research Centre, University of Manchester, http://www.sed.manchester.ac.uk/research/marc/news/seminars/latour/COSMOGRAM-INTER-GB_Spheres.pdf (accessed 30 January 2011).

[35] McAdie, "War Weather Vignettes," in *Alexander McAdie: Scientist and Writer*, comp. Mary R. B. McAdie (Charlottesville, Va., 1949), 296, 261; "Making the Weather," ibid., 325–6 (emphasis in the original).

[36] Frank L. Kluckhohns, "$28,000,000 Urged to Support M.I.T.," *New York Times*, June 15, 1947, 46.

[37] Fleming, *Fixing the Sky* (cit. n. 33).

from climatic differences, have been discussed as if they did. Military success and political arrangements have been and continue to be related to adaptation to particular climates. We believe that the importance of articles in this volume lies in recognizing that such claims have informed climatological thinking throughout history. As editors we feel privileged to bring attention to the richness of climate discourse, which is further buttressed by the richness of approaches presented here. *Klima* is thus an attempt to resurrect the many meanings of atmospheric environment through the different modalities of scholarship: from environmental history to intellectual biography, from history of science to historical geography, from the analysis of field notes and correspondence, through discursive and conceptual engagements, to studies of networks, places, and regions. If it serves to vex a more simplistic recent set of assumptions about climate, so be it; for the risks of not reading history are great.

CLIMATES INCARNATE

The incarnation of climate as social truth is one of the motivating themes in the contributions that follow. But so is the dialectic opposite of this process. Regardless of the time period or the concern at hand, the authors here demonstrate that social issues can be incarnated as natural threats. Climate is a discursive vehicle capable of naturalizing matters of social concern into matters of natural fact. Studying climatology is always about studying society, vicariously or not. While remaining a highly complex physical science, climatology has virtually always turned into a performative entity. Especially in the public realm and in current decision making, climate is made to perform acts of immense political magnitude and economic consequence. In these realms, climate has been clearly emancipated to become a fulcrum of social action.

Section 1: Natural Laboratories

We have divided the volume into four sections. In the first section, "Natural Laboratories," we include the articles dealing with the historical circumstances that gave rise to the complexities of modern climate discourse. We open with the contributions by Gregory Cushman, Deborah Coen, Sverker Sörlin, and Ruth Morgan because they share a common interest in exploring the intersections of knowledge and politics, space and concepts, experience and theory. They look at "natural laboratories" as sites of the production of climate knowledge: a Peruvian lake (Cushman), Turkestani steppes (Coen), Greenland glaciers (Sörlin), and Australian deserts (Morgan). These articles are clear about the varied mechanisms of coproduction of climate knowledge between naturalists, authorities, and local publics. They highlight the relevance of first-hand involvement with the materiality of climatic zones and geological features. They lead us to recognize that climatological ideas often derive from both somatic and social encounters with airs, waters, and places and depend on the ability of practitioners to extend and expand limited, partial, and small-scale data into general "truths" of climatology. Importantly, these articles reveal that the political meanings of climate and its social implications are due to the fact that the concept itself refers to a hybrid realm comprising land, water, air, living beings, people, and cultural institutions. Klima, in this sense, is paradigmatic in its binding of culture and nature that represents civilization as a result of materiality, contingency, and particularity of place. Climate discourse is environmentalism before there was an environment.

More specifically, Cushman discusses the political epistemology of climatic change stemming from land use and deforestation. His contribution demonstrates that the popularization of the modern belief that land use changed climate can be traced to Alexander von Humboldt's treatment of the Lake Valencia Basin in Venezuela and the desert coast of Peru. Cushman portrays these places as "natural laboratories" and sites of contact between geophysical and cultural agencies. In particular, he points out that Humboldt's treatment (and use) of desertification and climatic change drew heavily on his political stances against colonialism and plantation slavery. Such and similar preconceptions have long influenced how individuals and communities imagined their place within the climatic belts of Earth. Where Humboldt might have deplored the climatic effects of slavery, his contemporaries might have understood slavery as a necessary form of production in the tropics. Scholars and laypeople alike perpetuated anthropological and racial stereotypes about the "torrid zone." Coen further addresses the construction of regional climates in her account of Austrian scientists working on the spatial differentiation of climate as an element in regional economics. Imperial climatographers described regional climate in relation to human life, stressing its relationship with vegetation, agriculture, industry, and human settlements. The mountain climatology of Heinrich von Ficker and A. I. Voeikov, Coen argues, conformed to the patterns of the continental-imperial science of "regionalization" and thus embodied the (syn)optics and priorities of Austrian geographic identity early in the twentieth century.

Sörlin looks in more detail at the work of Swedish glaciologist and policy adviser Hans Ahlmann, author of the "polar warming theory." Sörlin interprets the career of Ahlmann's theory as an epitome of its author's idea that fieldwork should be conceived as a form of laboratory procedure involving networked data gathering and quantifiable demonstration. Ahlmann's ideas can be exemplified in what Sörlin calls the "instrumented glacier," combining the subject of investigation, the instrumental infrastructure, and the community of investigators and local informers. Morgan provides extraordinary evidence that the growth of scientific interest in the rainfall of the Southwest region of Western Australia during the twentieth century was "strongly influenced" by political, social, and economic concerns. Morgan provides an account of how, more recently, climate science became deeply enmeshed with party politics when the Labor Party took to market its green credentials, leading to the government's decision to link support for national research to the implications of the greenhouse effect. Climate science cannot be divested from the circumstances surrounding its production, as regionalism, cultural difference, and local senses of belonging define vectors of research and even the basic meaning of climate.

Section 2: Social Contexts

These issues lead to our second section, "Social Contexts." Rethinking climate in light of its meanings in social contexts takes us further away from its indexical status as an average. Memory, local knowledge, and expectations about physical surroundings have played important roles in thinking about climates, stable or otherwise. In this section, Brant Vogel, Mark Carey, and Georgina Endfield engage with the past tropes, stereotypes, and values at work in shaping the "climate dimension" of cultural experience. What did it mean to say that climate was

"wholesome," or "enervating" or "enfeebling," as was often the jargon of colonial naturalists? And what sort of relationship could there be between climate
and national identity? What did it mean to argue that a climate was changing at a
time when there were no reliable measurements of meteorological trends, local or
global?

The chronic difficulties in providing reliable explanations of climatic patterns and
the fears surrounding their future have defined the status of climatology as science.
At dynamic historical junctures in eras other than our own, when human activities
appeared so intensive as to encroach on providential order, concerns about the climatic outcomes gave rise to discussions on "anthropogenic climate change."[38] Vogel
shows that such discussions thrived even in the early modern period, when British
and American observers argued over how a reported warming trend corresponded
with the issues of land management and colonial enterprise. Following the arguments in an anonymous letter from Dublin, Vogel finds that the author's doubts as
to the cause of the reported warming made him an advocate of instrumental series
of measurements, which he hoped would decide the issue. For Carey, such views
informed nineteenth-century medical climatology that brought to attention unfamiliar pathologies in non-European latitudes. Carey, however, detects a substantial change in the perceptions of the Caribbean climate, from unhealthy to bracing, and follows the change in the writings of contemporary physicians, residents,
state officials, travelers, and missionaries. The emerging paradisiacal image of the
Caribbean did not result from a "rational" discovery of its healthiness but from an
assemblage of interests, Carey explains, negotiated among groups such as tourist
and transport organizations. The global swell of mass tourism testified to, among
other things, a rapid increase in the disposable income of (mostly) the European and
North American industrial bourgeoisie. Endfield's treatment of the work of British
climatologist and geographer Gordon Manley demonstrates the real possibility to
"reculture climate change discourses" by alerting us to his association of weather
with people, spaces, and places. Endfield argues that Manley understood climate to
be both culturally and spatially variable, layered with meaning and linked to cultural
habitus.

Section 3: International to Global

In our penultimate section, "International to Global," we move toward more recent
developments in the science of climate change as an anthropogenic phenomenon.
European and American industrialization, the growth of megacities, and the social
problems that came in their wake made many recognize that the environmental foot-
print of growth may well turn out to be the farthest reaching in its global consequences, not least in its effect on climatic patterns. Interest in gas physics and in the
atmospheric changes due to industrial emissions was accordingly on the rise in the
early decades of the twentieth century. The insights of Svante Arrhenius and Guy
Stewart Callendar's visionary claims about the relationship between the observed
global temperature rise and carbon dioxide emissions were followed, by the 1950s,
with more orchestrated and better funded research into the problem of anthropogenic

[38] James Rodger Fleming, *Historical Perspectives on Climate Change* (New York, 1998).

climate change.[39] But while the main protagonists of this research in the Anglophone world are well known and written about, the non-English-speaking world has not received enough attention. Maria Bohn redresses this through a fresh assessment of the work of Swedish climatologists. Bohn looks at the longer history of carbon dioxide measurements in Scandinavia and the Arctic before the Mauna Loa series, showing that the reasons why Swedish protagonists undertook them in the mid-1950s had to do more with local agendas than with the idea of testing the "greenhouse effect" hypothesis.

Adrian Howkins shows in his account of the climatology of Antarctica after the International Geophysical Year of 1957–8 that, regardless of scientific dimensions, the threat of climate change in Antarctica was politically opportune in reinforcing the great powers' exclusive domination in the territorial politics of the continent. Symptomatically, however, the political opportunity was so alluring that it stifled a fair assessment of scientific results and allowed for simplified interpretations and reductionist narratives. The politicization of climate discourse is further echoed in the contribution of Matthias Dörries, who underscores the need to pay more attention to climate change as a military defense entity whose imagery of "nuclear winter" penetrated deeply into the imagination of leading scientists and public figures. Whichever way one looked, "doing atmospheric research *was* politics," if for no other reason than because the consequences of a nuclear war could well mean the end of politics as the world knew it. Samuel Randalls argues for the importance of social science histories of climate change and presents an account of the role economics plays (or does not play) in the genre of contemporary science-policy histories of climate change. His focus is on how cost-benefit analysis from the 1970s to 1990s, as practiced by William Nordhaus and others, subtly altered debates about climate policy as heuristic economic models took on a prescriptive status.

Section 4: **Klima Redux**

In the postscript to the volume, Mike Hulme directly addresses the perennial issue of climatic determinism, marginalized in the first half of the twentieth century, yet resurgent in the twenty-first century as heightened anxieties about changes in climate and the hegemony of climate models foster a new "climate reductionism" regarding society and the future. Hulme argues that climate reductionism is exercised through "epistemological slippage," in which predictive authority is transferred from one domain of knowledge (physical climatology) to another (social science) without appropriate theoretical or analytical justification. The immense role of climate as a trope shaping communal perceptions has left an imprint on the science of climatology. Climatology is sometimes hailed as a tool for thinking about our future. But is it the only available expertise for the diagnosis, prognosis, and cure of the climate crunch? Climatology has taken the form of a planetary medicine: what the medical sciences do for the sick body, climatological knowledge can do for the sick planet. As a result of this repositioning, climatological modeling has ceased to act as a mere form of

---

[39] James Rodger Fleming, *The Callendar Effect: The Life and Work of Guy Stewart Callendar (1898–1964)* (Boston, 2007); Spencer Weart, *The Discovery of Global Warming* (Cambridge, Mass., 2003).

expertise and has become a normative, value-laden instrument for assessing the fate of economic and social worlds.

CLIMATE MATTERS

"Climate is a rather elusive entity," wrote Landsberg in 1950, as he sorted out some twenty or so competing definitions.[40] He went on to note that the greatest puzzle of climatology is why and how climates have changed (and will change). In addition to various physical and geographical approaches, climatologists framed their discourse as commensurable with political, ethical, and other master narratives. In the literature, as if counterpoised on a conceptual seesaw, the nebulous, portentous (some would say pretentious) concept of climate is paired by many authors with such macroconcepts as culture, society, civilization, time, life, literature, war, cosmology, evolution, comfort, diseases, landscape, architecture, capitalism, global survival, the British scene, and the energy of nations.

If the atmosphere—as a medium that shapes life in a most fundamental and most dramatic way—can no longer exist outside human past and future, we also suggest that climate, as a framework of the material possibilities of life, can no longer exist outside the temporality of the social world. Historically speaking, it is no longer viable to think of climate as a subject of climate science only, no matter how one wishes to define it or practice it. Reducing climate to climatology is like reducing language to linguistics. We rather speak of climate discourse as one of the historically evolving (perhaps devolving) frameworks of possibilities by means of which societies make explicit their experiences of a special kind of *Umwelt:* one enframed by the forces of latitude, season, weather, illnesses, clothing, housing, diet, status, and social class. For us, climate *discourse*—not to be reduced to climate *science*—is a framing device that makes explicit all social concerns arising from anxiety over the sensible and latent experiences of living in an atmosphere of hunger and satiation, disease and health, poverty and wealth, isolation and community, angst and hope.

We hope that the lesson to be taken away from this volume is not just about the role of "climate" in the narratives of risks associated with environmental contingency. While we recognize that Klima is a strange attractor of fears arising from the uncertainties of Earth's pulse, we also note that it has as frequently been a detractor in cases when human responsibility was the only explanation of social distress. To say that climate explains the shape of material and spiritual life would be to give it an undue prominence on the stage of history. When and where this has been done, climate was used as a subterfuge for social ills, especially at times when the complexities of social fabric exceeded human understanding or when the reality was too grim for acceptance of responsibility. Thus it sometimes seems more logical, as emerging planetary surgeons argue, to "fix" the sky technologically rather than address social ills, or to justify imperialism as a form of civilizing process by a race nurtured by bracing northern climates. Issues like famines and "climate refugees" are often much too complex to be thought of as a result of physical circumstances only.[41] Yet they

---

[40] Helmut Landsberg, "Climatic Analysis and Climatic Classifications," seminar talk before the Geophysical Research Directorate, Air Materiel Command, August 4, 1950, Helmut Landsberg Papers, University of Maryland, College Park.

[41] Mike Davis, *Late Victorian Holocausts* (London, 2000).

are sometimes reduced to geophysical systems and made amenable to disinterested scientific inquiry.

The articles presented here epitomize a historical agenda to problematize knowledge claims, particularize climatic experience, and pluralize the meanings of climatological expertise.[42] The subject is venerable, but we have only just begun to explore its rich historical complexity.

[42] David N. Livingstone, commentary, "Cultural Spaces of Climate" session, Royal Geographical Society annual meeting, Manchester, August 2009.

NATURAL LABORATORIES

Humboldtian Science, Creole Meteorology, and the Discovery of Human-Caused Climate Change in South America

by Gregory T. Cushman[*]

ABSTRACT

The belief that human land use is capable of causing large-scale climatic change lies at the root of modern conservation thought and policy. The origins and popularization of this belief were deeply politicized. Alexander von Humboldt's treatment of the Lake Valencia basin in Venezuela and the desert coast of Peru as natural laboratories for observing the interaction between geophysical and cultural forces was central to this discovery, as was Humboldt's belief that European colonialism was especially destructive to the land. Humboldt's overt cultivation of disciples was critical to building the prestige of this discovery and popularizing the Humboldtian scientific program, which depended fundamentally on local observers, but willfully marginalized chorographic knowledge systems. In creating new, global forms of environmental understanding, Humboldtian science also generated new forms of ignorance.

In February 1937, at the behest of Swiss-born botanist Henri Pittier, Venezuela established its first national park in order to protect a forested swath of coastal mountains between the Caribbean Sea and the fertile Lake Valencia plain (fig. 1). Over the next few years, the Venezuelan state created a forest service with twenty-five rangers at its command, established a botanical directorate to survey the biodiversity of government lands, began sponsoring Arbor Day celebrations, and signed the 1940 Hemispheric Convention on Nature Protection. Nearly all of these projects were placed under the administration of trained scientists intent on preventing the degradation of Venezuela's forests in order to protect the country's water supply.[1] In April 1940 in

[*] Department of History, University of Kansas, 1445 Jayhawk Blvd., Room 3650, Lawrence, KS 66045; gcushman@ku.edu.

Five individuals provided indispensable access and insights into primary source materials: Frank Baron, Luis González, Neil Oatsvall, Adam Sundberg, and especially Keri Lewis. Grants from the American Meteorological Society and the KU Center for Research enabled research for this project.

[1] Henri Pittier, *Consideraciones acerca de la destrucción de los bosques y del incendio de las sabanas* (Caracas, 1936), 1–4; Pittier, *Notas sobre la crisis de agua en la parte central de Venezuela* (Caracas, 1948), 13–5, 22–5; Manuel González Vale, *Un plan nacional forestal venezolano* (Trujillo, 1942), 3–4, 8.

Figure 1. Map of the Lake Valencia basin and coastal range of north-central Venezuela, with major conservation areas. Map by author; used with the author's permission. Based on "Parque Nacional copiado del mapa levantado por Dr. Alfredo Jahn Sr" (1937), overleaf insert in La protección de la naturaleza en las Américas, by Wallace Atwood (Mexico City, 1941); "Mapa del Lago de Valencia" (1955), overleaf insert in El desecamiento del Lago de Valencia, by Alberto Böckh (Caracas, 1956); Christopher Garrity et al., Digital Shaded-Relief Map of Venezuela (Reston, Va., 2004), electronic resource accessed through the University of Kansas Library.

Peru, a similar group of scientists, agronomists, and engineers formed the Comité Nacional de Protección a la Naturaleza. This private advocacy group focused most of its early efforts on the protection of native forests and establishment of forest plantations, particularly along Peru's arid coast. Their efforts led directly to the creation of a government Department of National Reforestation in 1944. One committee member, oceanographer Erwin Schweigger, a Jewish refugee from Nazi Germany, later proposed using recurrent El Niño rains to try to plant trees in the coastal Sechura Desert with the hope of permanently moistening the climate of this barren, sandy waste.[2] In the United States, Russian-born forester Raphael Zon convinced the New Deal government to begin installing a massive forest shelterbelt through the heart of the Great Plains as a remedy for the 1930s Dust Bowl.[3] After an intense period of drought during the late 1940s, Josef Stalin inaugurated an even more grandiose campaign to reengineer the landscape and climate of collective farms in central Russia by using trees to wall off this valuable agricultural region from the desiccated heart of Inner Eurasia. A few years earlier, Adolf Hitler had personally canceled a wartime plan to resettle thousands of Eastern European Jews to drain the wild Pripet marshes in order to prepare the way for eventual German settlement of the Ukraine, partly out of fear that this action would permanently desiccate the region and make German-ruled lands to the west vulnerable to dust storms blowing in from the eastern steppes.[4]

The widespread popularity of these plans to plant and preserve trees in order to affect the climate spanned the political spectrum. They derived from a century and a half of scientific opinion, dating back to the 1799–1802 visit by Prussian naturalist Alexander von Humboldt to northern Venezeula and coastal Peru, that *human land use* had caused the climate in these regions to become significantly drier over time. Humboldt believed that the progressive shrinking of Venezuela's Lake Valencia and the incredible aridity of Lower Peru provided "striking proofs of the justness" of what became the most influential dictum of scientific conservation of this epoch:

> By felling the trees, that cover the tops and the sides of mountains, men in every climate prepare at once two calamities for future generations; the want of fuel, and a scarcity of water.[5]

It is difficult to overstate the historical significance of the supposed discovery of human-caused climate change in northern South America and its apparent confirmation by Humboldt's disciples. Conservation advocates as well known as George Perkins Marsh directly cited this work as proof that deforestation and forest regrowth had produced measurable climate change over large regions. These findings reenergized international debate, dating back to earlier centuries, over whether large-scale

[2] *Boletín del Comité Nacional de Protección a la Naturaleza* 1 (1944): 30–2, 89–96, 99–104; 2 (1945): 68–80, 216–21; 5 (1948): 37–44; 7 (1951): 39–44, 69–71; 8 (1952): 31–8.

[3] Raphael Zon, "Shelterbelts—Futile Dream or Workable Plan," *Science* 81 (1935): 391–4; Donald Worster, *Dust Bowl: The Southern Plains in the 1930s* (New York, 1979), 220–3.

[4] Raphael Zon, "The Volga Valley Authority: The New Fifteen-Year Conservation Plan for the U.S.S.R.," *Unasylva* 3, no. 2 (1949), http://www.fao.org/docrep/x5349e/x5349e02.htm (accessed 19 November 2010); Paul Josephson, *Industrialized Nature: Brute Force Technology and the Transformation of the Natural World* (Washington, D.C., 2002); David Blackbourn, *The Conquest of Nature: Water, Landscape, and the Making of Modern Germany* (New York, 2006), chap. 5.

[5] Humboldt, *Personal Narrative of Travels to the Equinoctial Regions of the New Continent* (1818–29; repr., New York, 1966), 4:143.

deforestation had caused climate change on a continental scale in places like eastern
North America and central Asia. During the late nineteenth century, these findings
profoundly influenced the science of forestry and forest conservation policy in places
as diverse as India, Australia, the United States, and Pittier's Switzerland.[6] Related
ideas continue to fuel anxieties about desertification and its possible connection to
global warming.[7] One recent book even cites Humboldt's interest in "the general connections that link organic beings" as the fountainhead of American environmentalism
and the modern science of ecology.[8]

The idea that human activities have changed the climate has been deeply politicized for at least two hundred years. Yet we have given surprisingly little attention to
the origin—much less the validity—of these deep-seated beliefs. This is especially
true when it comes to evaluating the life and work of Humboldt (1769–1858)—a
scientific saint of such stature that the followers of Hitler and Stalin as well as greens
and gays have all used his persona to bolster their political movements. Scientists
and historians continue to embrace uncritically the notion that Humboldt is the
father of modern environmental thought, environmental science, and environmental
movements.[9]

This article will trace the history and ultimate validity of the belief that human
activities caused large-scale climate change in parts of Latin America, from Humboldt's first arrival on the Venezuelan coast in 1799 to the advent of paleoclimato-
logical techniques for reconstructing past environments in the late twentieth century.
It will employ travel diaries and correspondence to produce a "thick description" of
Humboldt's and his disciples' reading of desiccated landscapes in order to reveal the
intellectual context and epistemological basis for this discovery and its confirmation. Most interpretations of Humboldtian science portray it as a method or program
of scientific research,[10] or emphasize its political uses.[11] Few give detailed attention
to its development and findings. The following account will emphasize the political
motivations and social dimensions of Humboldtian science, especially Humboldt's
interaction with locals during his travels, his exuberant efforts to cultivate disciples,
and his disdain for Spanish colonialism. It demonstrates that the methodological preoccupation with travel and precision measurements associated with Humboldtian
science served to establish patron-client relationships between Humboldt and his dis-

[6] George Perkins Marsh, *Man and Nature; or, Physical Geography as Modified by Human Action*
(New York, 1864), 8–9, 145–6, 160–1, 183–4, 191–3, 200–5; James Rodger Fleming, *Historical
Perspectives on Climate Change* (New York, 1998), chap. 4; Michael Williams, *Americans and Their
Forests: A Historical Geography* (New York, 1989), 144–5, 307–9, 379–90, 401–3; Richard Grove,
*Green Imperialism: Colonial Expansion, Tropical Island Edens, and the Origins of Environmental-
ism, 1600–1860* (New York, 1995), 364–79, 427–31, 436, 443; Joachim Radkau, *Nature and Power:
A Global History of the Environment* (New York, 2008), 212–21.
[7] Reid Bryson and Thomas Murray, *Climates of Hunger: Mankind and the World's Changing
Weather* (Madison, Wis., 1979), pt. 3; Helmut Geist, *The Causes and Progression of Desertification*
(Aldershot, 2005).
[8] Aaron Sachs, *The Humboldt Current: Nineteenth-Century Exploration and the Roots of American
Environmentalism* (New York, 2006), 2.
[9] Nicolaas Rupke, *Alexander von Humboldt: A Metabiography* (New York, 2005); Gregory Cushman and Kent Mathewson, "Humboldt, Guano, and the Hermeneutics of Empire," in *Humboldt and
the Americas*, ed. Vera Kutzinski, manuscript (Vanderbilt University, Department of English), chap. 7.
[10] Susan Faye Cannon, "Humboldtian Science," in *Science in Culture: The Early Victorian Period*
(New York, 1978), 73–110; Michael Dettelbach, "Humboldtian Science," in *Cultures of Natural History*, eds. Nicholas Jardine, James A. Secord, and Emma C. Spary (New York, 1996), 287–304.
[11] Mary Louise Pratt, *Imperial Eyes: Travel Writing and Transculturation* (New York, 1992), pt. 2;
Rupke, *Humboldt* (cit. n. 9).

ciples that aggrandized the scientific reputations of both parties, even as it served a technocratic political program aimed at placing enlightened, globe-trotting scientists in positions of power.

This approach reveals that Humboldtian climatology sometimes willfully marginalized or ignored other competing explanations for phenomena.[12] Unlike much of Humboldt's other work, the discovery of human-caused climate change in Latin America owed surprisingly little to local understanding and sometimes overtly flew in the face of it.[13] This is particularly apparent with regard to a vibrant Creole tradition of meteorology in late colonial Lima focused on climate anomalies and the environmental determinants of human health. The term "Creole science" is often used to refer to systematic knowledge about the natural world produced by European settlers and their American-born progeny under colonial rule. It typically incorporated a hybrid of European, indigenous, or even African-derived ideas, along with novel elements, and often focused on the interaction between race and place. Like the term "Creole" from which it derives, this term has been used promiscuously to refer to a vast array of knowledges produced in the colonial and postcolonial world.[14] This article proposes narrowing usage of the term "Creole science" to refer to a specific geopolitical context in which systematic knowledge of the natural world provided a basis for Americans of European and mixed ethnicity to assert their own authority and dominance over regional environments and their residents while living under colonial rule. This distinguishes it historically from systematic forms of knowledge primarily intended to legitimate imperial rule or to strengthen the controllers of centralized postcolonial states—phenomena better referred to as imperial science or national science, respectively. Humboldt is rightfully famous for undermining Spanish rule in the Americas, but it is vital to recognize that he often belittled Creole achievements under colonial rule as well.

The broad influence of Humboldt's views on the human causes of climate change in South America ultimately hinged on three factors: one epistemological, one social, and one political. (1) Humboldt convincingly portrayed closed lake basins and watersheds as natural laboratories for understanding the interaction between geophysical forces and human endeavors. Paleoclimatologists operating under this very assumption eventually invalidated Humboldt's conclusion that Spanish imperialism was responsible for the desiccation of Lake Valencia and other regions of South America. (2) Humboldt achieved great success at recruiting disciples loyal to his scientific

[12] On the history of scientific ignorance in colonial contexts, see Londa Schiebinger, *Plants and Empire: Colonial Bioprospecting in the Atlantic World* (Cambridge, Mass., 2004).

[13] Jorge Cañizares-Esguerra, "How Derivative Was Humboldt? Microcosmic Narratives in Early Modern Spanish America and the (Other) Origins of Humboldt's Ecological Sensibilities," in *Nature, Empire, and Nation: Explorations of the History of Science in the Iberian World* (Stanford, Calif., 2006), 112–28.

[14] Key recent studies include Stuart McCook, *States of Nature: Science, Agriculture, and Environment in the Spanish Caribbean, 1760–1940* (Austin, Tex., 2002), chap. 3; Schiebinger, *Plants and Empire* (cit. n. 12); Judith Carney, *Black Rice: The African Origins of Rice Cultivation in the Americas* (Cambridge, Mass., 2001); Jorge Cañizares-Esguerra, *How to Write the History of the New World: Histories, Epistemologies, and Identities in the Eighteenth-Century Atlantic World* (Stanford, Calif., 2002); James Delbourgo and Nicholas Dew, eds., *Science and Empire in the Atlantic World* (New York, 2007); Daniela Bleichmar et al., *Science in the Spanish and Portuguese Empires, 1500–1800* (Stanford, Calif., 2008); María Portuondo, *Secret Science: Spanish Cosmography and the New World* (Chicago, 2009). For an overview of earlier diffusionist perspectives on science and colonialism, see Antonio Lafuente and María Ortega, "Modelos de mundialización de la ciencia," *Arbor* 142 (1992): 93–117.

program. These self-identified disciples demonstrated great skill at accommodating potentially conflicting observations to the Humboldtian episteme. (3) Humboldtian science proved to be quite useful for undermining existing forms of domination in order to replace them with new, technocratic forms of authority. In fact, it proved so useful for these purposes that many Creole scientists learned to embrace the Humboldtian program for their own ends.

THE BIRTH OF HUMBOLDTIAN SCIENCE

Humboldt and his ever-present travel companion, French botanist Aimé Bonpland (1773–1858), first set foot in South America on July 16, 1799, in the coastal district of Cumaná in the Captaincy-General of Venezuela. They spent the second half of the year's rainy season investigating this mountainous region, following in the footsteps of Linnaean botanist Peter Löfling's expedition of 1754–6. In late November, they relocated to the province of Caracas. This was one of the up-and-coming regions of Spanish-ruled South America as an exporter of cacao, indigo, and coffee and had recently come to enjoy unimpeded access to the lucrative markets of the British and Danish Caribbean and the United States.[15] While staying in La Guaira, the province's main port, Humboldt learned that this newfound economic liberty had ominous consequences for local health. La Guaira physician José Herrera, a member of the Royal Medical Society of Edinburgh, operated a small meteorological observatory where he sought to determine the relationship between changing environmental conditions and epidemic disease. He told Humboldt that yellow fever had jumped from Philadelphia to the island of Trinidad during the Great Epidemic of 1793, then over to Venezuela with the opening of free trade in 1797. Herrera also thought there was a correspondence between this outbreak and a rare tropical cyclone, which dumped rain on the region for sixty straight hours. As we shall see, these were hallmarks of the Hippocratic revival in eighteenth-century medicine. In Puerto Cabello, the main port for the Lake Valencia valley, Humboldt spent an enjoyable time with Gaspar Juliac, a physician and polymath who kept his house well adorned with birds, books, plants, animals, and mulatto girls. Juliac reported that nine thousand people had died of yellow fever in Puerto Cabello since 1793, and that the European born only needed to breathe putrid, miasmatic air in this locale to become sick and die. According to Caracas residents, winds from the south were particularly deadly because they brought miasmas from the distant swamps of the Río Negro. It relieved Humboldt to learn that he had arrived during the dry season when more healthful airs predominated.[16]

At least at this juncture, Humboldt recognized the dangers of travel to new climates and the importance of extreme events in defining the climate of a place. Nevertheless, he emphatically rejected the advice of the Baron de Montesquieu and other thinkers that the safest and most productive course for an individual was to stay put in the climate of his birth.[17] For Humboldt, the act of travel provided a powerful method

[15] Michael McKinley, *Pre-revolutionary Caracas: Politics, Economy, and Society, 1777–1811* (New York, 1985), chaps. 2–3.

[16] Humboldt, *Reise durch Venezuela: Auswahl aus den amerikanischen Reisetagebüchern* (Berlin, 2000), 175–82; this German translation is the only complete published version of Humboldt's manuscript field journals. All quotations in this article originating from non-English texts are the author's own translation.

[17] Fleming, *Historical Perspectives on Climate Change* (cit. n. 6), 16–7.

to test the limitations that climate supposedly imposed on an organism. His own body provided a moving laboratory and a finely calibrated set of instruments for examining these climatic boundaries. Humboldt therefore relished the "frenzy of pain" he experienced while scaling the lofty Silla de Caracas. He thought the "punishments" of scientific travel made "the thinking man . . . more aware of his condition when compared to his accompanying rabble"—a group that included the capital's most notable naturalist, Capuchin friar Francisco de Andújar (1760–1817), an outspoken advocate of making science and mathematics features of study for the priesthood in Venezuela. For Humboldt, mountain climbing acted as "a wonderful balm" that "heals the wounds of the physical organism . . . which shackle one's own and others' reason." According to this emerging paradigm, travel provided a means to liberate the body and the mind from inherited strictures, while attachment to place blinded a person to the true workings of nature. In his private writings, Humboldt loved making fun of Venezuelan Creoles' hesitance to travel, and he associated their resistance to looking beyond the horizon with an atavistic lack of interest in the future. Unsurprisingly, he had little good to say about Creole science in his published writings about Venezuela, with the marked exception of scientist Carlos del Pozo y Sucre (1743–1813), who operated a meteorological observatory equipped with a cabinet of electrical instruments "nearly as complete as our first scientific men in Europe possess," despite living in the country's isolated inland savanna.[18]

This prejudice against place-based knowledge blinded Humboldt to certain features of the Venezuelan landscape and climate—particularly those that were temporary. Humboldt was immediately struck by the apparent dryness of the central Venezuelan coast as compared to Cumaná. He made special note of changes in vegetation while scaling the Silla de Caracas and explicitly blamed these changes on seasonal burning of the lower slopes by cattle ranchers, rather than natural conditions at elevation. This indicates that Humboldt was predisposed to see humans as major agents of environmental change from the moment he arrived in South America. While walking between Caracas and Lake Valencia, he was struck by wanton destruction of forests, particularly in the vicinity of a copper mine where slaves used huge quantities of charcoal for "incompetent smelting." Humans were obviously at fault for the resultant soil erosion, which was progressively covering ore-bearing strata with silt. But the "thick vegetation" of intact seasonal forests in this area also failed to impress him, because he never saw a large tree with a full branch of leaves. Humboldt speculated that their trunks were too thick to maintain good sap flow, and that their great age and size was debilitating. Locals informed him that the landscape looked like this because of the "winter calm" and lack of rain, but Humboldt kept looking for a reason to blame humans. After reaching the green fields at the head of the Tuy Valley, Humboldt still found plenty to criticize: "A great part of the [lower] valley is vacant and dry, not from drought, but because water never reaches it because it is taken up by La Victoria's coffee, sugar cane, plantains, and wheat fields that surround it."[19] Like many northern travelers before and after, Humboldt struggled mightily to make sense of the unfamiliar ways of nature in the tropics. Generations of scientific

[18] Humboldt, *Reise durch Venezuela* (cit. n. 16), 179, 185, on 179; Humboldt, *Personal Narrative* (cit. n. 5), 4:343–4, 398–9; Yajaira Freites, "De la colonia a la república oligárquica (1498–1870)," in *Perfil de la ciencia en Venezuela*, ed. Marcel Roche (Caracas, 1996), 25–92, on 33–4, 42–3, 74.
[19] Humboldt, *Reise durch Venezuela* (cit. n. 16), 177–82, 186, 188, 190–4, 196, on 186, 188, 193, 196.

travelers—and environmental historians depending on their accounts—have allowed similar prejudices to shape their interpretation of semiarid landscapes, to the continuing detriment of our understanding of desertification.[20]

Travel was an essential component of Humboldtian science, in which subjective experience produced by bodily sensations was just as important as objective measurements produced by the precision instruments Humboldt and his servants lugged with them. According to historian Ovar Löfgren, Humboldt profoundly influenced the practice of modern travel: from tourists' propensity to measure, evaluate, and describe the scenery, peoples, and situations they encounter, to the binoculars, thermometers, and other scientific knickknacks found in tourists' luggage, to the popularity of travel to southern climes. Humboldtian travel also had a social component. It enabled the traveler to attain liberation from the everyday norms of existence, to expand horizons, experience new environments, attain new levels of companionship, achieve a higher social status, even to take on a different bodily form—if only for a brief holiday.[21] Humboldt did not invent this style of travel. Journeying to the seaside, hot springs, and other healthful climes emerged as a popular pastime among Europe's leisure class during the late eighteenth century.[22] Humboldt spent the year 1790 traveling around northwestern Europe with naturalist Georg Forster (1754–94), a veteran of Captain Cook's expeditions to the Pacific. This trip directly exposed Humboldt to revolutionary movements emanating from France and to the idea that exhalations from soil, plants, and water might influence rainfall on Tahiti, St. Helena, and other tropical islands. Forster was fascinated by the relationship between climate and politics and dedicated the rest of his life to bringing revolution to the German lands. Their journey together awakened Humboldt to the possibility of traveling to South America and the Pacific Ocean.[23]

Humboldt's interest in these subjects also derived from the "great climate debate" of the eighteenth century. The former superintendent of the Royal Botanical Garden in Paris, the Comte de Buffon (1707–88), had popularized the notion that Earth had once been wetter and had grown progressively drier over time as it cooled from its original molten state. Buffon seized on a report by Harvard physician Hugh Williamson to the American Philosophical Society in 1770 that forest clearance along the eastern seaboard of North America had caused the climate to become warmer and more amenable to civilization. Buffon deduced from this that Europe would be as cold as equivalent latitudes in Quebec and Labrador if their forests had not been cut, their marshes not drained, and their rivers not controlled in centuries past. (Humboldt later invented the isoline to map this exact phenomenon.) Buffon vigorously

[20] Nancy Leys Stepan, *Picturing Tropical Nature* (Ithaca, N.Y., 2001), chap. 1; Georgina Endfield and Sarah O'Hara, "Degradation, Drought, and Dissent: An Environmental History of Colonial Michoacan, West Central Mexico," *Ann. Assoc. Amer. Geogr.* 89 (1999): 402–19; Endfield and David Nash, "Drought, Desiccation and Discourse: Missionary Correspondence and Nineteenth-Century Climate Change in Central Southern Africa," *Geogr. J.* 168 (2002): 33–47; Karl Butzer, "Environmental History in the Mediterranean World: Cross-disciplinary Investigation of Cause-and-Effect for Degradation and Soil Erosion," *Journal of Archaeological Science* 32 (2005): 1773–1800.

[21] Ovar Löfgren, *On Holiday: A History of Vacationing* (Berkeley and Los Angeles, 1999).

[22] Alain Corbin, *The Lure of the Sea: The Discovery of the Seaside, 1750–1840* (New York, 1995).

[23] Georg Forster, *Ansichten vom Niederrhein von Brabant, Flandern, Holland, England und Frankreich im April, Mai und Junius 1790* (Berlin, 1791–4); T. C. W. Blanning, *French Revolution in Germany: Occupation and Resistance in the Rhineland, 1792–1802* (Oxford, 1983); Grove, *Green Imperialism* (cit. n. 6), 153–67, 326–32, 364–75.

promoted the idea that humans had emerged as a powerful force for environmental change during the recent "Epoch of Man," and he speculated that the primeval Americas had fallen behind the development of the Old World in a host of ways because they lacked these climatic improvements.[24] Humboldt's mentor at the Royal Saxon School of Mines in Freiberg, Abraham Gottlieb Werner (1749–1817), also contributed to this debate. Werner's belief that the world was becoming gradually cooler and drier provided the basis for his Neptunist theory of the formation of geological strata by water. He taught Humboldt to use the layering of the landscape to reconstruct its ancient history. According to this geognostic methodology, a scientific traveler could move forward and backward through geological time by climbing mountains, descending valleys, and exploring caves and mines. The published work of Horace-Bénédict de Saussure (1740–99) describing his exploration of the Alps taught Humboldt the value of precision field instruments and convinced him that mountain lakes could serve as natural laboratories for understanding environmental change within a montane basin. Changes in Lake Geneva, for example, played a special role in convincing Saussure that the Alps had become much drier over time thanks to human activities.[25]

On the eve of Humboldt's voyage to the Americas, he worked closely with French analytical chemist Louis Nicolas Vauquelin (1763–1829) in a series of experiments on pneumatic chemistry, the absorption of oxygen by soils, and the influence of electricity on plant germination and crop growth. This laboratory work fueled Humboldt's interest in the environmental significance of gaseous chemicals and their role in plant fertility and animal vigor—key obsessions of experimental chemists of the time.[26] On Humboldt's way to South America, he made contact with Paris-trained botanist Antonio José Cavanilles (1745–1804) and American botanical explorers at the Real Jardín Botánico in Madrid. The relationship between agriculture, deforestation, soil fertility, disease, and drought was a major preoccupation in late eighteenth-century Spain. Based on his own travels, Cavanilles was of the opinion that deforestation had caused detrimental changes to the Spanish province of Valencia and that wet rice cultivation, which had replaced fields of wheat and fruiting trees, had made the region vulnerable to tertian fevers. His intellectual circle (*tertulia*) bestowed its blessing on Humboldt's voyage—a critical act of political patronage.[27] Humboldt eventually obtained some useful tools for understanding the microcosmic diversity and communitarian nature of plant life in the Andes while botanizing in what is now Colombia with José de Caldas (1768–1816), a Creole scientist from the mining center of Popayán. Caldas, in turn, adopted some of Humboldt's geognostic methods for understanding

[24] Fleming, *Historical Perspectives on Climate Change* (cit. n. 6), chap. 2; Clarence Glacken, *Traces on the Rhodian Shore: Nature and Culture in Western Thought from Ancient Times to the End of the Eighteenth Century* (Berkeley and Los Angeles, 1967), chap. 14; Antonello Gerbi, *The Dispute of the New World: The History of a Polemic, 1750–1900* (Pittsburgh, Pa., 1973), 171–3, 404–17.

[25] M. J. S. Rudwick, *Bursting the Limits of Time: The Reconstruction of Geohistory in the Age of Revolution* (Chicago, 2005), 15–22, 84–97, 420; René Sigrist, ed., *H.-B. de Saussure (1740–1799): Un regard sur la terre* (Chêne-Bourg, 2001).

[26] Wolfgang-Hagen Hein, ed., *Alexander von Humboldt: Life and Work* (Ingelheim am Rhein, 1987), 155–9, 164–5.

[27] Luis Urteaga, *La tierra esquilmada: Las ideas sobre la conservación de la naturaleza en la cultura española del siglo XVIII* (Barcelona, 1987), chaps. 8–9; Miguel Ángel Puig-Samper, "Humboldt, un prusiano en la corte del Rey Carlos IV," *Rev. Indias* 59 (1999): 329–51; Antonio González Bueno, *Antonio José Cavanilles (1745–1804): La pasión por la ciencia* (Madrid, 2004), 16–7, 161, 166, 303.

the underlying structure of the landscape.[28] These interactions underscore the significance of interpersonal contacts and companionship in both hemispheres to the early
development of Humboldtian science. As we shall see, social relationships of this
sort also played a vital role in its popularization.

Humboldt used a combination of these ideas and methods to make sense of the dry-
season landscapes of north-central Venezuela. He tentatively concluded that the high
ridges of the coastal range blocked the entry of humid air into interior valleys from
the Caribbean Sea to the north and the Orinoco plain to the south. From his own experiences during the rainy season in Cumaná, Humboldt had observed that rainstorms
in the tropics seemed to result from local electrical processes, usually during the heat
of the day, unlike in northern regions, where storms migrated from place to place.[29]
These meteorological observations are critical to understanding how Humboldt arrived at his conclusions regarding Lake Valencia and human-caused climate change
in the region. His belief that tropical thunderstorms form in situ enabled him to treat
this lake basin and other tropical watersheds as closed systems, analogous to tropical
islands, that could serve as natural laboratories for understanding the physical relationship between the air, soil, vegetation, moisture, and human agency.

On February 13, 1800, Humboldt and his associates reached the head of the Tuy
Valley, then journeyed quickly west along the northern shore of Lake Valencia, before making a quick jaunt to Puerto Cabello over the coastal range, taking instrumental measurements at almost every turn. The broad inland valley occupied by Lake
Valencia was the heartland of irrigated cacao, coffee, sugarcane, cotton, tobacco, and
especially indigo cultivation in Venezuela. The Marques del Toro and other valley
plantation owners well read in agricultural works explained to Humboldt that indigo
rapidly depleted the land: "In four to five years, nothing will grow in the wasted
soil. . . . One cultivates the soil like one operates a mine." Humboldt identified three
possible factors that played into this wastage: American greed to make money from
the land, poisonous "excrement" produced by the indigo plant, and the ill effects of
clearing natural vegetation for farming, which opened the soil to the sun and eliminated the shade and moisture needed for humus to form and for oxygen, hydrogen,
and carbonic acid to nutrify plants.[30]

Locals insisted that Lake Valencia had been drying up rapidly since about 1740.
The Cabrera Peninsula and a long list of hills, points, and outcrops had all reputedly
been islands within the past century. The previous eight years had been particularly
dry, by most accounts, resulting in the appearance of three flat-topped islets in 1796
known as the Nuevas Aparecidas, which now stood a meter above the waves. Most
locals considered the clear blue skies and drying of the lake basin to be a good thing,
however, because they spared the valley from the deadly fever epidemics of the previous decade and created rich alluvial lands along the lakeshore.[31]

Humboldt had no reason to doubt these assertions, but took a far more negative view of the situation. In his opinion, the "misconduct" of civilized society had
"forcefully disturb[ed] nature's economy" within the lake basin to such an extent that
it threatened to destroy the productivity of the region forever. After rejecting local

[28] Cañizares, "How Derivative Was Humboldt?" (cit. n. 13); John Wilton Appel, *Francisco José de
Caldas: A Scientist at Work in Nueva Granada* (Philadelphia, 1994).
[29] Humboldt, *Reise durch Venezuela* (cit. n. 16), 186.
[30] Ibid., 199, 208–9, 220, on 208–9.
[31] Ibid., 215–9.

speculation that subterranean filtration was responsible, he identified four interlinked causes for the shrinkage of Lake Valencia:

1. The surrounding mountains had been mostly cleared of vegetation for wood, charcoal, plantations, and grazing over the past century. Every night after the sun set, Humboldt was mesmerized by the "spectacular theatrical effect" of dry-season fires used by ranchers to eliminate pests and keep the upland range open and nutritious for cattle.
2. As a consequence, there was little tall, leafy, woody vegetation to suck up water and shade the ground in the valley. This allowed the tropical sun to beat down on the ground, inhibiting the formation of nutritive humus and accelerating soil desiccation. There was also little vegetation to slow runoff after storms. Humboldt compared this situation to that of the German province of Franconia, which suffered from a severe wood famine.
3. Practically all of the rivers and streams that flowed into the lake had been diverted for plantation irrigation during the recent indigo boom. "Hardly a drop now enters the lake," Humboldt noted—again based on what he saw during the dry season. According to local informants, these alterations dated back to the late seventeenth century, when a rancher redirected the flow of the Río Paíto away from the lake's western shore toward the distant Orinoco. "This drove the plain into desert" in the vicinity, Humboldt concluded.
4. Cleared ground absorbed much greater quantities of heat from the intense sunlight of the low latitudes, which caused the whole lake basin to heat up, thus decreasing the humidity of the air and speeding up the rate of evaporation from the lake. Humboldt did not think the lake would ever dry up completely, but he was greatly concerned that these processes would make it too salty to drink, kill off the lake's fish life, and turn the whole basin into "a dry desert." He encouraged the Marques del Toro, the town of Valencia's most prominent sugar planter, to install granite limnometers in the lake to precisely quantify its rate of recession.

Humboldt explicitly compared this situation to Lake Geneva, thus revealing his intellectual debt to Saussure. This explanation exemplified Humboldt's emerging obsession with "the confluence and interweaving" of geophysical, organic, and social forces. Humboldtian science came into definitive being for the first time along the receding shoreline of Lake Valencia.[32]

Humboldt repeated this reasoning almost verbatim in his *Personal Narrative of Travels to the Equinoctial Regions*. By the time he wrote this best-selling work in the 1820s, Humboldt had come to believe that climate change of this sort was representative of a much broader tendency of European colonialism to abuse American lands and peoples: "These natural desiccations, so important to the colonial agriculture, have been eminently considerable during the last ten years [the 1790s], in which *all America* has suffered from great droughts."[33] Humboldt's visit to the Pacific shore of the New Continent was particularly important in convincing him of this.

[32] Ibid.; Dettelbach, "Humboldtian Science" (cit. n. 10).
[33] Humboldt, *Personal Narrative* (cit. n. 5), 4:129–230, on 151 (emphasis added).

THE ULTIMATE HUMAN-DEGRADED LANDSCAPE?

On September 18, 1802, under the watchful eyes of several Andean condors, Humboldt, his companions, and his porters began their final descent from the highlands of northern Peru after eighteen months exploring the northern Andes. He was amazed by the aridity and barrenness of Peru's coastal slope, even though it lay in the heart of the tropics. Great torrents of water had obviously flowed through its deep, bone-dry canyons, perhaps "at a time when the abundance of water was much greater around the globe." At first glance, this landscape seemed to confirm the Buffonian idea that Earth had grown progressively drier over time, and he imagined that Lower Peru had once been as green and leafy as the lowland rainforests of the Orinoco or Marañon. Humboldt was particularly interested in what appeared to be newer valleys dissecting Peru's coastal range. They were too remote from the highlands to be caused by seasonal runoff. Perhaps a massive earthquake had thrown great quantities of dust into the atmosphere and caused an enormous rainstorm. This is the sort of explanation that Georg Forster and other devotees of exhalationist meteorology might have proposed.[34]

Suddenly, in the midst of the brown coastal desert, Humboldt came upon a massive stone wall of great antiquity that appeared to be defending the lowlands from highland invaders, and below that the "sad remains" of canals and aqueducts. What had become of the ancient civilization that built them? He soon found out. On September 26, 1802, Humboldt finally made it to the shore of the Pacific Ocean near Trujillo. He made sure to take a sea-level barometer reading the moment he arrived at the seashore, so he could convert his barometric measurements on Chimborazo and other Andean heights into absolute measures of altitude. He also dipped a thermometer into the cold ocean surf. Humboldt and his companions soon turned their attention to the nearby ruined city of Chan Chan, the ancient capital of the Chimor Empire. Its mud-brick labyrinth, palaces, and pyramids partly fulfilled his dream of exploring Egypt. Humboldt even arranged to meet fat, burbling Chayhuac, the latest of a long line of indigenous nobles descended from the last Chimu king. The Inca Empire conquered the Chimu at the end of the fifteenth century, followed soon after by the Spanish in the 1530s. Like generations of his lineage before him, Chayhuac used gold and silver looted from the tombs of his ancient ancestors to pay his people's tribute to the Spanish. To Humboldt, Chayhuac's situation encapsulated the abusive nature of Spanish colonial rule. "A bad government destroys everything," Humboldt recorded in his journal. "Away from their own lands, Europeans are as barbarous as Turks, or more, since they are more fanatical." On the long walk south to Lima, Humboldt passed many signs of abandoned irrigation works, and he was especially struck by a barren field near the mouth of the Santa River covered with human bones, mummified body parts, and crushed skulls. He imagined this to be the site of a cataclysmic battle at the time of the Inca conquest. He wondered how green the Peruvian coast must have been before the invading Incas and Spanish destroyed the Chimu aqueduct network and cut off the thick, silt-laden water that once fertilized the Peruvian coast

---

[34] Humboldt, "Diario de viaje," in *Alexander von Humboldt en el Perú: Diario de viaje y otros escritos* (1802; repr., Lima, 2002), 31–88, on 71–3; Vladimir Jankovic, *Reading the Skies: A Cultural History of English Weather, 1650–1820* (Chicago, 2001), 27–8.

"like the mud of the Nile."[35] For many years, Humboldt considered the dry deserts of coastal Peru to be the ultimate *human*-degraded landscape.[36]

Humboldt was even less impressed with the state of intellectual and cultural life in coastal Peru. In an acidic letter written to the governor of the highland province of Jaén, he harshly criticized the "cold egotism" of the *limeño* elite and their supposed lack of patriotic interest in the glories and good government of their country: "In Lima, I learned nothing of Peru."[37] This commentary says far more about Humboldt's prejudices than it does about the status of scientific life in the capital of the Viceroyalty of Peru. Humboldt was introduced to a long list of notables during the two months he spent in Lima surrounding his observation of the transit of Mercury across the sun on November 9, 1802. This included several figures with close ties to the Sociedad Académica de Amantes del País, an organization dedicated to the improvement of the viceroyalty that was best known for publishing the intellectual journal *Mercurio Peruano* (1791–5). This was one of many patriotic societies of its type, which were signature institutions of the Enlightenment within Spain's global empire.[38] In contrast to Venezuela, Lima had a long-established community with an interest in scientific pursuits. Most prominent among this group was Creole physician Hipólito Unanue (1755–1833), a native of the southern port of Arica on the edge of the Atacama Desert. Unanue is known for a host of contributions to cultural life in late colonial Peru and his leadership of the newborn Peruvian Republic during the independence wars. He was also the leading exponent of a vigorous Creole tradition of medical meteorology.[39]

This tradition had deep roots in Lima and many similarities to the "patriotic astrology" produced by Creole science elsewhere in the Americas.[40] It took on a far more visible form as the result of a natural disaster. On October 28, 1746, one of the most destructive earthquakes of modern times struck the central coast of Peru. The Spanish viceroy responded to this crisis by appointing French astronomer Louis Godin (1704–60), a recent member of La Condamine's geodesic expedition to the equator, to chair an expert commission to plan the rebuilding of Lima. Jesuit mathematician Juan Rehr (1691–1756) received a similar call to travel 1,100 kilometers from a remote indigenous mission on the other side of the Andes to oversee the reconstruction

[35] Humboldt, "Diario de viaje" (cit. n. 34), 73–81, on 73, 81; Miguel Feijóo de Sosa, *Relación descriptiva de la ciudad, y provincia de Truxillo del Perú* (1763; repr., Lima, 1984), 1:25–6, 84–6, 158–9; 2:90–1.

[36] Humboldt, *Political Essay on the Kingdom of New Spain* (1811; repr., New York, 1966), 2:45; Humboldt, *Personal Narrative* (cit. n. 5), 4:143, 296; 5:346; 6:182; Humboldt, *Aspects of Nature in Different Lands and Different Climates* (Philadelphia, 1850), 45–6.

[37] Humboldt to Ignacio Checa, 18 January 1803, in *Humboldt en el Perú* (cit. n. 34), 214–5.

[38] Paul Rizo-Patrón, "Arrogance and Squalor? Lima's Elite," in *Alexander von Humboldt: From the Americas to the Cosmos; an International Interdisciplinary Conference, October 14–16, 2004* (New York, 2004), 69–81; Robert Schafer, *The Economic Societies in the Spanish World, 1763–1821* (Syracuse, N.Y., 1958).

[39] Adam Warren, *Medicine and Politics in Colonial Peru: Population Growth and the Bourbon Reforms* (Pittsburgh, 2010); Jorge Cañizares-Esguerra, "La utopía de Hipólito Unanue: Comercio, naturaleza, y religión en el Perú," in *Saberes andinos: Ciencia y tecnología en Bolivia, Ecuador y Perú*, ed. Marcos Cueto (Lima, 1995), 91–108; John Edward Woodham, "Hipólito Unanue and the Enlightenment in Peru," (PhD diss., Duke Univ., 1964).

[40] Jorge Cañizares-Esguerra, "New World, New Stars: Patriotic Astrology and the Invention of Indian and Creole Bodies in Colonial Spanish America, 1600–1650," *Amer. Hist. Rev.* 104 (1999): 33–68.

of Jesuit properties in the capital. After spending two decades in the Amazonian wilderness, this Prague native took over from Godin as royal cosmographer and took command of rebuilding Lima's main cathedral.[41]

Rehr used his position to make some potent political statements regarding modern science. He firmly believed that "everyone can see clearly in the theater of nature" and publicly declared that Galileo had proven "our movement."[42] He continued the astronomical observations, astrological prognostications, and calendrical tasks that had long been required by this office. But Rehr was dissatisfied with the idea that the movements of the heavenly bodies provided a good instrument for predicting the meteorological shifts that affected local agriculture and medical practice.[43] His predecessor as royal cosmographer, Creole virtuoso Pedro de Peralta Barnuevo (1663–1743), had lamented that a "rustic" could foretell rain by the sky's appearance, while he as a "doctor of astrology" found it impossible to predict one of Lima's rare rain showers or other "accidents" of nature.[44] To remedy this situation, Rehr initiated a systematic program of instrumental measurements, supplemented by qualitative observations he called "rustic astrology." In the 1756 edition of the almanac *El conocimiento de los tiempos*, Rehr published the first known systematic atmospheric measurements from Peru. This friar's simple act of setting up a barometer was a sure sign that the Enlightenment had arrived in Peru.[45]

Rehr's research program was motivated by an intense desire to understand the local environment as part of "the workings of the world machine." His once-a-day observations with a mercury barometer informed him that the "weight of the Atmosphere" over Lima increased slightly during the winter months, leading him to speculate that ambient pressure on the body's blood vessels and soft tissues was responsible for the prevalence of strokes and paralysis during this season. His observations with an alcohol thermometer were intended to establish daily maximum and minimum temperatures so he and his successors could decide whether the climate of a particular day or season was relatively hot or cold, so physicians could prescribe appropriate treatment to their patients.[46] These records confirmed what everyone already knew to be true: the climate and weather of Lima barely varied from day to day and year to year. Rehr's successor as royal cosmographer, Cosme Bueno (1711–98), discontinued publishing barometric measurements in 1761, since they almost never varied more than a couple of lines during the year. Barometers provided very little enlightenment to Lima's cognoscenti.[47]

---

[41] *El conocimiento de los tiempos* [hereafter cited as *Conocimiento*]. *Ephemeride del año de 1745* (Lima, 1744), final page; *Conocimiento. Ephemeride del año de 1750* (Lima, 1749), 3, 21; *Diccionario histórico y biográfico del Perú: Siglos XV–XX* (Lima, 1986), s.v. "Rehr, Juan"; David Block, *Native Tradition, Jesuit Enterprise, and Secular Policy in Moxos, 1660–1880* (Lincoln, Neb., 1994), 40, 114; Charles Walker, *Shaky Colonialism: The 1746 Earthquake-Tsunami in Lima, Peru, and Its Long Aftermath* (Durham, N.C., 2008).

[42] *Conocimiento. Ephemeride del año de 1754* (Lima, 1753), 6; *Conocimiento. Ephemeride del año de 1756* (Lima, 1755), 4–5.

[43] *Ephemeride de 1750* (cit. n. 41), 5–6, 16, 18–22.

[44] *Conocimiento. Ephemeride del año de 1733* (Lima, 1732), 2–5; Manuel Moreyra Paz-Soldán, "Peralta: Astrónomo," *Revista Histórica* 29 (1966): 105–23.

[45] *Ephemeride de 1750* (cit. n. 41), 18; *Conocimiento. Ephemeride de 1755* (Lima, 1754), 12–3; Jan Golinski, *British Weather and the Climate of Enlightenment* (Chicago, 2007), chap. 5.

[46] *Conocimiento. Ephemeride del año de 1758* (Lima, 1757), 13–4; *Conocimiento. Ephemeride del año de 1780* (Lima, 1779), 22.

[47] *Conocimiento. Ephemeride del año de 1762* (Lima, 1761), 9–10.

But in other matters, Bueno greatly extended the environmental investigations of Rehr. Bueno was born in the countryside of northeastern Spain and immigrated to Lima at the age of nineteen, where he took up the professions of pharmacy, then medicine. He considered himself a follower of Leiden physician Herman Boerhaave (1668–1738) and his disciple, Viennese court physician Anton de Haen (1704–76). They were Europe's leading exponents of a return to Hippocratic medicine with its emphasis on airs, waters, and places, modified by the modern materialist belief that physical factors in the environment were the principal determinants of human health. Boerhaave and Haen made careful empirical observation of their patients and were among the earliest advocates of the use of the thermometer as a tool for diagnosis.[48] Bueno applied insights from Boerhaavean medicine, Stephen Hales's work on exhalations, and local wisdom to attempt to explain why the Peruvian coast almost never experiences rainstorms, but instead is watered by heavy winter mists (*garua*) and runoff from the Andes.[49]

Bueno also made an immense contribution to geographical knowledge in South America by soliciting reports from district officials (*corregidores*) throughout the Viceroyalty of Peru, which he published in serialized form. Humboldt relied heavily on information of this sort, particularly when drafting his political essays criticizing Spanish rule in the Americas.[50] For Bueno, acquiring exact chorographical knowledge of the people, places, and natural resources of Peru had a vital role to play in public life—far more so than the old astrological prognostications expected of the cosmographer's office: "Honesty, . . . perseverance, . . . justice, . . . pity, . . . foresight, . . . zeal, . . . disinterest, . . . are the true *astros*, . . . the true influences under the political sky of he who operates a Good Government."[51] He asked his respondents to give special attention to "accidents of nature." The Peruvian-born *corregidor* of Trujillo, Miguel Feijóo de Sosa (1718–91), produced a model report of this sort. It concluded with an entire chapter on the catastrophic coastal rains and floods of 1701, 1720, and 1728—a treatise now recognized as one of the first scientific treatments of the recurrent El Niño phenomenon in Peru.[52] Like Rehr, Bueno was preoccupied with the physical phenomena that produced these rare disasters mainly because of their influence on human illness. For example, Bueno noted that the strange weather of 1720 corresponded with a horrific plague in the Andes lasting from 1719 to 1721.[53]

Hipólito Unanue was a disciple of Bueno and this Creole tradition of medical meteorology. Humboldt spent significant time interacting with Unanue in Lima and successfully recruited Unanue to extend some of his observational investigations in Peru. Humboldt later commented favorably on Unanue's 1802 vaccination campaign against smallpox and his "excellent physiological treatise on the climate of Lima,"

[48] Roy Porter, *The Greatest Benefit to Mankind: A Medical History of Humanity* (New York, 1999), 246–61, 344; D. W. McPheeters, "The Distinguished Peruvian Scholar Cosme Bueno, 1711–1798," *Hispanic-Amer. Hist. Rev.* 35 (1955): 484–92; Glacken, *Traces on the Rhodian Shore* (cit. n. 24), chap. 12.

[49] Cosme Bueno, "Dissertación Physico Experimental sobre la Naturaleza del Agua," in *Conocimiento. Ephemeride del año de 1759* (Lima, 1758), 46–80.

[50] For Humboldt's use of Unanue's "guide for foreigners," see *Political Essay on New Spain* (cit. n. 36), 3:240, 245.

[51] *Conocimiento. Ephemeride del año de 1778* (Lima, 1777), 3–5.

[52] Feijóo, *Relación descriptiva* (cit. n. 35), chap. 12.

[53] *Ephemeride de 1759* (cit. n. 49), 12; *Conocimiento. Ephemeride del año de 1766* (Lima, 1765), 2–9; *Conocimiento. Ephemeride del año de 1768* (Lima, 1767), 62; *Conocimiento. Ephemeride del año de 1769* (Lima, 1768), 19.

Observaciones sobre el clima de Lima (1806, 1815), but Humboldt ignored its main findings for many years.[54]

Hints of Humboldt's influence are scattered throughout this text, which like many Creole works from this era sought to refute Buffon's derogatory speculations about climate in the New World. Unanue recognized that Lima's environs had changed over time and that humans could change the environment for the worse. But he had no use for the notion that human abuse of the land had created Peru's peculiar coastal climate. Echoing patriotic Creoles elsewhere, Unanue believed that God's design had blessed the Peruvian Andes with "all the climates of the universe" and given coastal locales an eternal spring that made it akin to paradise.[55]

Unanue also had little use for the idea that tropical weather is produced in situ. In his view, the heat of the sun acted as the basic force stimulating "evaporation" from the sea, "transpiration" from all organic beings, and atmospheric electricity. The fact that Lima lies in an enclosed basin did seem to explain why the valley was particularly susceptible to fogs produced by "aqueous vapors" emanating from the ground during winter. But Unanue insisted that evaporation from the Pacific Ocean was ultimately responsible for all the rain and snow that fell in the high Andes—an idea corresponding with ancient indigenous beliefs. The coldness of these waters originated from even farther away. In the 1815 edition of *Observaciones*, Unanue inserted a note crediting the ocean current flowing north along the Peruvian coast for bringing "the frigidity impart[ed] by the waters of Cape Horn" to the region.[56]

For Unanue, changes in the wind determined the variability of precipitation within the Viceroyalty of Peru and enabled distant regions to influence local climes. For example, Unanue thought that cool, "boreal" winds during the summer months, blowing regularly from the ocean to the northwest, were capable of causing winter to arrive early and bringing drought to strike the highlands, as was the case in 1799, a year in which Unanue kept a complete weather diary. On the other hand, if these northwesterlies blew irregularly, moisture carried by southeasterly winds greatly augmented summertime rains in the Sierra—and, on rare occasions, reached all the way to the coast, bringing thunderstorms to the desert.[57]

Unanue was mainly interested in climate phenomena because of their relevance to human health. Anomalies were therefore of special concern. The chorographical work of his forebears enabled Unanue to recognize immediately that the summer of 1791 was the hottest on record and that extreme rains and floods along the northern coast that year bore a strong resemblance to those of 1720 and 1728.[58] Unanue also noted similarities between 1791 and the summer of 1803, which culminated in one of the most spectacular weather events of his lifetime: a strong nighttime thunderstorm over Lima with bright lightning. He credited this extremely rare electrical storm with

---

[54] Humboldt, *Personal Narrative* (cit. n. 5), 6:202; Humboldt, *Political Essay on New Spain* (cit. n. 36), 1:8.

[55] Unanue, *Observaciones sobre el clima de Lima y sus influencias en los seres organizados en especial el hombre*, 4th ed. (1815; repr., Barcelona, 1914), 12–4, 18, 20–3, 40, 49–83, 104; Gerbi, *Dispute of the New World* (cit. n. 24), 252–68, 302–5.

[56] Unanue, *Observaciones sobre el clima* (cit. n. 55), 25–8, 32, 34–5; Victoria Cox, *Guaman Poma de Ayala: Entre los conceptos andino y europeo del tiempo* (Cuzco, 2002).

[57] Unanue, *Observaciones sobre el clima* (cit. n. 55), 31–5, 175–81.

[58] Ibid., 33; *Mercurio Peruano* 1 (1791): 275–80; 2 (1791): 258; *Conocimiento. Ephemeride del año de 1792* (Lima, 1791), 6–7; Joëlle Gergis and Anthony Fowler, "A History of ENSO Events since A.D. 1525: Implications for Future Climate Change," *Climatic Change* 92 (2009): 343–87.

causing the immediate arrival of the winter mists, which fell in such great abundance that year that the coastal desert erupted with life. This reveals a glaring epistemological blind spot of Humboldtian science. Its obsession with average conditions made it difficult to detect and make sense of climate variability and extremes. Just one year after Humboldt walked down the Peruvian coast in 1802, disgusted by its aridity, it was wetter and greener than Unanue had ever seen it.[59]

On the other hand, Unanue did not possess a clear enough picture of average conditions to determine whether the climate of Lima was changing over time. The summer of 1804 in Lima was hotter than ever and caused coastal crops to flower and fruit far ahead of schedule, only to give way to one of the worst highland droughts in memory, greatly reduced water for coastal irrigation, and a severe rabies epidemic. Unanue heard that Hamburg, Paris, and Vienna also experienced anomalously warm winters in 1804. According to Buffon, the global climate was supposed to be getting cooler and drier. Unanue wondered whether the strange conditions of 1803–4 represented a reversion to the warmer, wetter "primitive order" of the past—with ominous implications for societal progress. Sharing perceptions of climate across a distance was a major difficulty for place-based knowledge systems and made it impossible for Creole meteorologists to recognize the phenomena we now associate with El Niño and La Niña as anything other than a regional curiosity.[60] (Historical climatologists now recognize 1791 and 1803 as strong El Niño years and 1801–2 as a strong La Niña.)[61]

Humboldt's freedom to travel, in contrast, enabled him to see enough of the Americas to become convinced that human activities were changing the climate for the worse on a hemispheric scale. His travels to Mexico in 1803 strongly reinforced this perception. Since the catastrophic floods of 1607 and 1629, the colonial government had been working to drain Lake Texcoco and other lakes in the Valley of Mexico. By the late eighteenth century, Felipe de Zúñiga y Ontiveros (1717–93), the head of the royal observatory at Tacubaya and keeper of some of Mexico's earliest meteorological records, had become aware that the climate of central Mexico was becoming noticeably drier, and food scarcities and epidemics more frequent. This pattern only seemed to get worse at the beginning of the nineteenth century. Humboldt was a bit shocked to learn that Mexico City, which had been founded on the island ruins of Tenochtitlan in 1520, was now 3.5 kilometers from the lake. In his *Political Essay on New Spain*, Humboldt directly blamed the conquering Spaniards' "hatred" of trees and their massive drainage works for overturning the natural order in the Valley of Mexico and converting what had been a verdant garden into a desiccated plain. He did not think it would be long before "the new continent, jealous of its independence, shall wish to dispense with the productions of old" and throw off the shackles of colonialism that had produced this disorder. The massive peasant rebellion known as the Hidalgo Revolt (1810–1)—the opening act of Mexico's wars for independence—made Humboldt look like a prophet regarding the political significance of environmental mismanagement.[62]

[59] Unanue, *Observaciones sobre el clima de Lima* (cit. n. 55), 36–8.

[60] Ibid.; Enrique Tandeter, "Crisis in Upper Peru, 1800–1805," *Hispanic-Amer. Hist. Rev.* 71 (1991): 35–71.

[61] Gergis and Fowler, "History of ENSO Events since A.D. 1525" (cit. n. 58).

[62] Humboldt, *Political Essay on New Spain* (cit. n. 36), 2:31–3, 45–6, 167–73, 403–6, 507–8, 529–31, on 31, 530; Nicolaas Rupke, "A Geography of Enlightenment: The Critical Reception of Alexander von Humboldt's Mexico Work," in *Geography and Enlightenment*, eds. David N. Livingstone

HUMBOLDTIAN DISCIPLESHIP AND THE CONFIRMATION OF CLIMATE CHANGE

Creole disciples of Humboldt played a central role in the creation of the Republics of Peru, Bolivia, and Gran Colombia, the last of which dissolved into Colombia, Ecuador, and Venezuela in 1830. These disciples also played an indispensable role in convincing the world of science that Humboldt's findings regarding the connection between climate change, colonialism, and human abuse of the land were correct.

Simón Bolívar (1783–1830), the most important military figure in the liberation of these territories, aspired to place scientific men in Humboldt's mold in prominent political positions. In 1805, Bolívar accompanied a team of explorers led by Humboldt to investigate the eruption of Vesuvius. To Humboldt's great amusement, Bolívar often sported a version of Napoleon's "legendary hat and grey frock coat" and began aspiring to follow Napoleon's example during this trip. In 1820, after establishing a strong foothold for independence forces in northern South America, Bolívar sent his second in command, Medellin-born botanist Francisco Antonio Zea (1766–1822), on a mission to Europe to seek diplomatic recognition, foreign loans, and a cadre of European-trained technocrats to help rebuild these war-torn territories. Bolívar later appointed Hipólito Unanue as his second in command within the Republic of Peru during his campaign of liberation in the southern Andes.[63]

On Humboldt's recommendation, Zea recruited a young Peruvian-born mining engineer, Mariano de Rivero (1798–1857), to organize this technocratic mission to Gran Colombia. Rivero was one of the first foreigners to graduate from the École Polytechnique and École des Mines in Paris and was one of a long line of young South Americans whom Humboldt took into his confidence. Rivero, in turn, recruited Jean-Baptiste Boussingault (1802–87), a talented French chemist who was employed at a mine that was rapidly consuming the forests of Alsace and who was eager to travel abroad. Rivero was an exemplar of the Creole intellectuals who looked to Humboldt and Europe for legitimation as they sought to build self-sufficient postcolonial states founded on white dominance in the Americas. Boussingault exemplified the European travelers and traders who descended in droves on the liberated territories of Latin America and the West of the United States to develop their mineral wealth, in the hope of converting the Western Hemisphere into a scene of industry and efficiency.[64]

Humboldt drafted a detailed list of instructions for these young disciples and

and Charles W. J. Withers (Chicago, 1999), 319–39; Andrew Sluyter, "Humboldt's Mexican Texts and Landscapes," *Geogr. Rev.* 96 (2006): 361–81; Louisa Schell Hoberman, "Bureaucracy and Disaster: Mexico City and the Flood of 1629," *J. Latin Amer. Stud.* 6 (1974): 211–30; Hoberman, "Technological Change in a Traditional Society: The Case of the Desagüe in Colonial Mexico," *Tech. Cult.* 21 (1980): 386–407; Arij Ouweneel, *Shadows over Anáhuac: An Ecological Interpretation of Crisis and Development in Central Mexico, 1730–1800* (Albuquerque, N.Mex., 1996), 72–100; Georgina Endfield, *Climate and Society in Colonial Mexico: A Study in Vulnerability* (Malden, Mass., 2008), chap. 6.

[63] Jean-Baptiste Boussingault, *Memorias* (1892; repr., Bogota, 1985), 3:11–2, 97–103; Gerhard Masur, *Simon Bolivar* (Albuquerque, N.Mex., 1969), 36–43, 191, 203, 254–5, 282–4, 288–9, 313; Fred Rippy and E. R. Brann, "Alexander von Humboldt and Simón Bolívar," *Amer. Hist. Rev.* 52 (1947): 697–703; Haraldur Sigurdsson, *Melting the Earth: The History of Ideas on Volcanic Eruptions* (New York, 1999), 120–4, 163–4.

[64] F. Urbani, "Mariano Eduardo de Rivero y Ustariz (1798–1857)," *Boletín Histórico de Geociencias* 46 (1992): 18–38; Monique Alaperrine-Bouyer, ed., *Mariano Eduardo de Rivero en algunas de sus cartas al Barón Alexander von Humboldt* (Arequipa, 1999), 13–8, 24; Boussingault, *Memorias* (cit. n. 63), 1:21, 32, 39–42, 56, 100–2, 149–51, 164–5; Pratt, *Imperial Eyes* (cit. n. 11), chaps. 7–8.

provided them with a set of precision instruments so they could retrace his travels through northern South America, thereby aggrandizing his earlier work. Humboldt hoped their efforts would establish the northern Andes as a center for training additional disciples who would fan out and create a network of meteorological and magnetic observatories across this liberated continent. In return, Humboldt enthusiastically promoted Boussingault and Rivero's activities within European scientific circles.[65]

Rivero and Boussingault arrived at the port of La Guaira, Venezuela, on November 22, 1822—again at the beginning of the dry season—and dutifully set up a meteorological observatory for hourly observations in the hotel patio. Rivero was shocked at the disorder of the region after more than a decade of war. After ascending the Silla de Caracas and exploring the ruined city of Caracas, which had been mostly destroyed by the great earthquake of 1812, they followed the Tuy Valley to Lake Valencia, where they spent several weeks exploring. They set up a meteorological observatory at Maracay, participated in the coffee and cacao harvests, observed indigo extraction, witnessed the burning of upland slopes for cattle grazing, looked for guano in the caves of San Juan, and, in Boussingault's case, began taking Spanish lessons from an attractive *señorita*—all the while fearful of meeting a violent end at the hands of royalist guerillas still prowling the area.[66]

Rivero and Boussingault also gave close attention to the lake level and signs of environmental change in the valley since Humboldt's visit two decades before. At least initially, Rivero was of the opinion that the lake had continued to shrink.[67] From Humboldt's disparaging description, Boussingault expected the region to be desiccated and infertile, but instead was "surprised" at its verdure. Other things were different as well. The Nuevas Aparecidas had become submerged. To the consternation of locals, the north shore road around the lake, the isthmus of the Cabrera Peninsula, and lakeside fields were now frequently flooded. The town of Valencia, a bustling metropolis of 7,000 when Humboldt visited, was mostly deserted by 1823. Years later, in an often-cited article first published in the *Annales de chimie et de physique* in 1837, Boussingault concluded that wartime chaos and the flight of freed slaves had caused regional depopulation, leading to a marked decline in irrigated plantation agriculture and cattle grazing, which allowed reforestation by "the invasive jungle of the Tropics." The *absence* of human activity appeared to have alleviated the geophysical and cultural causes for this closed lake basin's desiccation, allowing the lake level to rebound.[68]

Boussingault's interpretation of environmental change in the Lake Valencia watershed was reinforced by the observation that the water supply to the mines of Marmato in Colombia's Cauca Province declined rapidly during the mining boom of the late 1820s as the area's forests were denuded to produce charcoal for ore smelting.

[65] Humboldt to Boussingault, 5 August 1822; Boussingault to his parents, 27 August 1822; Vaudet to Boussingault, 8 June 1823, 27 July 1823, in Boussingault, *Memorias* (cit. n. 63), 1:167, 179–81; 3:19–20, 154, 156.

[66] Rivero to Humboldt, 7 December 1822, in Alaperrine-Bouyer, *Mariano Eduardo de Rivero* (cit. n. 64), 48–51; Boussingault, *Memorias* (cit. n. 63), 2:10–2, 24, 37–52.

[67] Rivero to Humboldt, 15 February 1823, in Alaperrine-Bouyer, *Mariano Eduardo de Rivero* (cit. n. 64), 53–4.

[68] Boussingault, *Memorias* (cit. n. 63), 2:35–7, 52; Boussingault, "Sobre la influencia de los desmontes en la diminución de las aguas corrientes," in *Viajes científicos a los Andes ecuatoriales*, by Boussingault and François Désiré Roulín (1849; repr., Bogota, 1991), 1–23, on 1–7, 15–18, 19, 24.

Boussingault's visit to coastal Peru appeared to clinch the argument that deforestation could result in regional-scale climate change. Immediately after his famous as-
cent of the Ecuadorian volcano Chimborazo, where he surpassed the altitude attained
by Humboldt, Boussingault made a trip to the whaling port of Paita in early 1832 to
see the desert coast of Peru. Boussingault was astounded by what he saw: "The surroundings are as arid as a person could imagine; not a plant, nor a stream, sand is
everywhere." According to locals, the town had not experienced a substantial rainstorm in seventeen years.[69]

Paita's treeless, rainless, goat-ridden wastes became Boussingault's ultimate scenario for illustrating his conclusions on the subject of deforestation and climate
change. He made a definitive statement of these in his wildly popular study, *Rural
Economy* (1843–4):

> 1st. That extensive destruction of forests lessens the quantity of running water in a
> country. 2nd. That it is impossible to say precisely whether this diminution is due to a
> less mean annual quantity of rain, or to more active evaporation, or to these two effects
> combined. 3rd. That the quantity of running water does not appear to have suffered any
> diminution or change in countries which know nothing of agricultural improvement.
> 4th. That independently of preserving running streams, by opposing an obstacle to evap-
> oration, forests economize and regulate their flow. 5th. That agriculture established in
> a dry country, not covered with forests, dissipates an additional portion of its running
> water. 6th. That clearings of forest land of limited extent may cause the disappearance of
> particular springs. . . . 7th and lastly. That in assuming the meteorological data collected
> in inter-tropical countries, it may be assumed that clearing of the forests does actually
> diminish the mean annual quantity of rain which falls.[70]

For George Perkins Marsh and many other readers, Boussingault appeared to have
decided the question for good, and Humboldt's efforts to advance Boussingault's
career once he returned to France only reinforced this perception. Boussingault's
views went on to exercise a special influence over conservation policy in Restoration France, where scientists became obsessed with determining how damage to forests during the French Revolution might have increased the intensity of droughts
and floods.[71] Mariano de Rivero, meanwhile, was appointed by Bolívar as the Peruvian Republic's first minister of mines and public education and worked closely with
two subsequent scientific travelers to Peru, Joseph Pentland and Johann Jakob von
Tschudi, to further a number of projects Humboldt set out for them.[72]

THE FATE OF HUMBOLDTIAN CLIMATOLOGY

Later in life, Humboldt returned to the question of human-caused climate change. In
1829, he and two companions journeyed across the vast expanse of the Russian Empire in central Asia, from the rich mines of the Ural Mountains, across the swampy

---

[69] Boussingault, "Sobre la influencia de los desmontes" (cit. n. 68), 17–20; Boussingault, *Memorias*
(cit. n. 63), 5:147–50, on 148.

[70] Jean-Baptiste Boussingault, *Rural Economy, in Its Relations with Chemistry, Physics, and Meteorology* (London, 1845), 673–90, on 688–9.

[71] Radkau, *Nature and Power* (cit. n. 6), 219; Caroline Ford, "Nature, Culture and Conservation in
France and Her Colonies, 1840–1940," *Past Present* 183 (2004): 173–98.

[72] Urbani, "Mariano de Rivero" (cit. n. 64); William Sarjeant, ed., "An Irish Naturalist in Cuvier's
Laboratory: The Letters of Joseph Pentland, 1820–1832," *Bulletin of the British Museum of Natural
History, Historical Series* 6 (1979): 245–319.

Barabinskaja steppe of western Siberia, to the Altai Mountains on the Chinese imperial frontier, then back again to the north shore of the Caspian Sea. Humboldt blamed the progressive desiccation of the Barabinskaja plain and the great saltpans of this region on human culture, and he posited that the shrinking Caspian and Aral seas may once have been a massive arm of the Arctic Ocean. He handed out thermometers to notables he met along the way, whom he hoped to recruit as participants in his project to determine geophysical causes for the "inflexion of isothermal lines" around the globe. On his way home, Humboldt beseeched the Academy of St. Petersburg to take the next step and establish an imperial network of meteorological observatories modeled after those in the United States. In his 1831 account of these travels, Humboldt restated his opinion, backed up by citing Hales and Saussure, that the removal of forests and foliage had the effect of intensifying the climatic tendencies of this region. But Humboldt gave far more credit to the remoteness of the ocean in creating this "continental climate," which he blamed for the unstable, despotic tendencies of Asian society. Meanwhile, new temperature data from the eastern United States suggested to Humboldt that the connection between deforestation and desiccation had been overblown—a finding that opened the door for others to propose that rain would follow the plow as European settlement expanded into the dry continental interior of North America.[73]

Humboldt applied this interest in the climatic influence of oceans to other regions as well. He noticed the peculiar coldness of the Pacific off Peru when he first dipped his thermometer into the ocean in September 1802.[74] Stimulated by hydrographic data from Louis Isidore Duperrey's 1822–5 expedition to the region for the French navy, Humboldt began to reconsider his entire opinion regarding the geophysical causes of the aridity of the Peruvian coast. He eventually embraced Unanue's belief that a cold ocean current originating in the Antarctic powerfully influenced the coastal climate of Peru and that anomalous winds were somehow responsible for the extraordinary rains and electrical storms of 1552, 1701, 1720, 1728, 1747, 1790, and 1803.[75] Eager to promote the accomplishments of Prussian oceanography, cartographer Heinrich Berghaus (1797–1884) sought to recognize him for discovering the cold "Humboldt Current." Humboldt objected to receiving credit for discovering something long known to South American mariners. Berghaus changed the name to the "Peruvian Current" in his *Physikalischer Atlas* (1849–52), but retained the commentary celebrating Humboldt's role in its discovery. The first name stuck. Humboldt never credited Unanue for any insights on the subject, and he perpetuated the preposterous claim that these cold temperatures had "remained unnoticed until my visit to the shores of the Pacific."[76] This story is typical of the way place-based knowledge has become incorporated into globalist versions of modern science. We rightly celebrate Humboldt for liberating this knowledge from local obscurity. Yet in the rush to credit him with discovery, we elided Unanue from this history and forgot

[73] Humboldt, *Fragments de géologie et de climatologie asiatiques* (Paris, 1831), 1:45–6, 90–6; 2:309, 453–8, 493–517, 556–64; Fleming, *Historical Perspectives on Climate Change* (cit. n. 6), 48–52.

[74] Humboldt, *Political Essay on New Spain* (cit. n. 36), 4:149; Humboldt, *Personal Narrative* (cit. n. 5), 2:66–75; 7:369–70, 419–27.

[75] Humboldt, "Memoria sobre la corriente fría," in *Humboldt en el Perú* (cit. n. 34), 223–38, originally published in *Länder- und Völkerkunde* (Berlin, 1831–6).

[76] Humboldt, *Aspects of Nature* (cit. n. 36), 110.

altogether that Humboldt once mistakenly blamed human activities for creating Peru's arid coastal climate.[77]

Postcolonial American scientists quickly learned that they could attain legitimacy for their own work by associating themselves with Humboldt—including in the United States.[78] Later Humboldtians also examined the connection between deforestation, agriculture, and climate change. In 1831, Italian-born military engineer Agostino Codazzi (1793–1859) took command of a chorographical commission charged with producing a detailed geographical survey of the newly independent Republic of Venezuela. As part of this work, he resurveyed the Lake Valencia basin. Colombian-born scientist Joaquín Acosta (1800–52) interpreted his work as indicating that the lake had resumed receding. Acosta had personal reasons for caring about this issue. On Boussingault's recommendation, Humboldt had taken Acosta under his wing when the state of Gran Colombia paid for Acosta to study military engineering in Paris during the late 1820s. Codazzi and Acosta both made elaborate overtures to dedicate their national maps of Venezuela and Colombia to Humboldt and receive his approval. Acosta remained close to Boussingault for the rest of his life.[79] But this did not prevent Acosta from questioning whether changes in the vegetation of the Lake Valencia basin were capable of causing the lake's recession. The postindependence boom in coffee cultivation had not denuded the landscape nearly to the extent of the indigo boom, but the valley was drying up anyway. "The question is therefore complex and requires more careful consideration," Acosta concluded in an easily overlooked footnote to Boussingault's work.[80]

Subsequent Humboldtians muddied the waters even more. Alfredo Jahn (1867–1940) was a Caracas-born, German-trained engineer who combined professional skill in making topographical observations with broad interests in meteorology, ethnology, and geology. From 1892 to 1895, he made the first systematic attempt to track changes in the level of Lake Valencia while working for the Great Railway of Venezuela. According to Jahn's initial survey, the lake level had fallen at least four meters from Humboldt's time—then promptly rose by 1.57 meters during his first two years of observation (fig. 2). Pluviometric measurements from the lake basin also showed marked annual variability. The first thirteen years of the rainfall series seemed to confirm Humboldt's old belief that the lake basin was becoming drier over time, but the record varied much more wildly thereafter. The lake level rose abruptly by four meters after bottoming out at a record low of 399.86 meters above sea level in summer 1932. This increase was roughly equal to the amount the lake was thought to have dropped between Humboldt's visit and Jahn's first measurements, and it called into question the entire belief that human altera-

---

[77] Heinrich Berghaus, *Physikalischer Atlas* (Gotha, 1849–52), vol. 2, map 2; Gerhard Kortum, "Humboldt und das Meer: Eine ozeanographiegeschichtliche Bestandaufnahme," *Northeastern Naturalist* 8, special issue 1 (2001): 91–108.

[78] Sachs, *Humboldt Current* (cit. n. 8); William Goetzman, *Exploration and Empire: The Explorer and Scientist in the Winning of the American West* (New York, 1966).

[79] Alberto Böckh, *El desecamiento del Lago de Valencia* (Caracas, 1956), 103–4, 131–4; S. Acosta de Samper, *Biografía del General Joaquín Acosta* (Bogota, 1901), 13, 109–18, 135, 184–7, 207, 210–2, 425–7, 432–6, 488–92; Freites, "De la colonia a la república oligárquica" (cit. n. 18), 73–4, 78–9.

[80] Boussingault, "Sobre la influencia de los desmontes" (cit. n. 68), 7.

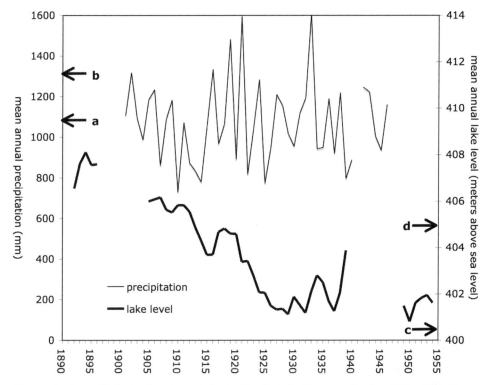

Figure 2. *Annual lake level and precipitation in the Lake Valencia Basin, 1890–1955. Precipitation averaged between Valencia, Maracay, and La Victoria. Lake levels:* a, *approximate level in 1800, 409.5 m (measured by Humboldt);* b, *approximate level c. 1835, 411.5–413 m (Agostino Codazzi);* c, *lowest historic mean annual level, 1976, 400.8 m (Ministerio del Ambiente);* d, *approximate level today, 405 m (International Lake Environment Committee). Figure by author; used with the author's permission. Based on Alberto Böckh,* El desecamiento del Lago de Valencia *(Caracas, 1956), tables 3, 8, 15; "Lago de Valencia," World Lakes Database, International Lake Environment Committee Web site, http://www.ilec.or.jp/ database/sam/sam-05.html (accessed 14 February 2011).*

tions within the lake basin were responsible for general tendencies in the regional climate.[81]

Another self-declared "follower of Humboldt" figured out a way to accommodate this anomaly to Humboldtian orthodoxy and, in the process, sparked the modern conservation movement in Venezuela. The career of Henri Pittier (1857–1950) exemplifies the long-term impact that Humboldtian science has had on environmental management and the institutionalization of science. Pittier became obsessed with tropical forests as a young engineering student living in central Europe after reading the travels of Humboldt and similar works. In Costa Rica, he organized a classic Humboldtian institution dedicated to combined topographical, geophysical, agricultural, and ethnographical investigations. Pittier settled permanently in Venezuela in 1917 to

---

[81] Böckh, *Desecamiento del Lago de Valencia* (cit. n. 79), 19–22, 42, 76; Yajaira Freites, "Auge y caída de la ciencia nacional: La época del Gomecismo (1908–1935)," in Roche, *Perfil de la ciencia en Venezuela* (cit. n. 18), 153–98, on 167–70.

protect his deteriorating health and received generous support from the authoritarian government of General Juan Vicente Gómez for the establishment of an agricultural experiment station at Maracay and a national museum. Pittier briefly managed Venezuela's National Astronomical and Meteorological Observatory and its network of meteorological stations, but he was run out of government service in 1933.[82]

Gómez died a year later, and the vast personal properties he had acquired with the nation's petrodollars returned to the ownership of the Venezuelan state. This included a huge forested tract known as Rancho Grande stretching across the mist-shrouded heights of the coastal range north of Lake Valencia. Pittier surveyed this estate for the new military government in 1935 and was horrified to see that squatters were already using the macadam highway Gómez had built through this reserve for illegal logging and farming. When Pittier reported his findings to Caracas, he chose to ignore that the region had just received record amounts of rain. Instead, he repeated the Humboldtian doctrine that deforestation, the annual burning of savanna, free-range cattle, and soil erosion had caused a general "deterioration of the climate" along Venezuela's coastal range—resulting in drier summers, irregular rainfall, and waterways that oscillated violently from raging torrents to an almost worthless trickle. If these changes continued, Pittier warned, then Venezuela's most densely populated and productive region would soon be doomed to near-constant thirst, with occasional destructive floods and landslides like those of 1934 that destroyed an opulent hotel and casino Gómez was building within Rancho Grande. Lake Valencia would turn into "a mere puddle." To reverse this disastrous trend, Pittier recommended rigidly enforcing existing laws intended to protect the country's forest and water resources, aggressively fighting forest and range fires, reforesting degraded areas important to the nation's water supply, and establishing a national system of forest reserves. To Pittier's surprise, General Eleazar López Contreras promptly established Rancho Grande as Venezuela's first national park. It now protects 107,800 hectares and has been renamed Henri Pittier National Park. Humboldt's long association with these ideas went a long way toward establishing their validity with government officials. Gómez's twenty-seven-year dictatorship helped out by sweeping away most traditional forms of political organization in Venezuela, which left his successors with a great deal of latitude to generate new forms of governance—most of which had a distinct technocratic orientation.[83]

Later governments did even more to fulfill Pittier's vision of a verdant Venezuela. In 1944, the national government built a concrete hydroelectric dam and reservoir in the Macarao Valley, and in the mid-1960s, it planted three million exotic pine and eucalyptus trees around the lake to protect the supply of drinking water to Caracas—all of which now lies at the heart of Macarao National Park (established 1973). In 1955–6, the authoritarian government of General Marcos Pérez Jiménez built a far more spectacular monument to Humboldt. On the mountains above northern Cara-

---

[82] Luis Alberto Crespo, *Henri Pittier: Caminante y morador de nuestro trópico* (Caracas, 1997), 26, 39–40, 48, 58; Marshall Eakin, "The Origins of Modern Science in Costa Rica: The Instituto Físico-Geográfico, 1887–1904," *Latin Amer. Res. Rev.* 34 (1999): 123–50; McCook, *States of Nature* (cit. n. 14), chap. 2.

[83] See n. 1 above; Humberto Ruiz Calderón, "La investigación científica en el gobierno, la universidad y el sector privado (1936–1958), in Roche, *Perfil de la ciencia en Venezuela* (cit. n. 18), 199–254, on 201–3; John Lombardi, *Venezuela: The Search for Order, the Dream of Progress* (New York, 1982), 211–51.

cas, German engineers erected a strikingly modern, fourteen-story spire connected to the city below by aerial cable car. The Humboldt Hotel now lies within El Ávila National Park (established 1958), which protects 81,000 hectares, including the lofty peak Humboldt scaled on New Year's Day 1800. Thanks to these efforts, almost five thousand square kilometers of Venezuela's coastal range lie under official protection today, including San Esteban National Park (established 1987), which protects 44,500 hectares just west of the original Rancho Grande preserve.[84] Since the late 1970s, government efforts to supply the cities and farms that surround Lake Valencia with water piped in from other watersheds have caused the lake to stop shrinking, but at the cost of severely polluting its waters with fertilizer, pesticides, and sewage.[85]

More recently, scientists have figured out ways of reconstructing environmental change within the Lake Valencia watershed by sampling layers of sediment formed in the lake bottom. These indicate that Lake Valencia has been drying up, more or less continuously, for 2,500 years. Similar studies using ocean sediments and cave stalagmites have demonstrated that Humboldt visited Lake Valencia at the tail end of the driest period for the coastal range of Venezuela since the end of the last ice age. Therefore, locals had good reason to be concerned about the desiccation of the region, although they cannot be blamed for its occurrence.[86]

Geological investigations have established that hyperarid conditions have existed along the Peruvian coast for at least fifteen million years, and perhaps much longer.[87] It is still an open question, however, whether deforestation and other human activities may have been responsible for climate change in northern South America during recent millennia. Over the past three thousand years, precipitation patterns have completely diverged from cyclic trends in solar radiation in northern Venezuela and many other regions, suggesting that human land-use patterns may be a factor in these changes. Climate simulations indicate that regional deforestation probably contributed to anomalously cool summers and autumns in the North American Midwest during the nineteenth century. Others have suggested that the regrowth of canopy forest in the tropical Americas in the wake of indigenous depopulation after 1492 may have played a role in generating the Little Ice Age, when Lake Valencia experienced its great drought.[88]

This article's examination of the origins and popularization of the belief that human land use is capable of causing large-scale climatic change also reveals some forms of ignorance inherent to the Humboldtian episteme. Most scientific voyages in the

[84] Information compiled from the respective national parks' profiles at Parkswatch, Venezuela, http://www.parkswatch.org (accessed November 2010).

[85] Robert Apmann, *Estudio ambiental del Lago de Valencia* (Caracas, 1979).

[86] Jason Curtis et al., "Climate Change in the Lake Valencia Basin, ~12 600 yr BP to Present," *Holocene* 9 (1999): 609–19; Luis González and Roger Gomes, "High Resolution Speleothem Climatology of Northern Venezuela: A Progress Report," *Boletín de la Sociedad Venezolana de Espeleología* 36 (2002): 27–9; Larry Peterson and Gerald Haug, "Variability in the Mean Latitude of the Atlantic Intertropical Convergence Zone as Recorded by Riverine Input of Sediments to the Cariaco Basin (Venezuela)," *Palaeogeography, Palaeoclimatology, Palaeoecology* 234 (2006): 97–113.

[87] Tibor Dunai et al., "Oligocene-Miocene Age of Aridity in the Atacama Desert Revealed by Exposure Dating of Erosion-Sensitive Landforms," *Geology* 33 (2005): 321–4.

[88] Stacey Rosner, "Stalagmite Based Paleoclimate Reconstruction, Northern Venezuela: A Record of Caribbean Holocene Climate Change," (MS thesis, Univ. of Kansas, 2006); Gordon Bonan, "Frost Followed the Plow: Impacts of Deforestation on the Climate of the United States," *Ecological Applications* 9 (1999): 1305–15; Richard Nevle and Dennis Bird, "Effects of Syn-pandemic Fire Reduction and Reforestation in the Tropical Americas on Atmospheric CO_2 during European Conquest," *Palaeogeography, Palaeoclimatology, Palaeoecology* 264 (2008): 25–38.

Humboldtian mode were too short to observe how a place develops over a season—much less over years, decades, or centuries. Humboldt recognized this problem, and this goes a long way toward explaining why he was so eager to cultivate disciples who would follow up on his journeys and measurements, ideally by systematizing them; why he gave credence to early Spanish chronicles, in order to provide his travels with historical depth; and why he was so obsessed with calculating the averages of climatic phenomena he encountered, at the expense of ignoring extremes. These considerations motivated some of the most significant efforts to observe geophysical phenomena of the nineteenth century, including the so-called Magnetic Crusade of the 1830s–40s.[89] In the process, "climate" lost many of its qualitative, chorographical properties and came to possess a much narrower meaning as "the long average of weather in a single place." This is typically defined by quantitative measurements of temperature, precipitation, wind, and the like—even for locales where mean conditions rarely exist.[90]

As historian Michael Dettelbach has pointed out, Humboldt and his disciples often took this fascination with the gathering of statistics and determination of the mean one step farther to assume that the economy of nature tends toward a steady-state equilibrium—unless humanity acts to disrupt it.[91] This belief lay at the heart of the Humboldtian political critique of Spanish colonialism and the "pristine myth" to which it is so closely attached.[92] It pervaded Boussingault's outspoken belief in a "balance of nature" and may have influenced the growing predominance of the counternotion during the late nineteenth century that climate is an environmental constant, rather than something that changes over human timescales. These steady-state ideas continue to haunt the sciences of climatology, ecology, and conservation, even today.[93] Meanwhile, American Creoles such as Jefferson, Bolívar, Unanue, Rivero, and Acosta appropriated these doctrines for their own use when advocating the overthrow of European rule and celebrating their ascendance as the rightful, technocratic stewards of the land. Interestingly, their actions as builders of centralized postcolonial states sometimes proved just as debilitating to place-based traditions of environmental knowledge as Humboldt's zeal for large-scale explanations for geographical phenomena.[94]

On the other hand, Humboldtian ideas on these subjects were responsible for inspiring some of the world's most important institutions for the conservation of forests and preservation of wild nature. Does it matter that we no longer accept the scientific reasoning that created them? Do we have to get the science right to reap benefits from policies designed to prevent catastrophic climate change?

[89] Christopher Carter, "Humboldt, Herschel, and the Magnetic Crusade," in *Alexander von Humboldt* (cit. n. 38), 509–18.

[90] *Oxford English Dictionary*, 2nd ed., s.v. "climate."

[91] Dettelbach, "Humboldtian Science" (cit. n. 10).

[92] Rupke, "Geography of Enlightenment" (cit. n. 62); William Denevan, "The Pristine Myth: The Landscape of the Americas in 1492," *Ann. Assoc. Amer. Geogr.* 82 (1992): 369–85.

[93] Jean Baptiste Dumas and Jean-Baptiste Boussingault, *The Chemical and Physiological Balance of Organic Nature* (New York, 1844); Fleming, *Historical Perspectives on Climate Change* (cit. n. 6), 50–3; Donald Worster, *Nature's Economy: A History of Ecological Ideas* (New York, 1994), pt. 6.

[94] Pratt, *Imperial Eyes* (cit. n. 11), chap. 8; Jankovic, *Reading the Skies* (cit. n. 34), 165–7; Frank Safford, *The Ideal of the Practical: Colombia's Struggle to Form a Technical Elite* (Austin, Tex., 1976); Julyan Peard, *Race, Place, and Medicine: The Idea of the Tropics in Nineteenth-Century Brazil* (Durham, N.C., 2000).

Imperial Climatographies from
Tyrol to Turkestan

by Deborah R. Coen[*]

ABSTRACT

This article argues for the importance of Europe's continental empires, Habsburg
and Romanov, to the emergence of a physical-dynamical model of the global climate before World War I. It begins to identify a set of questions and methods that
were distinctive of climatology as a continental-imperial science of "regionalization" with a global vision. The focus is on studies of mountain climatology by Heinrich von Ficker and A. I. Voeikov in the ecologically vulnerable regions of Tyrol and
western Turkestan. This continental-imperial context deserves historians' attention
because it suggests a new model for the globalization of knowledge: not simply a
matter of scaling up, globalization must be understood as a process of seeing across
scales, of recognizing causal connections between local, regional, and planetary
phenomena.

> "For a change, the material for a dissertation was procured
> mostly with the feet."
> Heinz Ficker, of his research on the *föhn* wind in Tyrol, 1906[1]

> "Science is something beautiful, especially when you can do it
> on horseback."
> Ficker in his journal in 1913, in a mountain pass in central Asia[2]

INTRODUCTION

This article is about an episode in the emergence of a physical-dynamical model
of the global climate. The setting is not an observatory in London, Paris, or Berlin,
but the mountains of Tyrol and Turkestan—two regions separated by roughly three
thousand miles, the Ural Mountains, and the Black and Caspian seas. Prior to World
War I, Tyrol lay near the western edge of the Austro-Hungarian Empire, straddling

[*] History Department, Barnard College, Columbia University, 415 Lehman Hall, 3009 Broadway,
New York, NY 10027; dcoen@barnard.edu.

Thanks to Lorraine Daston, Matthias Dörries, Jim Fleming, Vlad Janković, David Moon, and audiences at Colby College and the University of Pennsylvania for valuable suggestions. This research
was supported by the National Science Foundation (award no. 0848583).

[1] Ficker, "Die Erforschung der Föhnerscheinungen in den Alpen: Eine meteorologische Studie,"
Zeitschrift des deutschen und österreichischen Alpenvereins 43 (1912): 53–77, on 58. All translations
are my own unless otherwise noted.

[2] Heinrich von Ficker Nachlass, Staatsbibliothek zu Berlin-Preussischer Kulturbesitz, Berlin, Hand-
schriftenabteilung, acc. Darmst. 1924/5.

the Alps from Innsbruck in the north to the Italian-speaking region of Trentino in the south. Turkestan lay on the periphery of three empires: Russian, Anglo-Indian, and Chinese, with its western half annexed by Russia in the 1860s. Analogies between Tyrol and Turkestan may not come readily to mind today, but they did to the Austrian meteorologist and mountaineer Heinrich von Ficker (1881–1957), who frequently surveyed their landscapes side-by-side in his mind's eye.

The comparison was less fanciful than it might seem. The regions lay on opposite edges of two continental empires whose borders were contiguous, shifting each time they went to war over Polish and Ottoman lands, and whose fates were intertwined, as history would confirm in 1918. Moreover, it was a commonplace of turn-of-the-century geography that these empires were bound by a physical structure of critical geopolitical significance: the Eurasian steppe. As we will see, Russian and Austrian earth scientists traced a geological and climatic continuity from the Alps to the Pamirs, from the Hungarian plain to the steppe of Mongolia.

Today, the fates of Tyrol and Turkestan are more palpably entangled. Glacial retreat in these regions supplies striking evidence of global warming trends (anthropogenic and otherwise) since the nineteenth century. The loss of glaciers in the Alps and the Pamirs has the potential to cause floods, disrupt drinking water supplies and agricultural irrigation, and cripple the winter tourism industry in the Alps. Habsburg and Romanov scientists in the early twentieth century were in a unique position to begin to correlate evidence of such effects.

Historians have placed great weight on the roles of Europe's overseas empires in producing a global atmospheric science.[3] Yet it remains unclear how the numbers churned out by imperial observing networks before World War I related to global physical models of the atmosphere. In 1908 Napier Shaw complained of the "waste" of the British Met Office, of which he had recently retired as director. Data was streaming in, but to what end? What was needed was a "delicate adjustment between speculation and routine," between individual insight and standardized, distributed observation.[4] Shaw lamented as well the discontinuities of the imperial network— "four observatories and over four hundred stations of one sort or another in the British Isles . . . besides other ventures squandered abroad," from Gibraltar to British New Guinea, "and a couple of hundred on the wide sea."[5] In this sense, waste was a function of scientists' inability to mediate between local observations and global models. Likewise, Michael Reidy has recently shown that imperial British oceanography of the mid-nineteenth century failed to link its large-scale descriptive research to smaller-scale dynamical studies.[6]

Continental empires had one obvious advantage over the British: territorial continuity. The geographical breadth of these states' possessions and their technological infrastructures made it possible to follow atmospheric phenomena continuously across a continent. Equally significant, this article will argue, were the methods of geoscientific research that emerged in the two empires. Continental imperial scientists were charged with understanding the factors determining the spatial differen-

---

[3] Katharine Anderson, *Predicting the Weather* (Chicago, 2005); Fabien Locher, *Le savant et la tempête* (Rennes, 2008).

[4] On the latter see Lorraine Daston, "On Scientific Observation," *Isis* 99 (2008): 97–110.

[5] Napier Shaw, "Address of the President to the Mathematical and Physical Section of the BAAS," *Science* 28 (1908): 457–71, on 463, 464.

[6] Reidy, *Tides of History* (Chicago, 2008).

tiation of climate as the basic determinant of economic regions, as well as the ways in which local climates participated in a continental-scale regime of atmospheric circulation. This continental-imperial context deserves historians' attention because it suggests a new model for the globalization of knowledge: not simply a matter of scaling up, globalization must be understood as a process of seeing across scales, of recognizing causal connections between local, regional, and planetary phenomena. In this way, the geoscientific research of these empires circa 1900 speaks to urgent problems today—above all, the need for methods that can put global climate models and local knowledge into productive exchange and coordinate short-term, local reforms with long-term, international policies.[7]

CLIMATOGRAPHIES OF EMPIRE

The Russian and Austrian empires occasionally claimed to model themselves on western European colonialism, but their modes of governance and territorial visions were distinct. Politically, their territorial continuity made divisions between center and periphery, metropole and colony, far more ambiguous. Indeed, both states largely eschewed the term "colony," preferring to view most newly acquired lands as "natural" extensions of their rule. This pretension was in fact one ideological motivation for their support of the earth sciences. Like western European imperialists, Russian and Austrian elites viewed themselves as bringing civilization to the "primitive" peoples on their peripheries, yet it was far more difficult for them to point to the place on the map where civilization ended and backwardness began.[8] These states thus had an imperative to map and continually remap the empires' internal boundaries, social and natural, giving rise to what has been described as a scientific project of "regionalization," of which climatology was the foundation.[9]

Beginning in the eighteenth century, the Russian and Austrian empires were divided and redivided into climatic zones, as part of government projects of population management and agricultural development. Precedents for this regionalizing approach existed in central Europe in seventeenth-century local natural histories. As Alix Cooper argues, these detailed inventories of the plants or minerals of a town's environs were comparative and surprisingly cosmopolitan undertakings.[10] In Russia, immediate precedents can be found in the economic geographies of the liberal reformers associated with the Decembrists.[11] It was not, however, until the 1860s that the sciences of regionalization took mature form in each empire. In both states, that

[7] For instance, William E. Easterling and Colin Polsky, "Crossing the Divide: Linking Global and Local Scales in Human-Environment Systems," in *Scale and Geographic Inquiry: Nature, Society, and Method*, eds. Eric Sheppard and Robert B. McMaster (Malden, Mass., 2004), 66–85. On a regionalizing tendency in twentieth-century Australia, see Ruth A. Morgan in this volume.

[8] See, e.g., Mark Bassin, *Imperial Visions: Nationalist Imagination and Geographical Expansion in the Russian Far East, 1840–1865* (Cambridge, 1999); Nancy M. Wingfield, ed., *Creating the Other: Ethnic Conflict and Nationalism in Habsburg Central Europe* (New York, 2003).

[9] Nailya Tagirova, "Mapping the Empire's Economic Regions from the Nineteenth to the Early Twentieth Century," in *Russian Empire: Space, People, Power, 1700–1930*, eds. Jane Burbank, Mark von Hagen, and Anatolyi Remnev (Bloomington, Ind., 2007), 125–38.

[10] Cooper, *Inventing the Indigenous* (Cambridge, 2007), 86; see too Chenxi Tang, *The Geographic Imagination of Modernity* (Stanford, Calif., 2008), on the regionalizing perspective of Romantic geography, which likewise bridged multiple scales of observation.

[11] David J. M. Hooson, "The Development of Geography in Pre-Soviet Russia," *Ann. Assoc. Amer. Geogr.* 58 (1968): 250–72, esp. 254–5.

decade saw a liberalization of government and markets, the construction of railroads, and thus the growth of trade and industry. This was also a formative moment for nationalist movements in both monarchies. For all these reasons, the 1860s brought both Russia and Austria a growing awareness of regional economic disparities and of the political instability they might generate. This led in turn to a wave of imperial efforts to study "scientifically" the relationship between populations and land, with the aim of more efficiently exploiting each empire's natural and human diversity.[12] The regionalizing sciences of the 1860s were holistic enterprises intent on analyzing local differences in a larger, integrated context: "The dominant subject of study was no longer a simple description of diverse regions but an analysis of each region as part of an integrated whole."[13]

Among the products of this work were imperial "climatographies." As defined by the genre's undisputed master, the Austrian Julius Hann, climatography was a description of a regional climate in its relation to human life, "drawing attention to the geographic dependence of the natural plant cover and, particularly, of the region's agricultural and industrial circumstances, to the arrangement of the human settlements and to their way of life."[14] In these studies, regions mediated between local and global analysis.

By the 1890s, economic development and the spread of nationalism precipitated another common element of these imperial contexts: the rise of conservation movements. Like Chekhov's character Dr. Astrov in *Uncle Vania* (1896), many Russians believed their country's climate to have been "ruined" as a result of rapacious deforestation.[15] The landscape of the steppes, long seen as the physical manifestation of backwardness, began to appear in Russian art and literature as a thing of virgin beauty, identified in ambiguous ways with the Russian character. Scientists urged the creation of protected areas for the purpose of scientific research on this fertile "black earth."[16] In the Austrian case, analogously, a more ambivalently nationalistic German-Austrian identity became tied to the landscape of the Alps and the persona of the mountaineer.[17] Conservation societies emerged in both states to protect these environments, with natural scientists taking leading roles. As in European overseas colonies of the eighteenth century, conservationists in both states worried in particular about the climatic effect of deforestation.[18] Thus, across central and eastern

[12] E.g., Alexander Supan, *Österreich-Ungarn* (Vienna, 1889). For other examples, see David Moon, "Agriculture and the Environment on the Steppes in the Nineteenth Century," in *Peopling the Russian Periphery: Borderland Colonization in Eurasian History*, eds. Nicholas Breyfogle, Abby Schrader, and Willard Sunderland (New York, 2007), 81–105. On Mendeleev's scientific projects of imperial reform, see Michael Gordin, *A Well-Ordered Thing* (New York, 2004).

[13] Tagirova, "Economic Regions" (cit. n. 9), 131.

[14] Hann, *Klimatographie von Niederösterreich* (Vienna, 1904), 8–9. See also Deborah Coen, "Climate and Circulation in Imperial Austria," *J. Mod. Hist.* 82 (2010): 839–75.

[15] David Moon, "The Debate over Climate Change in the Steppe Region in Nineteenth-Century Russia," *Russ. Rev.* 69 (2010): 251–75, on 254.

[16] Christopher Ely, *This Meager Nature: Landscape and National Identity in Imperial Russia* (De Kalb, Ill., 2002); Douglas Wiener, *Models of Nature: Ecology, Conservation, and Cultural Revolution in Soviet Russia* (Pittsburgh, Pa., 2000); Moon, "Debate over Climate Change" (cit. n. 15), and "The Environmental History of the Russian Steppes: Vassili Dokuchaev and the Harvest Failure of 1891," *Trans. Roy. Hist. Soc.* 15 (2005): 149–74.

[17] Tait S. Keller, "Eternal Mountains, Eternal Germany: The Alpine Association and the Ideology of Alpinism, 1909–1939" (PhD diss., Georgetown Univ., 2006).

[18] Richard H. Grove, *Green Imperialism: Colonial Expansion, Tropical Island Edens and the Origins of Environmentalism, 1600–1860* (Cambridge, 1996).

Europe in the last third of the nineteenth century, climatologists embarked on imperial missions to map "natural regions" and "transition zones."

THE EURASIAN STEPPE, LABORATORY OF CLIMATE CHANGE

What connected the Austrian and Russian empires in a physical sense was the Eurasian steppe, running from the Hungarian plain in the west to northern China in the
east. Nineteenth-century geographers understood the Eurasian steppe as quintessen-
tially a land between empires, the scene of decisive battles for global dominance,
much as Afghanistan is seen today. It was the "heartland" in Harold Mackinder's
formulation, the scene of the "great game" in the parlance of British diplomacy, contested by the British, Russians, Austro-Hungarians, Ottomans, Chinese, and Japanese. Culturally, the steppe had even come to be seen by certain German and Anglo-
American scholars as the ur-homeland of the Aryan "race."[19]

It is not surprising, then, that the steppe took on global significance in the eyes
of physical scientists as well, nearly on a par with Antarctica and the oceans as an
iconic site of international geoscientific research.[20] The scientific study of the Russian steppe began in the eighteenth century under Russian patronage and German
leadership. From the German perspective, the study of the eastern steppe was part of
the Enlightenment project of defining Europe against Asia. In this vein, Alexander
von Humboldt identified the central Asian climate, with its extremes of heat and cold,
as "continental," in opposition to the "occidental" climate of western Europe. He thus
formulated in climatic terms the Enlightenment's cultural distinction between temperate West and wild East.[21] Perhaps the most fateful contribution of early German
surveyors of the steppe was their opinion that the region could indeed be cultivated
and settled by "Europeans."

By the 1830s, however, the first cracks in this optimism were visible. The drought
of 1832–4 sparked an international debate over the past and future climate of the
Russian steppes. Pessimists warned of desiccation due to deforestation. Optimists
retorted that the data on rainfall were too sparse, that theories of progressive climate change were unproven, that droughts and famines were the result merely of
climatic "oscillations," or that desiccation was a threat to the soil alone and could
be countered by improved irrigation. Russian geographers generally emerged from
these debates with their optimism about colonization intact. They sang the praises of
the "black earth," predicting bountiful harvests as Russian villages sprang up in the
steppes of "New Russia" (today's southern Ukraine) and later in Turkestan. In the
spirit of regionalization, soil scientist Vassily Dokuchaev mapped Russian Eurasia
into five "natural regions" and further "transitional" zones.[22] At the same time, he and
his colleagues stressed the global scientific significance of the Russian steppe, as part
of the "black-earth steppes of Hungary, Russia, Asia and America."[23]

---

[19] Peggy Champlin, *Raphael Pumpelly, Gentleman Geologist of the Gilded Age* (Tuscaloosa, Ala.,
1994), 165.
[20] Moon, "Environmental History of the Russian Steppes" (cit. n. 16). See also Adrian Howkins in
this volume.
[21] Alexander von Humboldt, *Asie centrale: Recherches sur les chaines des montagnes et la climatologie comparée* (Paris, 1843), 3:32. On Humboldt, see Gregory T. Cushman in this volume.
[22] Moon, "Debate over Climate Change" (cit. n. 15), and Willard Sunderland, *Taming the Wild Field:
Colonization and Empire on the Russian Steppe* (Ithaca, N.Y., 2004), 205.
[23] Moon, "Agriculture and the Environment on the Steppes" (cit. n. 12), 89.

The devastating drought and harvest failure of 1891 briefly shook the confidence of Russian scientists in the fertility of the steppes, but their response was vigorous. Dokuchaev won government funding for the construction of three research stations between the Dnieper and the Volga rivers, which would support the study of "agricultural meteorology," referring to an integrated analysis of soil, water, and air. This was a pioneering move to analyze climate on the small scale yet with a global comparative perspective, not to be matched in central Europe for another two decades.[24] Denying the charge that the climate of the steppe was growing drier through cultivation, Dokuchaev drew up a plan to "improve" the soil through forestry and artificial irrigation, an effort that has been judged an early example of sustainable development.[25]

Meanwhile, the steppes of Russia's southern and eastern edges were attracting influential central European earth scientists.[26] Eduard Brückner, who taught physical geography in Vienna from 1906 to 1927, focused the first of his highly influential studies of climate oscillations on the variability of water levels in the Caspian Sea.[27] Wladimir Köppen, whose father was sent by Tsar Nicholas I in the 1830s to investigate falling water levels in the Volga, began his own scientific career by studying climate and vegetation at his family's summer home in the Crimea. As Köppen assessed the situation in 1895, "Russia's conquests to its west have quickly expanded our knowledge. For the Russian soldier is followed with remarkable speed by the Russian [scientific] observer, and out of the impenetrable dens of thieves, which only a generation ago Vambery could enter only in the guise of a mendicant dervish, have for a long time now been appearing practicable, uninterrupted meteorological data series."[28] Köppen himself contributed fundamentally to the problem of defining the steppe (vis-à-vis deserts) in his seminal scheme of climatic classification. By the turn of the century, several central European scientists, along with a handful of Russian dissenters, had drawn international attention to the question of whether the climate of the Eurasian steppes was indeed deteriorating. Their critiques became part of a controversy over the past and future of central Asian civilization and the wisdom of European colonization in the region.[29]

TURKESTAN AND THE RUSSIAN CIVILIZING MISSION

By the turn of the twentieth century, the attention of central and eastern European geographers was shifting from the Caucasus to the newly settled territory in Turkestan. A series of international teams of geographers, geologists, and archeologists set out to uncover the environmental circumstances that might have caused the decline

[24] Moon, "Debate over Climate Change" (cit. n. 15); Deborah R. Coen, "Scaling Down: Mapping the Austrian Climate between Empire and Republic," in *Intimate Universality: Local and Global Themes in the History of Weather and Climate*, eds. James R. Fleming, Vladimir Janković, and Coen (Sagamore Beach, Mass., 2006), 115–40.

[25] Moon, "Environmental History of the Russian Steppes" (cit. n. 16), 171.

[26] Svetlana Gorshenina, *Explorateurs en Asie centrale* (Geneva, 2003).

[27] On Brückner, see Nico Stehr and Hans von Storch, *Eduard Brückner: The Sources and Consequences of Climate Change and Climate Variability in Historical Times* (Dordrecht, 2000).

[28] Wladimir Köppen, "Die gegenwärtige Lage der Klimatologie," *Geogr. Z.* 1 (1895): 613–28, on 615.

[29] Medical studies of acclimatization were also prominent on the negative side of the debate, with Russians themselves arguing that they, as "Europeans," were ill constituted for the "Asian" environment: Cassandra Cavanaugh, "Acclimatization, the Shifting Science of Settlement," in Breyfogle et al., *Peopling the Russian Periphery* (cit. n. 12), 169–88.

of past civilizations in the region. The Russian expatriate-anarchist-aristocrat and sometime geographer Peter Kropotkin argued that progressive desiccation had destroyed the early civilization of the high steppes, leading to the "great migrations and invasions of Europe" at the end of the Roman era.[30] In the hope of finding the "cradle of Aryan civilization," the American geographer Raphael Pumpelly led a Carnegie-funded expedition to central Asia in 1903–4. One of his companions, the American geographer Ellsworth Huntington, published his *Pulse of Asia* in 1907, arguing, on the basis of his own observations in central Asia, that desiccation was a fluctuating but unstoppable trend.[31]

Turkestan also emerged in the late nineteenth century as a focus of Russian economic development and one of the world's major cotton-producing regions. Yet its image as the empire's new promised land was tarnished by the evidence brought forth by Huntington and others that the region's climate had desiccated within the span of human history. Those claims were challenged by Russia's foremost climatologist, Alexander Voeikov (1842–1916). Voeikov had embarked on a life of world travel after his home university in St. Petersburg was closed in response to student agitation in 1861—a pivotal juncture in Russian history for reform, empire building, and geographical exploration.[32] Voeikov's voyages gave rise to ambitious theories of the global climate system and of the influence of climate on social life, but also to practical efforts to benefit Russian agriculture and tourism. His extensive inquiries into the human modification of the environment reflected the ambitions of a modernizing and expanding empire, not the aesthetic concerns of Western nature preservationists. He argued in particular that unfavorable environments could be improved through rational water use: "Dry countries with irrigation are the countries of the future."[33] Despite the claim by American conservationist George Perkins Marsh that the Aral Sea was sinking because of deforestation, Voeikov insisted that "a large Aral Sea is a *testimonium paupertatis*; it shows that man has not been able to use the immense quantity of water furnished by the melting of the mountain snows, and lets it evaporate without utility for him."[34] In Huntington's opinion, Voeikov had been blinded by his nation's imperial ambitions, but Huntington's own judgment was steeped in environmental determinism. It is worth noting that, in the 1960s, the Soviet Union not only acknowledged desiccation in central Asia but planned to reverse it, by redirecting the flow of Siberian rivers. The project's potential for environmental devastation was massive; but it stalled, fortunately, on account of tenacious resistance from Soviet environmentalists.[35]

The debate over desiccation in Turkestan coincided with the discovery of evidence of glacial retreat in the Austrian Alps. In the late 1890s, the Viennese physical geographer Albrecht Penck made detailed measurements of glaciers near the high-altitude Sonnblick observatory, from which he concluded that the Alps had experienced a

[30] Prince Peter Kropotkin, "The Dessication of Eur-Asia," *Geogr. J.* 23 (1904): 722–34, on 723.
[31] Champlin, *Pumpelly* (cit. n. 19); John E. Chappell, Jr., "Climate Change Reconsidered: Another Look at 'The Pulse of Asia,'" *Geogr. Rev.* 60 (1970): 347–73.
[32] Bassin, *Imperial Visions* (cit. n. 8); A. Kh. Khrgian, *Meteorology: A Historical Survey*, ed. Kh. P. Pogosyan (Jerusalem, 1970), vol. 1.
[33] Quoted in Ellsworth Huntington, review of Voeikov's *Le Turkestan Russe, Bulletin of the American Geographical Society* 47 (1915): 708. Voeikov had made this argument earlier in "De l'influence de l'homme sur la terre," pt. 2, *Annales de Géographie* 10 (1901): 193–215, esp. 193–5.
[34] Marsh, *Man and Nature* (New York, 1864), 299; Voeikov, *Le Turkestan Russe* (Paris, 1914), 116.
[35] Paul Josephson, *New Atlantis Revisited* (Princeton, N.J., 1997), chap. 5.

warming trend in the past quarter century.[36] The second (1897) edition of Julius Hann's *Handbook of Climatology* announced, "We live at present in a period of pronounced retreat of the Alpine glaciers," although Hann was optimistic that this process was about to reverse itself. Hann's widely read handbook turned directly from the issue of glacial retreat in the Alps to that of desiccation in Turkestan, implicitly posing the question of a link between the two phenomena.[37]

FICKER, FICKER, AND RICKMER RICKMERS

Heinrich von Ficker entered this debate in 1908 by way of a passion for mountaineering. Ficker had been a student of geology in Innsbruck when he first traveled to the southern periphery of the Russian Empire in 1903.[38] The conditions of this journey were spartan, and Ficker traveled not as an imperial scientist but as a mountaineer, determined to "conquer" the region's peaks. He and his sister Cenzi joined a party of central Europeans led by Willi Rickmer Rickmers, the man who introduced skiing to Tyrol. Cenzi scaled a 4,700-meter peak and so impressed the local aristocrat that he offered her the mountain itself, "as a gift."[39] Cenzi was soon known as the foremost female alpinist of her era. She returned east with Rickmers in 1907, recording meteorological observations that her brother Heinz would rely on for a study published the following year. In 1913 both siblings would follow Rickmers to Turkestan.

There is no space here for a full account of these journeys and of the themes they raise—such as the relative status of measuring instruments and personal observations and the complexities of communication among international scientific travelers. Here, we should merely note the development of Heinz Ficker's self-consciousness as a central European scientific explorer on imperial Russian territory. Ficker styled himself a *Naturmensch*, comparing himself to those who sought refuge from daily cares in the Alps: "Politics and the like were empty words for us." By contrast, he caricatured Russians as petty tyrants. On one occasion, an order for three horses fell one short, and the hiking party turned for help to a police officer, a "representative of the Russian imperial authorities." Ficker freely mocked the "giant in an officer's uniform," with his "pompous" attitude, "blaring" voice, and habit of wielding his whip against inanimate objects while addressing the native population. Yet Ficker was clearly grateful for the support of Russian officials against the local population, whom he found generally "greedy" and "untruthful." He and his comrades were not above imitating the manner of the Russian police if they thought intimidating the locals would be useful.[40] In short, the relationship between central European and Russian scientific ambitions in the steppes was alternately competitive and symbiotic.

---

[36] Penck, "Gletscherstudien im Sonnblickgebiete," *Zeitschrift des deutschen und österreichischen Alpenvereins* 28 (1897): 52–71; Reinhard Böhm, *Gletscher im Klimawandel* (Vienna, 2007), 100.

[37] Julius Hann, *Handbuch der Klimatologie*, 2nd ed. (Stuttgart, 1897), 364.

[38] On Ficker, see G. Oberkofler and P. Goller, "Von der Lehrkanzel für kosmische Physik zur Lehrkanzel für Meteorologie und Geophysik," *Veröffentlichungen der Universität Innsbruck* 178 (1990): 23–6; F. Steinhauser, obituary for Ficker, *Österreichische Akademie der Wissenschaften Almanach* 107 (1957): 390–402; Hans Ertel, "Heinrich von Ficker zu seinem 60. Geburtstag," *Die Naturwissenschaften* 47 (1941): 697–700. On tourism as a factor in climate research, see Sverker Sörlin and Mark Carey in this volume.

[39] Otto Marschalek, *Österreichische Forscher: Ein Beitrag zur Völker- und Länderkunde* (Mödling bei Wien, 1949), 124.

[40] Ficker, "Aus dem Kaukasus," *Deutsche Alpenzeitung* 3 (1904): 197–213, on 197, 198, 200, 202, 205; Alb. Weber, "Im zentralen Kaukasus," *Jahrbuch des Schweizer Alpen Club* 41 (1905–6): 206–27.

"BORDERLANDS" AND "TRANSITION ZONES"

Ficker's return from the Caucasus coincided with his decision to drop geology in favor of meteorology, a move inspired by the Innsbruck professor Wilhelm Trabert. It was Trabert who steered Ficker toward his dissertation research on the "*föhn* pause." Föhn is a warm, dry wind that strikes most often in the cold months, producing incongruously warm temperatures and allegedly unleashing a variety of human woes (from heart conditions to epileptic seizures). The "pause" referred to a sudden cold snap in the midst of föhn, which can cause wildly fluctuating temperatures, particularly on the north side of the Alps. Ficker soon ruled out the possibility that these cold snaps were lower atmosphere inversions—which result from the cooling of the ground at night, a strictly local radiation effect. By carefully adjusting the scale of observations, Ficker managed to explain the cold snaps as a nonlocal phenomenon, the effect of a cold air mass moving in from the north, filling the northern valley in a shallow layer before moving up the mountain. (Here, "air mass" [*Luftkörper*] refers to a large volume of air that acquires relatively uniform properties by virtue of having formed over a given surface of the earth.)[41]

Visible already in this study was a distinctive research style. According to private remarks by Trabert, Ficker took "a special interest in all that is climatologically important, and namely in *causes*."[42] Ficker was interested above all in understanding how orography constrains the vertical structure of the atmosphere, and it was this interest that would draw him to Turkestan. The high steppes of central Asia provided an "ideal case" for the study of mountain weather, one that was impossible to find in a mountain chain like the Alps, for they demonstrated "the influence of elevation in its purest form."[43] Ficker's explanations also bore the mark of the mountaineer, tending to describe changes in weather in terms of the modification of air masses as they traversed peaks and valleys. In this vein, during his research in Turkestan in 1913, Ficker often sketched the silhouette of a mountain in order to contemplate its meteorological effects. Likewise, he repeatedly contrasted landscapes that were "open" to air masses with those that were closed, going so far as to write of "invasions" of air into open valleys.[44]

These methods of research and description proved well suited to Ficker's contribution to the *Climatography of Austria*, announced in 1901 as "a monumental work, which will give a detailed picture of the climate of the so varied parts of our kingdom, for the benefit of all."[45] In his *Climatography of Tyrol*, Ficker used the Brenner Pass, the link between north and south Tyrol, to illustrate the monarchy's glorious natural diversity:

[41] Ficker, "Innsbrücker Föhnstudien," *Denkschriften der kaiserlichen Akademie der Wissenschaften, mathematisch-naturwissenschaftliche Klasse* 78 (1905): 84–163, and "Der Transport kalter Luftmassen über die Zentralalpen," *Denkschriften der kaiserlichen Akademie der Wissenschaften, mathematisch-naturwissenschaftliche Klasse* 80 (1907): 131–95. On the development of "air mass climatology," see, e.g., Erwin Dinies, "Luftkörper-Klimatologie," *Archiv der deutschen Seewarte* 50 (1932): 6.

[42] Trabert to W. Koeppen, 22 August 1909, Köppen Nachlass, Universitätsbibliothek, Graz.

[43] Ficker, "Zur Meteorologie von West-Turkestan," *Denkschriften der kaiserlichen Akademie der Wissenschaften, mathematisch-naturwissenschaftliche Klasse* 81 (1908): 533–59, on 557.

[44] Ficker, "Untersuchungen über die meteorologischen Verhältnisse der Pamirgebiete (Ergebnisse einer Reise in Ostbuchara)," *Denkschriften der Akademie der Wissenschaften, mathematisch-naturwissenschaftliche Klasse* 97 (1921): 151–255, on 157; see his line drawings in that publication.

[45] Wilhelm von Hartel, "Rede," *Österreichische Akademie der Wissenschaften Almanach* 52 (1902): 371–4, on 373.

> For the climatographer Tyrol is one of the most interesting regions of the monarchy. The climatic contrasts are stark and yet in their distinctive features mostly a blessing for the province. To assess these differences, to represent them in numbers, is an attractive task for the climatologist. Yet the contrast imprints itself in memory far more strongly when a spring journey of a few hours across the Brenner reveals to the senses the image of an abrupt change of climate like no other. The climatologist's burden is lightened in this way, since his numbers mingle effortlessly with the common ideas that every educated person almost intuitively associates with the concepts "north Tyrol" and "south Tyrol." Stark contrasts make the climatologist's work easier, particularly in regions that belong, thanks to their scenic beauty, to the most famous of the continent.[46]

It is the gaze of the imperial traveler that makes this particular climatic boundary so vivid. The quoted passage suggests continuity rather than mere contrast, demonstrating the imperial scientist's eye for "transition zones." Echoes of this passage would appear in Ficker's studies of Turkestan, where the contrast between the region's east and west, separated by the "borderland" (*Grenzgebiet*) of the Pamirs, seems to parallel that between north and south Tyrol.[47]

THE QUESTION OF ANTHROPOGENIC CLIMATE CHANGE

Ficker had never set foot in Turkestan when he first composed a portrait of its climate. The basis for his conclusions was the records of the imperial Russian meteorological stations in Turkestan, which had numbered twenty-seven in the 1870s, but had dwindled to twelve by 1913, allegedly because of "lack of interest" from the provincial government.[48] Only one was located in the high steppes (Pamirski Post, 3,640 meters). Ficker's argument was nonetheless unequivocal: Russian Turkestan was a "dying," "forsaken" land. This mountainous, arid region owed its fertility to runoff from melting glaciers, but these glaciers were shrinking, and, as a result, the region faced irreversible desiccation. Its present fertility was due solely to artificial irrigation, an unsustainable practice: "For the more water that I artificially divert to a region today, the more I deprive another region and deliver it up to destruction."[49]

Voeikov responded aggressively to Ficker, though he too knew Turkestan only secondhand. And, as he conceded, he was no mountaineer: when he finally journeyed to Turkestan in 1912, he traveled in a private wagon of the Trans-Caspian Railway, supplemented by coaches and automobiles.[50] Nor was Voeikov shy about defending Russia's imperial aims in central Asia. "The Russian rule of the past 35 to 40 years has given peace to this land; there are no longer feuds between the Khanats, no longer raids by nomads."[51] Agriculture was blooming, the canals on which it depended were being widened, and still the lakes of the region were rising. Turkestan's climate surely oscillated in Brückner's sense, but there was no evidence there of progressive

[46] Ficker, *Klimatographie von Tirol und Vorarlberg* (Vienna, 1909), 116.

[47] Ficker, "Pamirgebiete" (cit. n. 44), 152.

[48] Franz Xaver von Schwarz, *Turkestan, die Wiege der indogermanischen Völker* (Freiburg im Breisgau, 1900), 552. Ficker's other sources included Schwarz's book, the writings of the Danish explorer Ole Olufsen, and his sister Cenzi's observations.

[49] Ficker, "Zur Meteorologie von West-Turkestan" (cit. n. 43), 558.

[50] Voeikov, *Turkestan Russe* (cit. n. 34), vi.

[51] Voeikov, "Ist eine fortschreitende Austrocknung Turkestans vorhanden?" *Meteorologische Zeitschrift* 25 (1908): 567–8, on 568.

climate change.[52] Not that Voeikov denied the possibility of anthropogenic climate change at the regional scale. He had even warned that Bosnia faced desiccation and warming through the destruction of forests under Austro-Hungarian occupation (a reminder of the conflicting Russian and Austrian agendas in the Balkans).[53]

Several years later, Ficker countered that it was impossible to prove the *absence* of progressive climate change in Turkestan: "Even if we had thousands of years of precipitation measurements from Turkestan, and could find no reduction in precipitation in this span of time, nonetheless this period could have been an era of progressive desiccation for the valley, naturally setting aside occasional fluctuations."[54] According to Ficker's own theory, the relevant decrease in precipitation would have occurred at the end of the ice age, well before any desiccation was visible in the valleys, and well before written records began. In any case, Ficker insisted, his goal was not to settle the historical debate: "This account deals naturally only with the climatic possibility of such a development; as far as answering the question of whether this in fact took place, the meteorologist's opinion carries little weight."[55]

Not the past but the future was his principal concern, and so Ficker returned in his dispute with Voeikov to the issue of artificial irrigation. He argued that an empire could not afford to risk the economy of one region for the benefit of another: "The economic outlook of the country as a whole cannot be raised by privileging one region at the cost of another."[56] Thus, on both sides of their debate lay the perspective of a continental empire. To Voeikov, Turkestan was evidence of the benefits that enlightened rule and "wise use" policies could bring to Asia. To Ficker, Turkestan stood instead as a reminder of the interdependence of an empire's regions and of the imperative to sustain its natural diversity. It was a reminder, too, of the "many difficulties" a climatologist faced without firsthand experience of a region.

CHASING COLD WAVES

Ficker recognized that the Russian Empire, with its vast land mass and sharp diurnal temperature variability, was a particularly good laboratory in which to pursue his interest in cold waves; further west, their traces vanished beyond the Atlantic coast. While contemporaries like Vilhelm Bjerknes and Napier Shaw were studying atmospheric discontinuities with an eye to forecasting, Ficker framed his question principally in geographic terms: to "seek and delimit the region from which the cold waves develop" and the "borderlands [*Grenzgebiete*] . . . to which the cold air penetrates."[57] To that end, he worked through two hundred synoptic weather maps, tracking approximately fifty cold waves.

Ficker's method was drawn not from the tool kit of the weather forecaster but

[52] Ibid.; Voeikov, *Turkestan Russe* (cit. n. 34), 115.

[53] Voeikov, "De l'influence de l'homme" (cit. n. 33), 202.

[54] Ficker, "Pamirgebiete" (cit. n. 44), 254.

[55] Ibid., 253. See too Ficker, "Gegewärtige und eiszeitliche Vergletscherung in den westlichen Pamirgebieten," *Geografiska Annaler* 17 (1935): 300–5, on 304.

[56] Ficker, "Die Austrocknung Turkestans," *Meteorologische Zeitschrift* 26 (1909): 216–7, on 217.

[57] Ficker, "Die Ausbreitung kalter Luft in Rußland und Nordasien," *Sitzungsberichte der österreichischen Akademie der Wissenschaften, mathematisch-naturwissenschaftliche Klasse* 119 (1910): 1769–837, on 1770; Robert Marc Friedman, *Appropriating the Weather: Vilhelm Bjerknes and the Construction of a Modern Meteorology* (Ithaca, N.Y., 1989), 92–4, 169–74.

from that of the climatographer. What he traced on his Eurasian maps were not isotherms, the curves that observatories churned out in reams daily, but isallotherms,
lines of equal interdiurnal variability of temperature. For each case, Ficker drew lines
connecting the stations that reported sudden coolings on the same day and considered these lines as the vanguards of the cold wave—just as he had earlier done on a
smaller scale when connecting the stations at which a föhn pause appeared simultaneously (see fig. 1). Thus, nowhere on a typical weather map of the time were Ficker's cold waves visible. Moreover, Ficker refused to treat the station data statistically,
in typical meteorological fashion. Instead, he took a geologist's approach, identifying a small number of "typical" cold waves.[58] As Bjerknes recognized, a remarkable
aspect of Ficker's work was his ability to perceive a discontinuity where the weather
maps of the day all indicated continuity.[59] For Ficker, schooled in the regionalizing
gaze of the continental-imperial scientist, central Asia was fundamentally a land of
boundaries, of "great contrasts," where mountains confronted desert.

Prior studies had established that winter weather in Russia and central Asia was
dominated by the effects of a high-pressure center in Siberia, and Ficker therefore
expected to find that Siberia was the source of the cold waves.[60] To his surprise, their
"actual source" lay instead in the northern polar basin. By taking a similar approach
to warm waves, he determined that cold and warm waves followed each other, moving together as a system from west to east. From this combination of synoptic analysis and physical geography, a flurry of crucial insights followed. Most impressively,
he was able to sketch the temperature and pressure distribution that result when these
two types of air mass converge. He was thus able to show that when a warmer air
current flows over a colder one, there arises an isobaric depression—the center of a
cyclone. What he could not yet do was explain *why* this happens. That would be the
Bergen school's contribution: the full explanation of the development of the depression, or cyclogenesis.

Unlike Bjerknes, who dedicated himself to demonstrating the universal predictive power of his frontal method and exporting it throughout the world, Ficker—
even when discussing fundamental physics—was motivated by an interest in particular places and what he called their meteorological "peculiarities." He thus took
it to be "of significant meteorological and especially climatological interest" that
changes in weather at such widely separated points as southern Russia and Man-
churia could be tied to the same cold wave. Indeed, he identified as the principal
lesson of the study "that the cold waves originate not in northeastern Siberia but in
the northern Arctic, and that they follow particular paths in their travels."[61] As was
typical for a continental-imperial field scientist, he was particularly interested in the
geographical lesson that a boundary "between the cold polar [region] and the warm
southern region" one could locate on the map, with its northern border in European

[58] Wilhelm Trabert, "Gutachten über die Dissertation von H. Ficker," reproduced in Oberkofler and
Goller, "Von der Lehrkanzel für kosmische Physik" (cit. n. 38), doc. 8.
[59] "At that time Ficker caused a sensation with what we would now call the 'Polar Front.' However,
faced with the abundance of daily weather maps—which all seemed to point to continuity—even in
these signs of discontinuity we were able to see only interesting exceptions, and made no attempt to
follow up on this important clue." Bjerknes, 1933, quoted in Ertel, "Heinrich von Ficker" (cit. n. 38),
698.
[60] E.g., V. Kremser, review of J. Kiersnowskij, *Die Windverhältnisse im Russischen Reiche*, *Meteorologische Zeitschrift* 13 (1896): 485–8.
[61] Ficker, "Die Ausbreitung kalter Luft" (cit. n. 57), 1837.

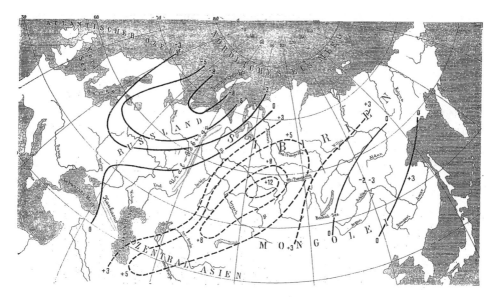

Figure 1. *An illustration of the movement of a warm wave across Eurasia by means of "isal-lotherms" (lines of equal interdiurnal variability of temperature). This was a climatological and geographic approach, in contrast to the more familiar "isotherms" (lines of equal temperature) used by weather forecasters of the day. In this way, Ficker was able to visualize a series of atmospheric discontinuities that remained hidden on forecasting maps. Reprinted from Ficker, "Fortschreiten der Erwärmungen" (cit. n. 62).*

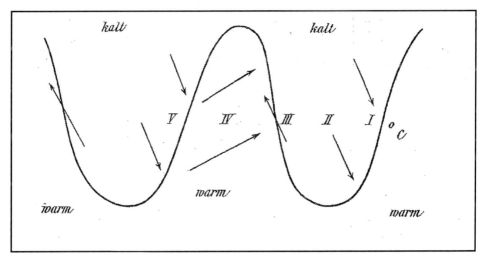

Figure 2. *Diagram of cyclone formation. Reprinted from Ficker, "Fortschreiten der Erwär-mungen" (cit. n. 62).*

Russia; a *"border zone . . .* that is influenced alternately by its *border regions."*[62] At the same time, he pointed to the significance of his discovery for an understanding of the general circulation of the atmosphere, or what we would call the global climate system.[63] Namely, the warm waves from the southwest appeared to be part of the circulation of warm air from the equator toward the North Pole, while the cold waves returned cold air from the pole to lower latitudes.[64] Unlike Bjerknes, then, Ficker framed his analysis of cyclone formation as a climatological insight—remarking, for instance, that warm waves are "climatically of equal importance" to cold waves. He refused to abstract his model of cyclone formation from its geographic context.[65]

As Ficker recognized, that context was ultimately global, but he was no less concerned with the regional implications of his account of cold waves. His work was not done when he had succeeded in scaling up to a hemispheric model. To the contrary, the real work of interpreting the model was just beginning, and again it was work to be done with the feet.

THE ALPS IN ASIA

In the summer of 1913 the German and Austrian Alpine Association sent Rickmers to the Alai-Pamirs of Bukhara (a Russian protectorate in western Turkestan), as the leader of a team of experienced climbers and scientists, including Ficker and his sister Cenzi. In order to justify this choice of destination, the team's geologist, Raimund von Klebelsberg, argued that the Alps themselves extended into Asia, and so, therefore, did the domain of central European science. "Through the entire Eurasian continent runs the system of the Alps, and only a small part of it are the Alps themselves." He went on to trace for his fellow alpinists a geological continuity from the Alps to the Carpathians, plunging under the Black Sea and up again through the Caucasus, down again through the Caspian Sea and up, finally, into central Asia, where the trace became harder to follow. There was no question, though, that "all these mountains meet in the highlands of the Pamir," and therefore that "the system of the Alps is inextricably connected to that of central Asia." "Only on the basis of comprehensive knowledge will one be able to draw further conclusions and approach closer to the grand questions that the Alps and the mountains of central Asia have in common," including those of climate and glaciation. "In short, an important chapter in the evolution of the earth as a whole is tied to our problem."[66] The expedition's perspective was thus defined as holistic, even hemispheric, but the path to such a comprehensive vision led through detailed, comparative experience of an unlikely pair of regions three thousand miles apart.

Rickmers and his followers found many points of reference for their exotic surroundings in their beloved Alps, from "relatives of Alpine plants" to mountain forms

---

[62] Ficker, "Das Fortschreiten der Erwärmungen (der Wärmewellen) in Russland und Nordasien," *Sitzungsberichte der Kaiserlichen Akademie der Wissenschaften* 120 (1911): 745–836, on 829, my emphasis.

[63] Gisela Kutzbach, *The Thermal Theory of Cyclones: A History of Meteorological Thought in the Nineteenth Century* (Boston, 1979), 216.

[64] Ficker, "Das Fortschreiten der Erwärmungen" (cit. n. 62), 831.

[65] Ibid., 745.

[66] R. v. Klebelsberg, "Die Pamir-Expedition des Deutschen und Österreichischen Alpenvereines vom geologischen Standpunkt," *Zeitschrift des deutschen und österreichischen Alpenvereines* 45 (1914): 52–60, on 52–3.

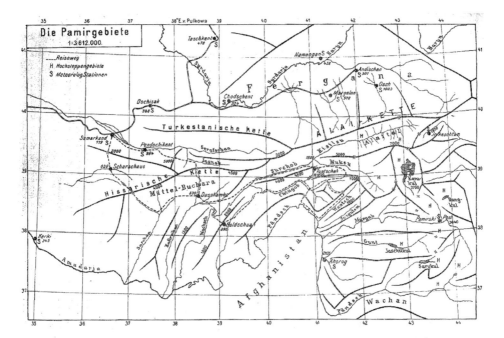

Figure 3. *Map of the route of the Pamir expedition of the German and Austrian Alpine Association in 1913. Reprinted from Rickmers, "Vorläufiger Bericht" (cit. n. 67).*

that reminded them of specific Tyrolean peaks. Yet comparison was only one facet of the expedition's methods. More important was their regionalizing perspective. They approached the exploration of Turkestan as a problem of spatializing natural (and human) difference in terms of boundaries and transitional zones. "The population is now growing increasingly Tajikish," remarked Rickmers in his journal, interpreting the population in terms of an ethnic transition zone.[67] Analogously, Ficker organized his climatological observations by tracking borders—of glaciers, vegetation, crops, even human settlements. For the climatologist, he argued, fixing such boundaries could potentially "replace averages of long series of observations," by indicating displacements of climate zones and thus climate change over time.[68]

Equally important, the perception of boundaries could bring into focus phenomena at multiple scales at once. "Extraordinary insights occur when we perceive relationships across and between boundaries."[69] Consider, for instance, Ficker's judgment that climate zones in central Asia could more effectively be defined according to average cloud cover than average rainfall. Tracking cloud cover brought into focus a boundary between east and west Turkestan: the former was drier, with sparser precipitation, yet cloudier than the latter. Having redrawn the climate zones, Ficker was in a position to track the phenomenon responsible for this geographic contrast. The

[67] Willi Rickmer Rickmers, "Vorläufiger Bericht über die Pamir-Expedition des Deutschen und Österreichischen Alpenvereins 1913," *Zeitschrift des deutschen und österreichischen Alpen-Vereins* 45 (1914): 1–51, on 20.

[68] Report on Ficker's lecture to the Gesellschaft für Erdkunde in Berlin, *Die Naturwissenschaften* 2 (1914): 827–8, on 828.

[69] Mitchell Thomashow, *Bringing the Biosphere Home* (Cambridge, Mass., 2002), 98.

cause of the summertime maximum in cloud cover in Mongolia and Tibet "cannot be sought in local phenomena," he argued. It was due instead to monsoons in east and south Asia. West Turkestan, on the other hand, was protected from monsoons by its mountain barrier.[70] In this way, Ficker's realignment of regional boundaries drew his attention to atmospheric relations on a continental scale. In this example, as in Ficker's research on the föhn pause, his investigation proceeded as an effort to distinguish between effects that were truly local in origin and those that merely appeared to be local, from too narrow an observational horizon.

Regionalizing also meant defining Turkestan and Tyrol against each other as climatic regions, whence one could generalize about the climatic effects of "high steppes" versus "mountain chains." To this end Ficker offered a sophisticated analysis of the multiple competing effects of mountains on the temperature of their enveloping air: the warming effect due to conduction of heat from ground to air and to wind shielding by neighboring peaks; the cooling effect due to the blackbody radiation of heat away from the ground (which increases more rapidly with elevation than does the warming effect of irradiation); the loss of heat by air currents forced upward by the mountain; and the convective effects as cooled air sinks down the mountainside while warmed air rises higher into the atmosphere. Simplifying, Ficker reduced this to a problem of weighing against each other, and tracing the trajectories of, "the warm and cool air that the mountain makes."[71] Resorting to a geological style of argument and to the visualization of mountain forms, he contrasted the climatic effects of two types of elevation, the first being "characteristic" of the Pamirs (see fig. 4). As Ficker pointed out, it was because of this orographical effect that temperature changes in Russia and north central Asia, like those he had tracked in his study of cold waves, were so sudden and sharp: the high plains acted like a dam to the flow of cold air across Eurasia.

Ultimately, then, Ficker's attempt to define Tyrol and Turkestan against each other as "ideal type" regions produced more than a comparative perspective. The regionalizing framework led him to try to disentangle local, stationary effects due to radiation from nonlocal, dynamical effects due to the movement of air masses. Further, the act of defining regions in relation to each other drew his attention, in a final comment, to a potentially global framework: "Naturally, there must then be regions of the earth's surface whose influence on the temperature of air masses compensates for the influence of mountain masses. With that however one arrives at questions that lead far beyond the limits of this investigation."[72] The power of the sciences of regionalization was that they brought regional differences into focus within a wider observational horizon.

CONCLUSION

As the Alpenverein party stressed in 1913, Turkestan was a land between empires. Scientifically, it fell into the interstices between the networks of Russia and British India. By the time full accounts of the expedition's findings could be published, the Russian Empire was no more. Nonetheless, the habits of imperial science lived on.

[70] Ficker, "Untersuchungen über Temperaturverteilung, Bewölkung und Niederschlag in einigen Gebieten des mittleren Asiens," *Geografiska Annaler* 5 (1923): 351–400, on 388.
[71] Ficker, "Pamirgebiete" (cit. n. 44), 197.
[72] Ibid., 198.

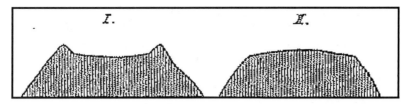

Figure 4. *Diagram of two types of high plains, concave and convex (reprinted from Ficker, "Pamirgebiete," cit. n. 44): type I, with narrow valleys, acts as a receptacle for cold air in winter, resulting in a reversal of the vertical structure found in the Alps. Instead of temperature inversions in which mountain peaks are warmer than neighboring valleys, type-I elevations are unusually cold in winter. Type II, on the other hand, allows the cold air to flow down in winter, and thus more closely resembles a "free peak" with its inversions.*

Ficker himself had been trapped between empires in World War I, taken prisoner by the Russians early in 1915 after an attempted escape by balloon from the besieged Galician fortress at Przemysl. Interned in Kazan, Ficker was permitted, thanks to an intervention by the Swedish king, to consult Russian meteorological publications. He was thus able to continue his analysis of the station data for Turkestan. He learned Russian and was even elected a member of the Soviet Academy of Sciences after the war. In 1919 Ficker raised the question of when the new Russia would take up the scientific traditions of the former empire. For now, he stressed, the network of western Turkestan represented "the only connection between the European and Indian networks."[73] Analogously, back home in Vienna, Ficker and his colleagues were busy negotiating the dismemberment of the Habsburg network among the empire's successor states. Yet Ficker's research was still driven by the concerns and language of imperial geography, with its regionalizing impulse and its attention to transitional zones: "The preceding studies are intended to be nothing more than a contribution to the effort to connect the old Russian network to the Indian-East Asian one, and to explore the transition zone, since the borders between climatic regions do not always follow the great mountain chains."[74]

The last section of his paper gathered together the many signs of desiccation, from the shrinking of forests to the retreat of glaciers. He was more convinced than ever that Turkestan was a dying land. But now Ficker saw Turkestan's problems in a wider frame: "In no case however will what man destroys in this region be replaced by nature; indeed, how quickly under such conditions whole regions can become permanently void of trees is clearly shown by the Karst region and by many stretches of the southern Alps."[75] To Ficker it seemed that Turkestan was following a course of environmental devastation that had already wreaked isolated havoc in central Europe. Thus Ficker did not stop at cautioning Russians against overexploitation of their natural resources; he also hinted at the need for conservation in regions of Austria—in Bohemia and south Tyrol, specifically. Instead of a contrast between enlightened West and uncivilized East, he saw two regions suffering a related fate, linked by a global climate system.

---

[73] Ibid., 152.
[74] Ficker, "Untersuchungen über Temperaturverteilung, Bewölkung und Niederschlag" (cit. n. 70), 392.
[75] Ibid., 246.

As an explorer in the service of a continental empire, Ficker recorded his meteorological observations through a regionalizing lens. He worked in a comparative framework, reasoning back and forth between Tyrol and Turkestan, in order to define each against the other. Defining a region was a holistic enterprise that involved tracking borders of various kinds, natural and human. It was necessary to pursue each phenomenon at a different scale of observation—from the local scale on which one could witness the death of a juniper forest, to the regional scale on which desiccation in Turkestan as a whole became visible, to the hemispheric scale on which Ficker followed cold waves across Eurasia. The regionalizing approach thus offered field scientists (1) a malleable, intermediate scale between local and global analysis, (2) a ready framework for comparative analysis, and (3) an integrated approach to the natural and human sciences.

EPILOGUE: CLIMATIC EXPRESSIONISM

Ficker presented his results on the Pamirs in 1919, the year of the founding of the League of Nations—a moment that, in retrospect, heralded the systematic internationalization of science and the end of the heroic age of individual exploration. Against this background, his conviction that the meteorological travel narrative "will become a very popular [genre]" may seem incongruous. Of course, what he could not have foreseen was the rise of computers and satellites, with their unprecedented power to synthesize data and support global climate models. But even before the computer age, the globalization of atmospheric science would mean the abstraction of meteorological knowledge from the people and places that produced it. Bjerknes's "polar front," a concept that began its rise to fame in 1919, is a case in point. Ficker and Bjerknes both identified a hemispheric discontinuity between polar and tropical air and recognized its role in generating cyclones. But Ficker focused on the climatological—one might even say ecological—implications of this finding. The implications he stressed were of the connection between distant places and the dependence of weather on a finely balanced global system. Seeking the origin of cold waves in Eurasia was, for Ficker, part of a broader quest to understand the connections between fragile climates like those of Tyrol and Turkestan. Bjerknes, by contrast, hailed the polar front more narrowly and instrumentally as the key to storm prediction.

With the triumph of Bjerknes's interpretation in the 1920s came a loss of meaning. Gone was the incentive to draw out the implications of this phenomenon at a regional, even local scale. The age of weather prediction was dawning, demanding that meteorological observations be abstracted and reduced to the neat numbers that calculators (human or mechanical) could manipulate. Meteorologists and physical geographers would soon part ways. Ficker's report belonged to a different age, brimming as it was with observations and digressions that resisted abstraction and quantification. His climate science remained a geographical enterprise in the style of a continental empire; hence his decision to pack in as much as possible—instrument curves, fine print, footnotes, and all.

Or not quite all: "I have refrained from printing the travel observations in their entirety," he wrote, "not only in consideration of the limits of the investigation, but also because it would mean publishing a great deal of unnecessary deadweight." Left in an archive, however, are a dozen pages of unpublished, partly illegible notes, apparently from the Pamir voyage of 1913.[76] Their themes and style evoke the literary and philosophical circle around Ficker's brother Ludwig and his journal *Der Brenner*, founded in 1910 as a forum for cultural criticism, expressionist lyrics, and Tyrolean landscapes. Ficker's journal entries shift between descriptions of landscapes and the moods they evoke and telling details of what Rickmers called the "piquant little amusements" that the party indulged in.[77] Here there is space only for the briefest discussion of one entry, titled "In Almalik (Pass in Central Asia/Persia)":

> . . . The Peter the Great chain. A proud man, who was only allowed to occupy
>   proud mountains . . .
> Science
> is something beautiful, especially when one can do it like us
> on horseback.
> Before we reached the central mountains of Almalik,
> In the middle of a blooming meadow
> Seated securely at saddle-height, we forgot
> Science.

The historian of science, eyes straining with effort, might be tempted to stop reading at the point where the author "forgot science." But the question of what comes next, of what lies beyond science, is too alluring.

> Science led us here.
> But was not at the beginning.
> If I had only known exactly, only something amusing, cheerful than [*sic*]
> The desire for new knowledge? . . .
> So beautiful, so indescribably beautiful the mountains are!

Now the historian seems to be committing an indiscretion, watching as the scientist swoons before a summit he describes as "powerful, commanding," as a "Tamerlane of the mountains":

> There is something more beautiful, far more beautiful than great Science:
> Great mountains.
> In the shade of a tree stood a young girl, as if
> Petrified, with large eyes lightly clothed.
> In an unsettling snowfall her cheeks glowed and, helpless, she gazed with brown,
>   gentle eyes.
>
> Like an Auerhahn chick[78] the beautiful child stood before me and raised her gaze
>   and lowered it again.
> The girl on the Almalik! My heart grows so warm when I think of that moment
> A wish, though it seemed to me there was no desire in me,
> My cheerful, sinful desire.

Perhaps, the historian considers, it is time to replace the manuscript in its folder and return to serious research. But the question tugs at us as it did at the author: What

---

[76] Ficker, "Pamirgebiete" (cit. n. 44), 152.
[77] Rickmers, "Vorläufiger Bericht" (cit. n. 67), 2.
[78] Known to hunters as a particularly wary bird.

bearing does an experience of a single moment in a single place, a sensual, speech-less moment, have on "science"?

What was mere science, mere mountains?

Here Ficker echoes Carl Dallago's formulation in the second issue of *Der Brenner*: " . . . science, when it is authentic; but then it is no longer mere science."[79] The pas-sage works reflexively to underline the loss of meaning entailed in shifting from a personal to a scientific register.

These lines also speak poignantly of Ficker's intuition of historical transition. Looking back on their journey of 1913 from the late 1920s, Rickmers would point to a sudden shift between two eras, "from the voyage of discovery to the voyage of workmanship, from the scout to the mobile office, from the hussars to the tanks of science."[80] Such a contrast could emerge only in retrospect, yet already in 1913 Rick-mers had hinted that the role of the explorer was in jeopardy. A large expedition with a strict "division of labor" left little room for personal freedom. To Rickmers, it was Ficker, "the Alpine scout," who represented freedom from such constraints:

> The geologist needs freedom in order to seek, like a restless spirit, those stony lines out of which he assembles his great picture of nature. In many ways the best-off is the Alpine scout, who forages in all corners and is able to taste the joy of discovery. Ficker took over this task with much skill and great success.[81]

Ironically, Ficker's next and last trip to the Russian East came not as an Alpine scout but as a prisoner of war. His research during his three-year captivity would be re-stricted to calculations based on published data. Thus, in Ficker's postwar publica-tions, calculation acquired a taint of imprisonment, while firsthand observation was all the more firmly associated with a lost age of freedom and heroism.

Yet already in the Pamirs in 1913 Ficker seemed conscious of the passing of the imperial age. This was the underlying theme of his musings on desiccation and civili-zational decline in Turkestan, past and present. His references to Peter the Great and Tamerlane in his journal are tinged with irony, as the mountains dwarf the spatial and temporal scale of human life. Ficker himself struck a blow at Peter's legacy when he determined that the "Peter the Great range" was not one but two mountain chains and proposed to apply the tsar's name only to the western one. For the eastern chain he suggested the name Catherine the Great—an appropriate choice, given that his com-panions included two female mountaineers. Even his pleasure in the peasant girl on the Almalik pass seems to be tinged by the fear of impotence—of "a wish" but "no desire."

At one level, then, Ficker's sketch of Almalik links the fading imperial age to a more capacious understanding of science. By the same token, the very act of record-

[79] Dallago, "Nietzsche und der Philister," *Der Brenner* 1 (1910): 25–31, on 26.

[80] Willi Rickmer Rickmers, *Alai! Alai! Arbeiten und Erlebnisse der Deutsch-Russischen Alai-Pamir-Expedition* (1930), quoted in a post by the historian of central Asia Thomas Loy on the Web site "Tethys: Central Asia Everyday": "Willi Rickmers und die deutsch-sowjetische Alai-Pamir Ex-pedition von 1928," December 19, 2008, http://www.tethys.caoss.org/index.php/2008/12/19/willi -rickmer-rickmers-und-die-deutsch-sowjetische-alai-pamir-expedition-von-1928 (accessed 24 No-vember 2010).

[81] Rickmers, "Vorläufiger Bericht" (cit. n. 67), 27.

ing this scene in a private diary was a concession to the modern demand to divide the personal from the scientific. And yet his journal attests that what lies "beyond science" is far more than "dead weight." He was, in effect, preserving observations that were unpublishable in his own day for a time when the sciences themselves might be better able to reconcile the physical and human dimensions of the environment.

The Anxieties of a Science Diplomat:

Field Coproduction of Climate Knowledge and the Rise and Fall of Hans Ahlmann's "Polar Warming"

by Sverker Sörlin*

ABSTRACT

In the decades between the world wars there were several attempts to document and explain perceived tendencies of atmospheric warming. Hans Ahlmann, a seminal figure in modern glaciology and a science policy adviser and diplomat, constructed a theory of "polar warming" using field results from glacier melting in the Arctic. This article aims to link the rise and fall of "polar warming" with Ahlmann's style of fieldwork. In Ahlmann's view, fieldwork should (1) enhance credibility of polar climate science by emulating laboratory methods and (2) secure knowledge in remote places through collaboration with local residents and fieldworkers. The bodily nature of this style of knowledge production turned out to be an asset in establishing Ahlmann's theory of polar warming but ultimately proved nonresilient to theories of anthropogenic climate change, which became influential from the 1950s onward.

PRECISION, TRUST, AND GLACIOLOGY

The field sciences occupy a special position in the "social history of truth,"[1] not least with regard to the way knowledge is shaped at the intersection of theoretical understand-

* Division of History of Science and Technology, Royal Institute of Technology, SE-100 44 Stockholm, Sweden; sorlin@kth.se.

I would like to thank the participants at the Colby College "Climate Anxieties" workshop in April 2009, including the editors, for helpful comments on earlier versions of this article. I would also like to acknowledge support from the Bank of Sweden Tercentenary Foundation for a grant to Michael T. Bravo and myself at the Scott Polar Research Institute, Cambridge, which hosted our project "Polar Field Stations and IPY History, Culture, Heritage, 1882–Present" (the project's Web site is at http://museum.archanth.cam.ac.uk/fieldstation/). Thanks are also due to all the participants in the project who shared views on the role of field science installations at seminars in Cambridge during 2006 and 2007. I thank as well the participants in the project "Colonialism, Empire, Environment: A History of Arctic Science in the 19th and 20th Centuries," which was part of the BOREAS program funded by the European Science Foundation in 2007–9, and in particular its coordinators, Ron Doel at Florida State University and Urban Wråkberg at the Barents Institute, Kirkenes, Norway. I also acknowledge financial support from the Swedish Research Council for my project "Ice and Credibility" and the Swedish Research Council Formas for my project "Models, Media, and Arctic Climate Change." Finally, thanks to Gudmundur Hálfdánarson, University of Iceland, for comments on Icelandic language intricacies.

[1] Stephen Shapin, *A Social History of Truth: Civility and Science in Seventeenth-Century England* (Chicago, 1994).

ing, bodily practice, and field narrative.[2] In this article I will follow the development of a culture of precision in glaciological fieldwork through the long and successful career of glaciologist and science diplomat Hans Wilhelmsson Ahlmann (1889–1974), who became professor of geography at Stockholm University in 1929, turned to science advice in the 1940s, and served as Sweden's ambassador to Norway from 1950 to 1956.[3]

As a scientist Ahlmann was one of the first to provide empirical documentation of ongoing climate change in the Arctic and the North Atlantic region in the 1920s and 1930s, and he pursued an intense research program to that effect. He also developed his own particular theoretical understanding of climate change, a phenomenon he spoke of as a "polar warming." This notion, however, did not fare well over the long run, as it did not, even in the slightest way, become a part of the theory of human climate forcing, although both theories acknowledged the warming trend. During the 1950s Ahlmann's idea faded into the background.

This is not just an irony, or a tragedy, that can serve as material for a narrative of the ever-changing conjunctures of climate science. More important, it raises questions about how we can move away from looking at major science issues in a genealogical fashion and instead locate them in the complexities of contemporary science practice and politics. What did Ahlmann's innovative research fail to achieve? Why did his groundbreaking work fail to provide a comprehensive climate-change science?

These questions are even more pertinent because by most contemporary standards Ahlmann was, in fact, right. He became one of the world leaders of his field: he was invited to write the foreword to the first issue of the new International Glaciological Society's journal in 1947, and he crowned his career with the position of president of the International Geographical Union from 1956 to 1960.[4] In Stockholm he formed a group of well-trained followers to carry his torch. He secured professional networks and built institutional stability for Arctic and climate science not just in his own country but in other countries as well, notably in Norway. He even worked closely with local Arctic and Scandinavian residents, paying tribute to inherited field skills and traditional knowledge. But his scientific brainchild, polar warming, was in the late 1950s already as doomed as the dodo, and with the advent of global change theories and computer-based modeling in the following decades it became all but forgotten, surviving only in a rudimentary form as global warming's "Arctic amplification," which has recently been established as also having a human origin.[5]

POLAR WARMING AND AHLMANN'S ANXIETIES

In early September 1939, at the outbreak of World War II, Hans Ahlmann was returning from glaciological fieldwork in northeast Greenland. Following North Atlantic

---

[2] Michael T. Bravo and Sverker Sörlin, "Narrative and Practice: Introduction," in *Narrating the Arctic: A Cultural History of Nordic Scientific Practices*, eds. Bravo and Sörlin (Canton, Mass., 2002), 3–32. See also, for somewhat different perspectives, Henrika Kuklick and Robert E. Kohler, "Introduction," in *Science in the Field*, eds. Kuklick and Kohler, *Osiris* 11 (1996): 1–13.

[3] For general information on Ahlmann, see Sverker Sörlin, "Hans W:son Ahlmann, Arctic Research and Polar Warming: From a National to an International Scientific Agenda, 1929–1952," in *Mundus Librorum: Essays on Books and the History of Learning* (Helsinki, 1996).

[4] Ahlmann, "Foreword," *Journal of Glaciology* 1 (1947): 3–4. In 1962 Ahlmann was named honorary member of the International Glaciological Society; see *Ice* 9 (1962): 8–9, and 19 (1972): 16.

[5] Nathan P. Gillett et al., "Attribution of Polar Warming to Human Influence," *Nature Geoscience* 1 (2008): 750–4.

studies on Svalbard and Iceland, he had investigated the Fröya Glacier at 74 degrees north. His Greenland results confirmed data that he had been assembling for more than three decades in Scandinavia, Iceland, and Svalbard. They all pointed in the direction of ongoing, rapid polar warming, but also of very different melting (ablation) regimes under different geographical and meteorological conditions.[6]

The outbreak of war meant that Ahlmann's research program came to an abrupt halt, at least temporarily. Plans and ideas for a more extensive network of sites and measurements had to be aborted or postponed. The situation was unfortunate, but Ahlmann was still actively promoting his research plans and used the war years to sum up and present his results thus far.[7]

In a series of lectures delivered in the Geographical Auditorium at Oslo University in November 1940, Ahlmann claimed that both his own research, from his first studies on the Horung Massif glaciers in Norway starting in 1918 to the latest results from northeast Greenland in 1939–40, as well as glaciological work throughout the world universally demonstrated the same pattern: a gradual retreat of glaciers, reflecting warmer temperatures in higher latitudes, indicating a process of *polar* warming. In the third and last of his lectures Ahlmann discussed possible causes of deglaciation, stressing particularly the role of strong winds, which could be the reason why glaciers retreated more rapidly when located close to the coasts.[8] He had previously described other features of his theory, such as "increased cyclone activity," "intensified circulation of the atmosphere," and, as a consequence, "enhanced maritimity" (*ökad maritimitet*).[9]

At this time no single theory could explain what Ahlmann would call the "embetterment of climate [*klimatförbättringen*] during the years 1920–1940."[10] The knowledge was vague, patchy, at times contradictory. The observational data seemed solid enough, but the time series were limited, only a few decades long. Ahlmann was not even sure that the warming was a permanent process; it could soon reverse, he said, which indeed it did in the 1950s and 1960s. The milder climate in cold areas, such as the Arctic and the North Atlantic, he maintained, was not accompanied by any similar climate change elsewhere. In a lecture in 1943, however, he assumed that polar melting occurred in Antarctica as well, which could explain the observed rise

[6] Ahlmann, "Ablation: Physico-Geographical Researches in the Horung Massif, Joutunheim, [part] IV," *Geografiska Annaler* 9 (1927): 35–66; Ahlmann, "Studies in North-east Greenland, Part III: Accumulation and Ablation on the Fröya Glacier; Its Regime in 1938–39 and 1939–40," *Geografiska Annaler* 24, nos. 1–2 (1942): 1–22, esp. 19–22; Ahlmann, *Glacier Variations and Climatic Fluctuations*, Bowman Memorial Lectures 3 (New York, 1953). A list of publications from Ahlmann's glaciological field investigations in Norway, Svalbard, and Iceland is included in Ahlmann, "The Styggedal Glacier in Jotunheim, Norway: Its Regime, Its Variations in Size and Their Climatological Causes, and Some General Remarks on These," *Geografiska Annaler* 22 (1940): 95–130, on 129–30.

[7] E.g., Ahlmann, "Le régime des glaciers: Ses elements, ses variations," *Revue de Géographie Alpine* 29 (1941): 337–56.

[8] The Norwegian press reported this news: "Breene i Norge går med underskudd," *Bergens Tidende*, November 29, 1940; "Norges og Öst-Grönlands breer går tilbake med femti procent," *Morgenbladet*, November 29, 1940; "Vindhastigheten har stor indflydelse på breenes tilbakegang," *Aftenposten*, November 30, 1940.

[9] E.g., "Glaciärerna som uttryck för den pågående klimatändringen" [Glaciers as evidence of the ongoing climate change], *Ymer* 59 (1939): 51–8, on 57 (all translations in this article are my own, unless otherwise indicated). This publication was originally presented as a lecture to the Swedish Society for Anthropology and Geography on December 9, 1938.

[10] Ahlmann, "Den nutida klimatfluktuationen: Det varmare vädret i Norge och på Svalbard," *Ymer* 61 (1941): 11–24, on 12.

in sea levels since the nineteenth century.[11] However, a broad summary of his North Atlantic results for an international audience did not appear until 1948.[12] At that point he was even making some concession to the possibility of climate change being a phenomenon spread beyond the polar regions, albeit then caused by variations in solar activity:

> If we find in the Antarctic similar evidence of the present climatic fluctuation as has been found in other parts of the world, we shall be justified in concluding that the present fluctuation is a world-wide phenomenon and probably the result of variations in solar activity which, slow as they may be to take effect, are actually resulting in an improvement in the climate of our world.[13]

During the 1930s and 1940s Ahlmann's moderate view of climate change—as ongoing, but limited to polar areas—concurred with mainstream research on historical climate. However, there were other theories that circulated among the scientific community, including the one on human climate forcing through emissions of carbon dioxide (CO_2) that was proposed by Guy Stewart Callendar, the British steam-power engineer, well versed in combustion theory and the radiative properties of atmospheric gases.[14] Callendar had sophisticated training from London's Imperial College, did synthetic work on radiation, atmospheric chemistry, and dynamics, and published a string of papers in important journals, the first in 1938.[15] Callendar argued that global weather data and other indicators, including retreating glaciers, supported his idea that warming was ongoing and global, and he said the cause was anthropogenic.[16]

Ahlmann's and Callendar's approaches to climate change were in some ways in agreement; both identified a warming trend, and both thought of higher temperatures as a largely beneficial climate "embetterment." They disagreed fundamentally, however, on the root cause of climate change, and they did so in ways that were later recognized as defining the main dividing line of the "climate issue": Ahlmann always denying human impact, Callendar always advocating it. Their career paths, and the careers of their respective hypotheses, differed too. Callendar had a less impressive institutional platform, working for the Royal Air Force in relative obscurity and with little or no opportunity for international travel or large-scale research projects.[17]

---

[11] Lecture at the Swedish Association for Anthropology and Geography, March 31, 1943, printed as "Is och hav i Arktis," *Kungl. Svenska vetenskaps-akademiens årsbok* 41 (1943): 327–36. A translation into English by Karin A. Gleim, "Ice and Sea in the Arctic" (manuscript), exists; one copy, sent in March 1947 to the American Navy Department by leading war meteorologist Carl-Gustaf Rossby, is in the Vilhjalmur Stefansson Collection at Dartmouth College, Hanover, N.H., Stef MSS 242(1).

[12] Ahlmann, *Glaciological Research on the North Atlantic Coasts*, Royal Geographical Society Research Series 1 (London, 1948).

[13] Ahlmann, "The Present Climate Fluctuation," *Geogr. J.* 112 (1948): 165–93, on 192–3.

[14] Callendar, "Infra-red Absorption by Carbon Dioxide, with Special Reference to Atmospheric Radiation," *Quarterly Journal of the Royal Meteorological Society* 67 (1941): 263–75.

[15] Callendar's first and seminal paper on the issue was "The Artificial Production of Carbon Dioxide and Its Influence on Temperature," *Quarterly Journal of the Royal Meteorological Society* 64 (1938): 223–40. James R. Fleming, *The Callendar Effect: The Life and Work of Guy Stewart Callendar (1898–1964), the Scientist Who Established the Carbon Dioxide Theory of Climate Change* (Boston, 2007).

[16] Callendar, "Artificial Production of Carbon Dioxide" (cit. n. 15).

[17] He lived modestly with his family in southern England and made in his lifetime only two research-related trips abroad, to Germany in 1930 and to the United States in 1934; Fleming, *Callendar Effect* (cit. n. 15).

Although he was often met with respectful reviews and gradually won influence among a few leading climate scientists, he had little success to begin with in convincing the world of the effects of CO_2 emissions. Not until the 1950s could Callendar reap in full the fruits of his seminal papers in the preceding decades, and by then he exerted a significant influence on leading proponents of the CO_2 hypothesis such as Gilbert Plass, Roger Revelle, and David Keeling. Ahlmann, in contrast, controlled the climate science establishment in Scandinavia from the 1930s and rose in the immediate post–World War II years to become a leading and very visible international authority, widely cited, with extensive personal and professional networks, and reviewed positively in international journals[18]—a social standing in the world of science that Callendar never enjoyed.

Ahlmann held this strong position, scientifically and institutionally, not just in Scandinavia but also in the British glaciological and meteorological communities, where after the war he would use his contacts in preparation for one of his seminal projects, an international glaciological expedition to Antarctica, which ultimately had three participating nations, Norway, Britain, and Sweden, and was known as NBSX.[19] In this period he was already drawn deeply into polar policy and diplomacy. It suited him: he was, as a person, diplomatic, careful, modest, and a good listener, yet firm and quite stubborn, with a clear internationalist agenda based on the belief that scientific cooperation fostered peace. This included bridge-building efforts to connect Western scientists with their Soviet counterparts. In 1934 he visited Leningrad to lecture in the Arctic Institute, the first of what was to become a series of visits to the Soviet Union over a quarter century.[20] Diplomacy, moving cautiously, resonated with deeper traits of his personality. He was, even during his years as a world authority on glaciers and Arctic climate, insecure, anxious, and cautious to the brink of cowardliness. He was also fundamentally impractical, with his "thumb right in the middle of his hand," as the saying goes in Sweden. But these were private and personal shortcomings, little known to others besides his closest associates and hidden behind a public façade of success and splendid leadership.

Ahlmann's moment of fame was short. The tides changed dramatically. In the 1950s and 1960s, when human climate forcing—Callendar's at first improbable-seeming

[18] E.g., he received several reviews in the 1940s and 1950s in *Annals of Geography*.

[19] Lisbeth Lewander, "The Norwegian-British-Swedish Expedition (NBSX) to Antarctica 1949-52— Science and Security," *Berichte zur Polar- und Meeresforschung*, vol. 560, ed. Cornelia Lüdecke (Munich, 2007), 123–41.

[20] *Geograficheskoe obschestvo za 125 let* [Geographical society during 125 years], ed. S. V. Kalesnik (Leningrad, 1970), 361. Ahlmann cofounded the Society for the Promotion of Cultural and Economic Relations between Sweden and the Soviet Union in 1935; see the society's Web page (in Russian), http://www.russwed.ru/pubs.php?num=15&content=ar&lan=rus (accessed 31 August 2008). See also Sovjetunionens vänner [Friends of the Soviet Union] to Ahlmann, 21 December 1934, Ahlmann Collection, Royal Swedish Academy of Sciences, Stockholm (hereafter cited as KVA); other letters in the archives suggest that he was already in touch with the German society Gesellschaft für Kulturelle Verbindung der Sovjetunion mit dem Auslande in 1929. Further information on Swedish-Soviet intellectual relations can be found in Statens Offentliga Utredningar [Sweden's Official Public Inquiries], *Övervakningen av "SKP-komplexet*," vol. 93 (Stockholm, 2002), 60–1. In the summer of 1945 Ahlmann visited Moscow and Leningrad as part of a high-level Swedish delegation, reinforcing his Soviet networks and picking up useful intelligence on Soviet Arctic science. See Sverker Sörlin, "The Global Warming That Did Not Happen: Historicizing Glaciology and Climate Change," in *Nature's End: History and the Environment*, eds. Sörlin and Paul Warde (London, 2009), 93–114, on 100–1.

idea—gained increased recognition, Ahlmann's view faded from prominence, and glaciological field science left the limelight of global science politics.[21]

ELITE AND POPULAR FIELD SCIENCE

Ahlmann's claims about climate may seem modest, but they were in fact quite far-reaching and empirically problematic. He felt that his data suffered from both spatial and temporal deficits—they only covered limited areas for brief periods of time. His anxieties—concerning data, the precision, credibility, and quality of fieldwork, and often also bad weather, field dangers, and looming disaster—were understandable. How did he deal with them?

Ahlmann's preferred path was to emulate the laboratory and exact sciences. Eager to use the latest equipment, he introduced automatic instruments and used statistical methods whenever he could. Compared to most of his colleagues, Ahlmann gained recognition as an unusually meticulous and innovative field glaciologist. In a review brimming with praise of "the Scandinavian workers," British geographer Gordon Manley described at length the sophisticated and ingenious field methods developed by Ahlmann and his students and concluded that in understanding climate variations we can "particularly . . . admire the steady, lively, coherent and fruitful series of investigations initiated by Ahlmann," following old traditions among "the men of the north" in "art, skill, craftsmanship and science."[22]

Ahlmann himself consistently contrasted his work with that of his less rigorous colleagues. He may not have been the first polar glaciologist—and he was certainly not the first polar field scientist—but he was aware that field endeavors would only be relevant if polar scientists pursued specialized science that would yield universally valid knowledge. "With the increased demands of precision and thoroughness that are nowadays raised on natural science," a polar expedition must not be a comprehensive undertaking of all possible subjects. It can only study "a certain group of phenomena," as he dictated in a broad, programmatic article, timed for the second International Polar Year (IPY) 1932–3.[23] His methodological gospel drew heavily on his experience of a public inquiry into the disastrous attempt by fellow Swede S. A. Andrée to reach the North Pole by balloon in 1897.[24] Andrée had been an intrepid engineer but at heart an adventurer, Ahlmann thought. He cited more recent examples of mishaps, such as Umberto Nobile's expedition of 1928, which claimed the life of Norwegian hero Roald Amundsen during the rescue, or Hubert Wilkins's aborted adventure with the submarine Nautilus in 1931, intended to reach the North Pole under the ice.[25]

[21] Ahlmann's view thus has been met with little interest in the seminal literature on the history of climate change science as well; see, e.g., Spencer Weart, *The Discovery of Global Warming*, rev. ed. (Cambridge, Mass., 2008), available in a continuously updated hypertext version on the Web site of the Center for History of Physics of the American Institute of Physics, http://www.aip.org/history/climate/index.html (accessed 28 December 2010); Gale E. Christiansen, *Greenhouse: The 200-Year Story of Global Warming* (New York, 1999); Tim Flannery, *The Weather Makers: The History and Future Impact of Climate Change* (Melbourne, 2005).

[22] Manley, "Some Recent Contributions to the Study of Climatic Change," *Quarterly Journal of the Royal Meteorological Society* 70 (1944): 197–218, on 218.

[23] Ahlmann, "Polarforskningens värde & berättigande," *Ord och Bild* 41 (1931): 195–207, on 198.

[24] Edward Adams-Ray, trans., *The Andrée Diaries: Being the Diaries and Records of S.A. Andrée, Nils Strindberg and Knut Fraenkel . . .* (London, 1931).

[25] Ahlmann, "Polarforskningens värde och berättigande" (cit. n. 23), 198. See also Simon Nasht, *The Last Explorer: Hubert Wilkins, Australia's Unknown Hero* (New York, 2005).

With time, such advocacy became his second nature and appeared as a rhetorical style in his papers, clearly visible in the last of his series of East Greenland reports. There he described the "Devik ablatograph," a device to measure glacier mass balance invented by polar geophysicist Olaf Devik, one of Ahlmann's many Norwegian friends, "which functioned irreproachably."[26] He went on at length about field details: time intervals between observations and the design of the punching drill. He gave time, place, names, material, frequencies, statistics, and tables to enhance credibility, and as if he needed to persuade the reader: "The ablation values obtained in this manner at 4-hour intervals on ice of a specific gravity of 0.9 are quite reliable."[27]

This emphasis on laboratory-like precision in the field is to be expected from someone trying to argue from few and scattered observations of very large and complex phenomena. For Ahlmann it became a field research style.[28] This tendency became more pronounced in the 1930s, when he sensed he was on the cusp of establishing "climate fluctuation," which was a major issue in the Scandinavian countries and the North Atlantic region.

However, Ahlmann also pursued another, less obvious strategy: to work with local people as field assistants. So not only were his field campaigns always international and Nordic, they were also multiclass and multicultural: he worked with Lapland Sami, Greenland Inuit, Norwegian trappers and hunters, and Icelandic farmers. He was able to utilize local skills and technologies in securing field data through what was in essence a field-based coproduction of climate change knowledge. This was obviously nothing new to natural history, nor even to the geological sciences, where local informants and assistants had for a long time played a role in the field.[29] However, for Ahlmann the close collaboration with local people was also a way of reaching out to nonelite collaborators to secure a popular and democratic dimension for the scientific enterprise.

His ideal of field science practice was developed over a long period and in tandem with his wider outlooks on science and politics. Ahlmann had experiences in Jotunheimen, where he was given essential assistance by local scientists from Bergen and their young Norwegian students and fieldworkers, that made him prone to see the advantages of cooperation for success in the field. The expeditions to Svalbard in 1931 and 1934 were different, with, by definition, no native population or knowledge sensu stricto at hand. Still, he used local knowledge from Scandinavia for transport, logistics, and housing, which he contrasted with amateurish British methods—man-hauled sledges, no dogs, zero sense of glacier seasonality—as embodied by the Oxford expedition to North East Land in 1924, which had, as a consequence, failed to reach any significant scientific results. His own meticulously prepared journey, with dogs, sleds, and comfortable equipment, was completely different, even with regard to results: "Our observations were so good that a first overview was secured of both the topographic conditions and the physical nature of the inland ice sheets."[30] The

[26] For the function of the Devik ablatograph, see Ahlmann, "Variations of Glaciers and Measurements of Ablation," in *Association internationale d'hydrologie scientifique, Congrès d'Edimbourgh, 17–26 septembre 1936* (Paris, 1936), 417–29.

[27] Ahlmann, "Studies in North-east Greenland, Part III" (cit. n. 6), 4.

[28] Sverker Sörlin, "Narratives and Counter-narratives of Climate Change: North Atlantic Glaciology and Meteorology, c. 1930–1955," *J. Hist. Geogr.* 35 (2009): 237–55.

[29] See, e.g., Jane R. Camerini, "Wallace in the Field," in Kuklick and Kohler, *Science in the Field* (cit. n. 2), 44–65.

[30] Ahlmann and Sigvard Malmberg, *Sommar vid Polhavet* (Stockholm, 1931), 66–70, on 105.

Scandinavian field expertise that he claimed repeated of course the famous contrasts between Roald Amundsen (dogs, skis, fur clothes like those of Arctic peoples) and Robert Scott (misplaced machinery, snowshoes, wool and cotton British garments), a heritage that Ahlmann silently but certainly knowingly adhered to; he always maintained, in full congruence both with his antiheroic program and with his personal angst, that suffering in the pursuit of science was not a virtue.[31] That was precisely the opposite of the "sacrifice for science" ethos that had been so prevalent among previous Anglo-American explorers, from John Franklin to Robert E. Peary and Frederick A. Cook.[32]

On his Svalbard expeditions in 1931 and 1934 Ahlmann continued to refine his treatment of the field as laboratory with high-quality sleds, exquisitely crafted field tents including a "daily office" tent, precision measurements, large quantities of data collection, and stupendous documentation.[33] He favored traditional techniques and domestic animals, along with the latest high-tech research equipment. On the Isachsen Plateau in 1934 he worked with Norwegian oceanographer Harald Ulrik Sverdrup, who would remain a lifelong brother in arms on the Arctic diplomatic front. They were visited in the field by Helge Ingstad, a lawyer and polar scholar, who arrived in style, commanding a large sled and dog party. Affinities between the men ran high, and two years later Ingstad, then in Bergen, provided dogs for Ahlmann's Iceland expedition.[34] The Svalbard field campaigns were labor-intensive operations, and although he put pressure on his field teams, he always took on a great deal of the work himself, making a point of upholding a principle of equality. He consciously used the credentials he had built as a member of the Swedish and Norwegian elites, working contacts and funding opportunities in both countries. He was also in essence harvesting trust sown by his earlier cooperation on the governmental Andrée commission in Sweden in 1930 and by many years of exchange with Norway, which was also the home country of his wife Lillemor, the daughter of a well-to-do Bergen shipping family.

His most demanding expedition turned out to be the one in Iceland in the late spring and early summer of 1936.[35] Ahlmann's book from the expedition, *På skidor*

---

[31] Ahlmann, "Polarforskningens värde och berättigande" (cit. n. 23), 198.

[32] Rebecca M. Herzig, *Suffering for Science: Reason and Sacrifice in Modern America* (New Brunswick, N.J., 2005), esp. chap. 4, "Explorers," 64–84.

[33] Robert Marc Friedman, *The Expeditions of Harald Ulrik Sverdrup: Contexts for Shaping an Ocean Science* (La Jolla, Calif., 1994), 23–6.

[34] Ahlmann, *Land of Ice and Fire: A Journey to the Great Iceland Glacier*, trans. K. and H. Lewes (London, 1938), chap. 2.

[35] The scientific results were published in *Geografiska Annaler* as the series "Vatnajökull: Scientific Results of the Swedish-Icelandic Investigations 1936–37–38," in eleven chapters: chap. 1, Ahlmann and Sigurður Þórarinsson, "Object, Resources and General Progress of the Swedish-Icelandic Investigations," chap. 2, Þórarinsson, "The Main Geological and Topographical Features of Iceland," chap. 3, Ahlmann and Þórarinsson, "Previous Investigations of Vatnajökull: Marginal Oscillations of Its Outlet-Glaciers, and a General Description of Its Morphology," chap. 4, Ahlmann, "Vatnajökull in Relation to Other Present-Day Iceland Glaciers," *Geografiska Annaler* 19 (1937): 146–231; chap. 5, Ahlmann and Þórarinsson, "The Ablation," *Geografiska Annaler* 20 (1938): 171–233; chap. 6, Ahlmann and Þórarinsson, "The Accumulation," *Geografiska Annaler* 21 (1939): 39–66; chap. 7, Ahlmann, "The Regime of Hoffellsjökull," chap. 8, Þórarinsson, "Hoffellsjökull, Its Movements and Drainage," chap. 9, Þórarinsson, "The Ice Dammed Lakes of Iceland with Particular Reference to Their Values as Indicators of Glacier Oscillations," *Geografiska Annaler* 21 (1939): 171–242; chap. 10, Ahlmann, "The Relative Influence of Precipitation and Temperature on Glacier Regime: Hoffellsjökull in 1936–38," *Geografiska Annaler* 22 (1940): 188–205; chap. 11, Þórarinsson, "Oscillations of the Iceland Glaciers during 250 Years," *Geografiska Annaler* 25 (1943): 1–54.

Figure 1. *Ahlmann working in his daily office tent on the 1934 Svalbard expedition. Source: 1934 Norsk-Svenska Spetsbergs-expeditionen [Norwegian-Swedish Spitsbergen Expedition], Ahlmann Collection, KVA, vol. 55. All figures are reprinted with permission from the Royal Swedish Academy of Sciences.*

och till häst i Vatnajökulls rike, published in Stockholm before Christmas the same year, and translated as *Land of Ice and Fire* (1938), is one long tribute to local knowledge and to the people of Iceland, whom he admired unreservedly, a not uncommon position vis-à-vis the people perceived as the true heirs of Norse history. The attitude was rooted in Nordicist or Scandinavianist sentiments, which ran strong as an alternative to nationalism in the mid-nineteenth century but survived after Norway's secession from Sweden in 1905 mostly among liberal and progressive elites of the kind Ahlmann belonged to.[36] Already at the outset of the Vatnajökull climb, when eight men "from the area" lifted and hauled the heavy sledges up the first ice fall on the morning of the second day, Ahlmann admitted that this kind of work would not be possible without a science that was firmly tied to the local community. His description of these men in *Land of Ice and Fire* was almost biblical. They were simply dressed, with no adornment, their feet were clad in shoes made of thin sheepskin, and they worked silently and loyally.

There was a high-minded political dimension to Ahlmann's passion for local

[36] Ruth Hemstad, *Fra Indian Summer til nordisk vinter: Skandinavisk samarbeid, skandinavisme og unionsupplösningen* (Oslo, 2008). Ahlmann was also active in the Norden Association (Föreningen Norden) and published in 1943, in the midst of war, a textbook on the cultural and physical geography of Norway which demonstrated his deep sympathy for the country.

Figure 2. *Field work on East Greenland, 1939. Kåre Rodahl (left) with Inuit collaborators Petrus and Jonatan. Source: 1936 Island [Iceland], Ahlmann Collection, KVA, vol. 56.*

knowledge. It was rooted in a respect for artisanship and vernacular ingenuity. Popular knowledge was as important as the scientific for the end result, which was a deeper, more nuanced understanding. This knowledge was a civic asset, and in shaping it all social classes, and certainly all groups of people, had their role to play. This was an *ethnopolitical* argument for local knowledge rather than mainstream ethnographical knowledge, which in his view was retrospective and at worst reactionary. In Scandinavia the "Lapp shall remain Lapp" philosophy had been proposed by conservative elites and advocated in Sweden by Karl Bernhard Wiklund, professor of Finno-Ugric languages at Uppsala University.[37] Essentialist understandings of non-Western peoples were still common among anthropologists, sometimes still with a racist undercurrent.[38] Ahlmann had no in-depth contact with the ethnographic literature. Rather, his political and moral compass, also expressed in his emphatic anti-Nazism and his growing sympathies for Social Democracy, concurred with his admiration for popular savoir faire.[39]

---

[37] Lennart Lundmark, *Stulet land* (Stockholm, 2008); Gunnar Eriksson, "Darwinism and Sami Legislation," in *The Sami National Minority in Sweden,* ed. Birgitta Jahreskog (Stockholm, 1982), 89–101.

[38] See, e.g., Gretchen E. Schafft, *From Racism to Genocide: Anthropology in the Third Reich* (Urbana, Ill., 2004); Elazar Barkan, *The Retreat of Scientific Racism: Changing Concepts of Race in Britain and the United States between the World Wars* (New York, 1992).

[39] On Ahlmann's anti-Nazism, see his interview in 1972 by Wilhelm Odelberg; transcript in Secretary's Collection, KVA, 25a:1.

Finally, on the last stop of his long North Atlantic odyssey on the Fröya Glacier in August 1939, he worked with Stockholm meteorologist Backa E. Eriksson and the Norwegian medical student Kåre Rodahl, who took observations on the glacier and also conducted diet studies involving vitamin A. They were assisted by Norwegian hunters, Henry Rudi and Schölberg Nilsen, and a third Norwegian by the name of Helland who served as factotum (*altmuligman*). Two other assistants, Petrus and Jonatan, were Inuit and served Rodahl, and thus indirectly Ahlmann, during his winter of observation in 1939/40.[40] Ahlmann wrote about the field assistants, "I owe them many great thanks for all the help and friendship they without compensation showed to me and the entire expedition."[41]

FIELD COPRODUCTION OF CLIMATE KNOWLEDGE— THE CASE OF VATNAJÖKULL 1936–8

There was, however, also a pragmatic and nonidyllic aspect to Ahlmann's cooperation with local residents. This is clear from a closer look at how the coproduction of climate knowledge in the field unfolded during and after the 1936 expedition to Iceland's Vatnajökull Glacier. The field campaign started in late April of 1936 and was finished in June. The weather was dreadful; several meters of snow fell in two weeks, and the team, camping on the glacier, almost had to give up. Ahlmann, true to his nature, pointed out the systematic method of measuring annual snow accumulation and ablation. He referred to his "snow holes," carefully dug by the altmuligman Jón fra Laug, as their "field laboratory," no less precise than any indoor equivalent. The assembly point for the expedition was the small farmstead Hoffell in the far southeast of Iceland, where the scientists could get food, shelter, treatment of injuries, and help with logistics.

Hoffell was a site of knowledge in its own right. The principal of the farm was Gudmundur Jónsson, a local sheep farmer, amateur natural historian, mining entrepreneur, and mineralogist who in 1909 had discovered a sizable deposit of spar on his own land. The spar was used for, among other things, ornamental purposes in the main hall of the University of Iceland in Reykjavik, officially opened in June 1940 during the British occupation.[42] Gudmundur was the father of Leifur, who helped Ahlmann solicit field support from a number of other men from surrounding farms. The most significant of these men were Helgi Gudmundsson, Leifur's brother who had famously and somewhat boldly undertaken a 500-kilometer glacier walk with two friends to North Iceland and back in 1926; Jón fra Laug, the altmuligman; Skarphjedinn, a glacial observer and transport man; Sigfinnur Pálsson, a horse-riding orphan who was raised on Leifur's farm; and a Swedish ski and sled expert, Mac Lilliehöök, who incidentally committed suicide shortly after the expedition, a fact that did nothing to ease Ahlmann's anxieties for what the field could do to a person, although there was never any suggestion that the Iceland journey caused

[40] Kåre Rodahl, "Med 'Polarbjörn' til Grönland og med fallskjerm til Norge," *Polar-Årboken 1945* (Oslo, 1945), 52–66; Rodahl, "Hans W:son Ahlmann: Polarforskeren og Norgesvennen," *Magasinet*, nos. 41–2 (1946): 22–6.

[41] Ahlmann to Gerard De Geer, 7 September 1939, Ahlmann Collection, KVA, vol. 5.

[42] Gudmundur Jónsson wrote about this in his autobiography, *Skaftfellskar þjóðsögur og sagnir ásamt sjálfsævisögu höfundar*, ed. Marteinn Skaftfells (Akureyri, 1946), 39–48. See also Páll Sigurdsson, *Úr húsnaedis- og byggingarsögu Háskola Íslands* (Reykjavik, 1961), 1:209–15.

Häst med Nansen-släde

Figure 3. Field coproducers of climate change knowledge in Iceland, 1936. Jón fra Laug, altmuligman (left), and Skarphjedinn, glacial observer (right). Photo by Sigurður Þórarinsson. Source: 1939 Nordostgrönland [North East Greenland], Ahlmann Collection, KVA, vol. 57.

the young man's melancholy.[43] Also important were the Icelandic horses, in particular the remarkable Skúmur,[44] which Sigfinnur rode on his ventures into the field (horses had already been tested by Alfred Wegener on Vatnajökull in 1912 and were commonly used).

This local knowledge was also gendered according to deeply rooted social traditions, but in ways that did not exclude anybody. At Hoffell the women, with Gudmundur's wife Valgerður Sigurdardóttir as their sharp-eyed principal, prepared the food, brought comfort, and catered to the injured, notably after a severe freezing incident. All in all, these local actors were indispensable to Ahlmann's enterprise. Perhaps most important, they reduced Ahlmann's unease. He was pursuing advanced science, but he was among caring and absolutely capable fellow humans. His dependence reverberates on every page of his book.

The science team had four members: Ahlmann, Carl "Calle" Mannerfelt, a trained athlete and member of the Swedish national track-and-field team, Jón Eythórsson, Iceland's state meteorologist, and his countryman Sigurður Þórarinsson, Ahlmann's

---

[43] Gösta Lilliehöök to Ahlmann, 6 December and 23 December 1936, Ahlmann Collection, KVA, vol. 12; Mac Lilliehöök's obituary, 22 November, ibid., vol. 34.
[44] *Skúmur* is Icelandic for a bird, the great skua (*Stercorarius skua*).

young favorite student, also a gifted guitar player and poet, nicknamed "Skallagrim" after the ninth-century Icelandic poet. Þórarinsson made sure, along with Ahlmann, that the glaciological research environment in Stockholm was generously mixed with culture, including lecture tours in workers' unions and social democratic circles and editorial projects to bring Icelandic poetry to a Swedish readership.[45] The scientific work was carried out in the same mode of expedition qua station as the previous ones in Svalbard, and the data collection in the field was, as usual, impressive and zealous despite the unspeakable weather conditions, which on more than one occasion almost caused Ahlmann to abort the mission and return, as Mannerfelt later recalled.[46]

THE ECONOMICS OF FIELD WORK—CLASS, NETWORKS, AND HIERARCHIES

To Ahlmann the measurements on Iceland were of the utmost significance for the general understanding of North Atlantic climate change. However, the measurements would be of no use if they could not continue after the expedition. This part of the story unfolded in a Stockholm-Reykjavik-Vatnajökull network of hierarchical and quite asymmetric relations, amid frequent and cordial expressions of friendship, but also with a sprinkling of strain and commandeering behavior that followed social hierarchies.

Eythórsson's loyalty to Ahlmann dated back more than fifteen years. In August 1920 he and Ahlmann attended the geophysical meeting in Bergen, which was then a leading center of geophysical and meteorological research. Ahlmann gave a paper that caused Eythórsson to note in his small black book, "Limits of glaciation are a function of summer temp[eratures], precipitation, and whether there is . . . a relation between 'mountain forms' and 'glaciation,'" an early summary as short and accurate as any of Ahlmann's emerging ideas of glaciation.[47] Now Eythórsson served as Ahlmann's field science broker, making sure that the workers in the field did their job and provided the data.[48]

At Hoffell Gudmundur Jónsson and his sons had been in repeated action since the expedition left in June 1936. Others, including Steinthor Sigurdsson (who was mapping the eastern parts of Vatnajökull) and Skarphjedinn (who did ablation measurements), had already presented their data late in the summer, as usual through Eythórsson, who incidentally was also asked to manage the finances of Mannerfelt's journey to Sweden after the field season and to arrange for a suitable bookstore in Reykjavik to receive a prospectus of Ahlmann's coming book.[49]

Ahlmann kept his Icelandic team busy for three consecutive years. In December 1938 he was still demanding more data on ablation at Hoffell for the full year of 1938. He also asked Eythórsson to make sure that "Gudmundur or some of his sons measure the lower poles at Hoffellsjökull between Christmas and New Year's" and to

[45] Letters from the Swedish national radio service (boxes 40, 42) and from Nordic cultural associations (boxes 37, 39, 41), University of Iceland Archives, Reykjavik.

[46] Carl Mannerfelt, interviews with the author, Stockholm, 26 July and 9 August 2008.

[47] Eythórsson, noteboook 1919–20, Jón Eythórsson Collection, National Library, Reykjavik, box marked "Dag- og minnisbaekur." On their Joutunheimen collaboration, see Ahlmann, "Researches on Snow and Ice, 1918–1940," *Geogr. J.* 107, nos. 1–2 (1946): 11–25, on 11.

[48] Ahlmann to Eythórsson, n.d. [probably December 1936], Jón Eythórsson Collection, National Library, Reykjavik, box marked "Vatnajökulsför. Bréfaskipti vid H. Ahlmann" [The Vatnajökull investigation, exchange of letters with H. Ahlmann] (hereafter cited as JE Vatnajökull).

[49] Ahlmann to Eythórsson, 5 October 1936, JE Vatnajökull.

cable these and other new data in early January, since the ablation chapter was due to the printers that month. Þórarinsson's address in Stockholm was provided since Ahlmann and his wife Lillemor, *comme d'habitude*, would spend the vacation period in her native Bergen. He also asked that copies be sent to Karin Gleim, a former student of Ahlmann's, in Boston, where she was helping with translations (as she would later do with Ahlmann's 1943 paper; see n. 11 above).[50]

Two Icelanders were, thus, finally responsible for the production of the Vatnajökull results: Jón Eythórsson for the primary production of the data in the field through subaltern nationals and Þórarinsson for final preparations of the finished manuscript. Eythórsson, however, was not swift enough, and Ahlmann pushed him again in January 1939 to send measurement data for December, telling him not to bother about all the details, as changes could be made in the proofs. He repeated that Karin Gleim needed to get the manuscript soon for her translation into English, talked of Gudmundur's poles—couldn't he just continue his measurements until early summer?—and asked again for Eythórsson's "meteorological chapter" (it was never delivered).[51]

This network chain of command represents a way of organizing labor in time and space that did not come without a cost. In reality the scientific hierarchy, obvious already at the time of the expedition, had a corresponding "economics" that was applied in the many months of continued work on the field infrastructures. At the top of the economic chain, which was closely linked with the regular scientific reward system, was Ahlmann, who sought and received grants and stipends for his project and his students, notably for Þórarinsson from the Wenner Gren Foundation.[52] In the next position below came Eythórsson, who performed some services for which he received compensation but principally collected data from the observers, a job that Ahlmann simply expected him to do as part of their long friendship. Eythórsson received immaterial rewards such as a celebratory telegram from the Swedish Society for Anthropology and Geography, signed by Lennart von Post, the geologist;[53] Ahlmann had a strong position in that organization. At the lowest end of the chain of command were the local informants, indispensable for the basic data and the overall result, but with no interest in publishing or analysis. They were the ones who actually did get paid for their efforts. The single individual who would emerge empty-handed from this little potlatch would be the most junior member of the network, Sigfinnur, who worked for Þórarinsson in the field. Sigfinnur received no funds, or perhaps he counted as a farmhand on Gudmundur's Hoffell and was fed by him.[54]

Were relations between the different levels of this chain of services and exchanges therefore strictly businesslike? Not at all. The "economy" was more complicated. The remuneration was usually inadequate. Ahlmann did receive grants for his

[50] Ahlmann to Eythórsson, 8 December 1938, JE Vatnajökull.

[51] Ahlmann to Eythórsson, 4 January 1939, JE Vatnajökull. See also the series of publications from the enterprise by Ahlmann and Þórarinsson, "Vatnajökull" (cit. n. 35), and Ahlmann to Eythórsson, 27 December 1945, JE Vatnajökull.

[52] Wenner-Grenska Samfundet to Þórarinsson, 5 September 1940 and 20 March 1946, Sigurður Þórarinsson Collection, University of Iceland Archives, Reykjavik, boxes 38 and 40. The kind of legwork performed by Ahlmann is also visible in his correspondence with the foundation (in the Ahlmann Collection, KVA) and with the Swedish state antiquarian Sigurd Curman, 11 November and 16 December 1939, in the Curman Collection, National Board of Antiquities Archives, Stockholm.

[53] Telegram, 21 November 1936, JE Vatnajökull.

[54] Letters from Sigfinnur to Þórarinsson (10 December 1936, 16 March 1937, 13 November 1938, and 26 November 1938) give evidence of their working relationship. Sigurður Þórarinsson Collection, University of Iceland Archives, Reykjavik, box 40.

expedition; in fact he had a reputation for his grant-earning skills, even in the hard times of the 1930s. Still, the grants did not allow for any excesses. Ahlmann's total budget for the Iceland journey was 20,000 Swedish kronor, or $3,500 in contemporary currency, which barely made ends meet, and frugality reigned throughout. After his return he had just about enough funds left to pay his postexpedition field staff in Iceland. Services in Hoffell had been free. Like many other field scientists Ahlmann was dependent on the loyalty of his observers. Ideally they were so interested and engaged in the investigation that they were motivated for that reason alone to collect the data. That was, of course, not always the case. Pedagogical instructions to informants were crucial, and these were facilitated by good personal relations—even friendship became in this context an economic factor. Ahlmann cultivated relationships and friendships systematically. He had a friendly style of communication: he calmed, soothed, praised, and talked softly, gently, and convincingly. He was almost never openly irritated or angry. This was, again, part of his diplomacy, but it was also a survival strategy for a field scientist operating across vast distances on a multiscalar network of relations.

In Stockholm, Ahlmann, ever the constructive diplomat, was busy setting up a small foundation for stipends to Icelandic students.[55] Þórarinsson repeatedly brought up this subject, which further deepened their shared radical "Nordism." This philosophy also meant keeping a certain distance from both NATO and the Soviet Union in the emerging cold war, but still claiming a role in building bridges between the two.[56]

POLAR WARMING—THE GLORY YEARS

After his expeditions to Iceland in 1936 and Greenland in 1939 Ahlmann had to take an involuntary pause from his field glaciology because of the war. In the meantime he started launching his polar warming theory, for which he had by then built a firm empirical foundation. He also focused on the big-picture politics of science. It was at this time that he opened a southern front for his polar warming work, hoping in earnest for an Antarctic campaign to test his warming hypothesis there. Aerial photos of the Drygalski Mountains in the German Neu Schwabenland area, taken by a German expedition in 1939 but published only in 1942, caught his eye in 1943 and served as a useful rhetorical device that helped him suggest ongoing climate change in the Southern Continent as well. He first publicly indicated his ideas of Antarctic glacier melting corresponding to that in the Arctic in a lecture in Stockholm in March 1943. The theme came up at about the same time in his correspondence with British and Scandinavian colleagues and was intense over the following years.[57]

---

[55] Ahlmann to Eythórsson, 2 March 1937, JE Vatnajökull.

[56] There was also an element of good advice to Western institutions to stay in touch with the important (according to Ahlmann's firsthand reports from June and July 1945) polar and climate science in the USSR. See Ahlmann, "Geography in the Soviet Union," trans. M. Burnett, *Geogr. J.* 106 (1945): 217–21, and a more candid take on the need to shape up the Scandinavian polar presence and research in *Polar-Årboken* (Oslo, 1945). In this respect he was, ultimately, leaning to the West, just like his friend and colleague Halvard Lange, who was Norwegian foreign minister 1946–65.

[57] See n. 11 above on the lecture. Early correspondence with Andrew Croft and James Wordie in England: Wordie to Ahlmann, 6 September 1944; see also Olaf Devik to Ahlmann, 16 November 1944, both in the Ahlmann Collection, KVA, vol. 8; Ahlmann to Croft, 17 March 1943, ibid., vol. 3. The German Antarctic Expedition's report and some of its photographs were published by the expedition commander Alfred Ritscher, *Wissenschaftliche und fliegerische Ergebnisse der Deutsche Antarktische Expedition 1938–1939*, vol. 1 (Leipzig, 1942). Ahlmann's Antarctic correspondence with Wordie, J. G. Andersson, and others is in the Ahlmann Collection, KVA.

The Antarctic expedition was an attempt to further test the theory of polar warming in a southern context, including building observational networks and linking stations of meteorological and climatologic interest.[58] Less publicly, but probably more important when it came to funding and state support, the project was part of strengthening Nordic scientific and strategic cooperation and safeguarding Norwegian territorial interests in the Atlantic sector of Antarctica. Ahlmann's skills in "selling" field-based earth science were different from but in no way less efficient than those of his American counterparts, stressing military and strategic interests less than peaceful and small-state territorial interests.[59] The expedition dovetailed nicely with growing concern in the United States and around the world about warming tendencies in the polar areas that might affect both military security and global sea levels. Ahlmann was repeatedly cited on this in the international media, partly as an effect of a visit he paid to the United States in the early summer of 1947, which included a stop at the Pentagon, where possible warming in the Arctic was of huge strategic interest.[60]

Ahlmann did not, however, aspire to play any major part in the NBSX, although as late as the late fall of 1948 he said he would join the vessel going down and spend the first Antarctic summer there.[61] At some point during 1949 he changed his mind, and in 1950 he assumed the position of ambassador to Oslo, a move that confirmed his role as a policy actor but at the same time put him on the sidelines of polar research in Sweden. Indeed, beginning in the late 1940s his impact was larger in Norway, where he played a major role in forming a modern polar organization, largely through his influence on Sverdrup and his many friends in Norway's new post–World War II political leadership.[62] Among these was Halvard Lange, who was one of Ahlmann's assistants in Jotunheimen and then went into (Social Democratic) politics and advanced to the position of foreign minister, in which he worked closely with Ahlmann during his years in Oslo.

Ahlmann also kept worrying about the correspondence between field observations and his warming theory. He needed secure, long-term production of field data that could serve as the empirical backbone of his work, and indeed of his entire polar warming school of thought. He had by now started growing a handsome collection of unusually gifted PhD students. Þórarinsson and Mannerfelt were the first, and later to come was Valter Schytt, who would soon be the chief heir to Ahlmann's legacy in Sweden. The lasting monument in this regard—apart from Ahlmann's Bowman Lecture in 1952 (which I discuss in the next section)—was the *Festschrift* for Ahlmann's

[58] Ahlmann, "Nutidens Antarktis och istidens Skandinavien," *Geologiska Föreningens Förhandlingar* 66 (1944): 635–52; Ahlmann, "Researches on Snow and Ice" (cit. n. 47); Ahlmann, "Den planerade norsk-svensk-brittiska Antarktis-expeditionen," *Ymer* 68 (1948): 241–67, esp. 247, 256. See also Peder Roberts, "Neither Explorers nor Adventurers: Inventing the Antarctic Specialist," paper presented at the Division for History of Science and Technology, Royal Institute of Technology, Stockholm, 18 November 2007; Roberts, "When the Polar Scientist Became Professional," translated into Swedish by Margareta Eklöf and published as "När polarforskaren blev professionell," in "Polarår" [Polar years], eds. Gunhild Rosqvist and Sverker Sörlin, special issue, *Ymer* 129 (2009): 129–50. See further Adrian Howkins in this volume.

[59] Fae L. Korsmo and Michael P. Sfraga, "From Interwar to Cold War: Selling Field Science in the United States, 1920s through 1950s," *Earth Sci. Hist.* 22 (2003): 55–78.

[60] Gladwin Hill, "Warming Arctic Climate Melting Glaciers Faster, Raising Ocean Level, Scientist Says," *New York Times*, May 30, 1947; Ronald E. Doel, "Polar Melting When Cold War Was Hot," *San Francisco Examiner*, October 3, 2000, A15. See also Sörlin, "Narratives and Counter-narratives" (cit. n. 28).

[61] Ahlmann, "Den planerade norsk-brittisk-svenska Antarktisexpeditionen" (cit. n. 58), 262.

[62] Robert Marc Friedman, "Å spise kirsebaer med de store," in *Norsk Polarhistorie*, vol. 2, *Vitenskapene*, eds. Einar-Arne Drivenes and Harald Dag Jölle (Tromsö, 2005): 331–420.

sixtieth birthday in 1949, *Glaciers and Climate*, a special volume of the journal *Geografiska Annaler*. This volume was edited by Mannerfelt and Schytt, along with Lawrence Kirwan of the Royal Geographical Society, coordinator of the U.K. contribution to NBSX, and Carl-Gustaf Rossby, the Swedish-born meteorologist who had played an important role in the American war effort and was now back in Sweden to improve training of meteorologists there, thanks to efforts by the Swedish government, the Swedish Air Force, and Ahlmann.[63] The Stockholm glaciologist, the editors claimed, had "recognized that all such climatic changes and fluctuations must ultimately be considered as manifestations of changes in the circulation of the earth's atmosphere."[64]

In retrospect that sounds like a bold statement, but at the time it seemed quite safe. It was corroborated by other contributions in the volume, notably that of C. E. P. Brooks, who affirmed that "the recent great retreat [of glaciers is] due . . . to variations in the strength of atmospheric circulation between equator and pole."[65] Human climate forcing was not even an issue, either in Brooks's contribution or in any other chapter of the 400-page volume. At this period, at least among this circle of geophysicists, polar warming was at the forefront, whether or not that concept was explicitly used, whereas the CO_2 hypothesis, and thus Callendar, belonged to the past. Brooks expressed this in even clearer terms a couple of years later: "[Svante] Arrhenius and [Thomas] Chamberlin saw in this [i.e., increased atmospheric carbon dioxide] a cause of climatic changes, but the theory was never widely accepted and was abandoned when it was found that all the long-wave radiation absorbed by CO_2 is also absorbed by water vapour." Callendar, Brooks went on, was simply mistaken: "This theory is not considered further."[66]

Ahlmann-style polar warming for a while seemed to be the triumphant paradigm for thinking about climate change. It resonated well with common sense, it was in line with empirical observations, and it sat well with local knowledge on changes in climatic conditions going back generations. It also fit the strategy of field precision that was the modus operandi of Ahlmann's research, and under this umbrella of consensus he could move forward with NBSX and his Nordic institutional ambitions. With the right of age—he was approaching sixty, which was old at that time for an Arctic field scientist—and with the intention of building a long-term site for his successful enterprise, he also returned closer to home, in Swedish Lapland. With funding from the newly founded Natural Science Research Council, he set up a mountain glacier research station at Tarfala in 1946 with his students and colleagues, again in close collaboration with local residents as well, in this case the Sami.[67]

[63] Rossby's move to Sweden was initially proposed at the end of the war by sources in the Air Force and later advocated by Ahlmann in a report with his co-investigator, Uppsala professor Harald Norinder, "Sakkunnige rörande organisationen av den meteorologiska forskningen och undervisningen" [The expert committee concerning the organization of meteorological research and education], 28 January 1946, Swedish National Archives, SE/RA/321102. More on the report to the government by Ahlmann and Norinder is in Sörlin, "Narratives and Counter-narratives" (cit. n. 28).

[64] L. P. Kirwan, C. M:son Mannerfelt, C. G. Rossby, and V. Schytt, "Glaciers and Climatology: Hans W:son Ahlmann's Contribution," in *Glaciers and Climate*, chief ed. C. M. Mannerfelt, special volume, *Geografiska Annaler* 31 (1949): 11–20, on 12.

[65] Brooks, "Post-glacial Climate Changes in the Light of Recent Glaciological Research," in *Glaciers and Climate* (cit. n. 64), 21.

[66] C. E. P. Brooks, "Geological and Historical Aspects of Climate Change," in *Compendium of Meteorology*, ed. Thomas F. Malone (Boston, 1951), 1004–18, on 1016.

[67] Per Holmlund and Peter Jansson, *Glaciological Research at Tarfala Research Station* (Stockholm, 2002), indicate the importance of the station for the continuity of monitoring and research

CONCLUDING DISCUSSION—THE FRAGILITY OF FIELD KNOWLEDGE

Around 1950 Ahlmann was at the center of the increasingly widespread notion of an ongoing, yet enigmatic, process of warming, and this made him occasionally a figure in the mass media, in Scandinavia, Europe, and the United States. He was invited to give the prestigious Bowman Lecture at the American Geographical Society in New York in 1952 and chose, of course, his quintessential topic, "glacier variations and climate fluctuations."[68] The outlook on the consequences of warming was still by and large optimistic.[69] With time, however, the sinister aspects of warming would become hegemonic, and very soon after the Bowman Lecture a public discussion emerged of the possible negative effects of fossil fuel combustion on the climate, with ramifications that would certainly be most marked around the poles but might extend to other parts of the planet. The scientific ground had been broken by Gilbert Plass in a seminal 1956 paper and by what we now realize was a sizable number of scholars in the United States, the United Kingdom, and Scandinavia.[70] Carl-Gustaf Rossby, who had long taken an interest in the subject, emphasized it in one of his last publications: "Mankind . . . is performing a unique experiment of planetary dimensions by now consuming during a few hundred years all the fossil fuel deposited during millions of years."[71] In the context of the International Geophysical Year (IGY) 1957–8, climate change had already become a research issue in its own right. A popular presentation of the IGY mentioned the "greenhouse effect" (in quotation marks) as a possibility and stated that humankind's outpouring of CO_2 was likely to "cause significant melting of the great icecaps and raise sea levels," an idea Ahlmann would certainly not have supported.[72]

Bearing in mind this development, which was only to be reinforced in the final decades of the twentieth century, how should we regard the seemingly single-minded care with which Ahlmann pursued his polar warming hypothesis and his scrupulous fieldwork program? I have already noted above that up until the early 1950s his ideas resonated quite well with the geophysical orthodoxy, in which indeed for many years he occupied a leading position. But in order to understand his anxieties we need to also look at the competitive and changing intellectual landscape where things

---

programs. For a brief contextualization, see my contribution to an IPY 2007–8 research project on the history, culture, and governance of field stations: http://museum.archanth.cam.ac.uk/fieldstation/members/sorlin.

[68] Ahlmann, *Glacier Variations and Climate Fluctuations* (cit. n. 6).

[69] The Ahlmann Collection, KVA (vols. 34–7 in particular), contains dozens of articles from newspapers around the world that overall present a picture of Ahlmann as popular because of his results indicating a warming trend in northern parts of the world, which was seen as welcome in the British Isles and the Scandinavian countries, among others.

[70] Plass, "Carbon Dioxide and the Climate," *Amer. Scient.* 44 (1956): 302–16; Weart, *Discovery of Global Warming* (cit. n. 21), 23–6; Fleming, *Callendar Effect* (cit. n. 15); for Scandinavia, see Maria Bohn in this volume.

[71] Rossby, "Aktuella meteorologiska problem" (1957), translated by members of Rossby's own International Institute of Meteorology in Stockholm as "Current Problems in Meteorology," in *The Atmosphere and the Sea in Motion: Scientific Contributions to the Rossby Memorial Volume*, ed. Bert Bolin (New York, 1959), 9–50, on 14. It is instructive to compare this volume with the Ahlmann *Festschrift* a decade earlier, which still had a nice mix of meteorologists, glaciologists, and other earth scientists; by 1959 meteorology and formal methods have achieved full hegemony, and Callendar is sometimes cited, Ahlmann never.

[72] National Academy of Sciences, *"Planet Earth": The Mystery with 100,000 Clues* (Washington, D.C., 1957), 27.

were going on that were potentially threatening to his views, as the cited literature suggests.

Right when Ahlmann's period as a world authority began in the late 1930s, the modern rise of the theory of human-forced climate change also started with the publication of Callendar's 1938 paper and a string of publications in the 1940s that gradually drew more and more interest among insightful scientists, including Plass and Keeling. Ahlmann was aware of Callendar's ideas, but quite obviously he did not believe in them.[73] Rather, as we have seen, he always attributed climate change—including global climate change, if there indeed was any, beyond polar warming—to nonanthropogenic causes: warm tropical winds, and later also variations in solar radiation, which he had at first rejected.[74]

On the other hand, he was aware that it was an extreme challenge for field science to establish reliable knowledge on historical climate change, not to speak of making statements about the future climate. It was mandatory to vastly improve the data collection, to widen the geographical network of monitoring sites, and to establish long-term and reliable time series of geophysical and glaciological data. This was a reality that Ahlmann sought to remedy during the war years. Just as he skillfully used the geopolitical situation to make the Antarctic expedition a priority in order to satisfy the necessity of widening the data-gathering network, he identified the need for a major field monitoring location more easily accessible to his growing crew of colleagues, PhDs, and students at Stockholm's Högskola; this became the Tarfala Research Station.

Why this obsession with the field? Why this quest for long-term presence? The shortest possible answer is that the object of study was so fleeting. Ahlmann needed a *longue durée* of observation, which in turn helps explain the longevity and stubbornness of his theoretical understanding. That is also why the reliable production of knowledge in the field became central for him and his theory. In that sense there is an interesting relationship between polar warming as a theory and the practice and experiences of the field. Ahlmann's climate change knowledge was paradoxical. He wished it to be detached, instrumental, objective, impersonal, but it could not escape its embeddedness in the field site. Despite Ahlmann's claim to the contrary, his knowledge took on what we may call a bodily nature. It required heavy investment in physical work, installations, and instrumentation of a fixed and long-term character. The chief protagonist in his program was ultimately the glacier itself, a main source of information in what Gordon Manley, the British geographer, significantly termed the "instrumental" period in the study of glaciers and climate.[75] In Latourian language, the glacier became an *actant* that served as the ultimate voice of truth, speaking from nature itself, but on behalf of Ahlmann, who had instrumented it. This image of the glacier as physical "truth spot,"[76] building authority in the field, speaks to Michael Polanyi's idea of the body as an instrument in itself and scientific instru-

[73] Gerald Seligman, "Glacier Fluctuations," *Quarterly Journal of the Royal Meteorological Society* 70 (1944): 22, quotes Ahlmann as saying that Callendar had misrepresented Ahlmann's view.

[74] Ahlmann, "Present Climate Fluctuation" (cit. n. 13).

[75] Manley, "Some Recent Contributions to the Study of Climatic Change" (cit. n. 22), esp. 205–10; Manley, "The Extent of the Fluctuations Shown during the 'Instrumental' Period in Relation to Postglacial Events in NW. Europe," in "Post-glacial Climatic Change: Discussion," *Quarterly Journal of the Royal Meteorological Society* 75 (1949): 165–71.

[76] Thomas F. Gieryn, "Three Truth-Spots," *J. Hist. Behav. Sci.* 38 (2002): 113–32; Gieryn, "The City as Truth-Spot: Laboratories and Field-Sites in Urban Studies," *Soc. Stud. Sci.* 36 (2006): 5–38.

ments as an extension of it.[77] The "instrumented" field station version of the glacier should be seen as not just an extension of the laboratory into the field, which has been argued for the biological sciences,[78] but also an extension of the mind of the scientist into the glacier, a hybrid category, combining the natural object, the scientific instrumentation, and the community of investigating bodies.

This field version of what has been called "science incarnate" had two aspects, which stand in an interesting tension with each other just as the scientific instrument and the physical field laborer do. With the instrumental approach the field became a powerful carrier of data, experiences, and research. This was of course because the field was where the primary seeds of knowledge were harvested, through physical observation. But it was also because the field contained the sites, research stations, equipment, and other monuments that served the function of uniting the group of scientists—in Ahlmann's case, everyone who shared in the experience, including local assistants and guides.[79] The research infrastructure thus served as both safeguard of precision and boundary object, establishing the claim that what took place on the glacier was a production of knowledge. At the same time there was no way to escape the heavy involvement of the body in the experience. The body of the scientist was an intimate part of the fieldwork, and as he researched the glacier he also turned it into a "body," with a temperature, a mass, a balance, a life, and a career, words that were all in the language Ahlmann used to speak about glaciers. Ultimately the glacier emerged as a versatile, nondeterministic entity, a "sphere of a multiplicity of trajectories,"[80] with an openness to future possibilities that suited a cautious man.

This is where we find the roots of the anxieties of a science diplomat searching for trust and credibility. Locating the production of malleable knowledge in an elusive field through the extension of a vulnerable body was the same as admitting that this was a knowledge that suffered from all the shortcomings of the ephemeral, the changing, and the local, with an object that was melting, moving, or even disappearing and an observer who must rely on his senses and physical labor far more than in the laboratory sciences, using coarse tools such as spades, skis, and measuring rods.[81] The strategy deployed by Ahlmann was to embrace the contradictions: striving for precision derived from instruments to enhance the credibility of his results, but at the same time inviting the more bodily experience, first by making himself, his coworkers, and his boundary objects into field props, and second by inviting those local residents who worked primarily through experiential, bodily knowledge to participate in the knowledge production with point measurements. What might have been a threat to credibility he could thus turn into a virtue. His field staff secured with their permanence, their experience, and their local savoir faire the universal nature of the

[77] Polanyi, *The Study of Man: The Lindsay Memorial Lectures* (Chicago, 1959); see Stephen Shapin and Christopher Lawrence, "The Body of Knowledge," in *Science Incarnate: Historical Embodiments of Scientific Knowledge*, eds. Shapin and Lawrence (Chicago, 1998), 1–20, on 6–7.

[78] Robert E. Kohler, *Landscapes and Labscapes: Exploring the Lab-Field Border in Biology* (Chicago, 2002), esp. chaps. 1 and 2.

[79] The classical study is S. L. Star and J. Griesemer, "Institutional Ecology, 'Translations' and Coherence: Amateurs and Professionals in Berkeley's Museum of Vertebrate Zoology," *Soc. Stud. Sci.* 19 (1989): 1907–39.

[80] Doreen Massey, *For Space* (Thousand Oaks, Calif., 2005), 119.

[81] Mark Carey, "The History of Ice: How Glaciers Became an Endangered Species," *Environ. Hist.* 12 (2007): 497–527; Garry K. C. Clarke, "A Short History of Scientific Investigations on Glaciers," special issue, *Journal of Glaciology* 33 (1987): 4–24.

knowledge that the transient field scientist could not achieve on his own. In passing, and without articulating the issue in those terms, Ahlmann thus also responded to Cuvier's classical objection to the *naturaliste voyageur*, who could only follow his own thin trace in space and time and was never able to get the full drama of the physical reality.[82] Ahlmann thought he got both.

Building a broader community in the field had strategic advantages. Characteristic of his style of fieldwork is the informality with which the labor was mobilized and organized and the significance of officially "unskilled" labor. In practice, building cryospheric knowledge in Ahlmann's fashion thus went far beyond the more austere and puritan precision approach that was the norm and that he himself advocated in his programmatic texts. However, he did not reveal in his scientific publications the full extent to which he relied on informal labor, a reflection of long-standing colonial and class distinctions in the history of field science. He did not seem sure of how to handle this information; he did mention some of the cooperation, and he gave more detail in his popular work, most evidently in *Land of Ice and Fire*. In his scientific knowledge claims and in his rhetoric of science and geopolitics, he did not have use for local residents, so they were omitted. There was an element of pragmatism that tempered his ethnopolitical ambition.

In a deep sense this had to do with the anxieties that were tied to the body as a category in relation to truth and evidence. Whose authority was to reign? How would he secure the universal character of the knowledge? One answer to questions like these was his food-chain approach to organizing the glacier knowledge enterprise. He gave everyone positions with incentives and rewards but also navigated so that his own position as the ultimate custodian of knowledge was securely elevated into the realms of pure science and the judgmental balance of diplomacy. He always remained discretely superior, a role that was again socially congruent with his diplomatic authority and scientific leadership.

I do not mean to say that all fieldwork was the same. There was a clear role for the field science specialist. The field "monuments," and the monitoring instruments in particular, also served as symbolic markers of the professional. Some of those who made knowledge claims lacked instruments and also field presence, which was increasingly a drawback as far as trust was concerned. The outsider, weakly connected to professional networks and not based in a science institution, would not be able to collect data in the same way as the established scientist. The obvious example in the history of climate forcing was Callendar, who turned out to have an extremely good hypothesis but was not widely recognized. Ahlmann, on the contrary, held on to a theory that stopped short, but was acclaimed and popular among his peers.

We may wish to regard Ahlmann's entire research effort as a major, comprehensive act of science diplomacy, all geared toward building trust and confidence and easing the anxieties of uncertainty in a complex social coproduction of ephemeral field knowledge. All his various science-related activities supported each other: his obsession with data and precision, his untiring efforts to build networks of confidence among his leading peers, his political and institutional ambitions, his careful cultivation of the media both as an expert in news reports and increasingly as an essay contributor to the leading Stockholm newspapers, and his growing professional

---

[82] Dorinda Outram, *Georges Cuvier: Vocation, Science and Authority in Post-revolutionary France* (Manchester, 1984).

role as a diplomat and policy actor. In this picture one can also locate his cooperation with local residents. It had obvious pragmatic purposes, but at the same time he widened the base of his knowledge claim by linking his findings and methods not just to the polar scientist as a "specialist" but also to collaborators with experiential knowledge. In his programmatic article in 1931 he in no way opposed local knowledge or vernacular experience; he just disregarded them, having not yet discovered their full potential, and he kept disregarding them in texts where he spoke to his peers. What he attacked was never local knowledge but the presumptuous "sport" version of science that wanted to break records and reach poles and peaks and that went hand-in-hand with media sensationalism. In that sense his ideal was the opposite of what has been described as the faith in "measurement by means of 'muscular exertion'" that was characteristic of strands of Victorian geophysical field science.[83] Ahlmann's field science was a hybrid, where body was subaltern to instrument.

Although methodologically and theoretically a pivotal effort, Ahlmann's program had clear defensive dimensions. Climate was infinitely complex. His disbelief in Callendar was obvious, but did he, after all, have any solid evidence himself? A truly open and less anxious mind would perhaps have broadened the research program to allow for competing ideas. Ahlmann did not. He was not trained in modeling climate or thinking big about atmospheric change. He was tuned to measurement and stayed on the singular path of polar warming. And he was cautious, not just because it was a scientific virtue but because he was fundamentally uncertain. Deep inside he was anxious, and in some moments his diplomatic façade broke and revealed a pernicious hope of revelatory empirical rescue. One such moment was on October 11, 1936, when Ahlmann was still polishing the last chapter of his *Land of Ice and Fire* manuscript. He brought up the issue of the measuring poles at Hoffellsjökull, saying that they needed to be moved up the glacier and that measurements should continue through the winter, and in a mysterious postscript he added, "Is there any value for the ablation during the night *Pourquoi-pas* [a ship] capsized?"[84] The question can only be understood in relation to his theory that warm and strong winds could explain much of the ablation. So what about this perfect storm? Would the strong winds that night be able to corroborate his deglaciation theory? In some moments even a meticulous scientist resorts to faint hope rather than systematic evidence.

Ahlmann's defeat came quite unexpectedly and from a very close source. As more and more evidence over the course of the 1950s suggested that there was good reason to take a new look at CO_2 in the tradition of Arrhenius and Callendar, his old friend and close colleague Rossby was, as always when important new things were going on, right at the heart of the developments.[85] Rossby and his team in Stockholm as well as his colleagues in the United States brought computers to bear on issues of weather predictions and climate change and subsequently introduced CO_2 into their equations. This whole line of work, taken up around the world in the decades to come, would ultimately prove devastating to the idea of polar warming. This was not easy to take. In fact, some members of the Ahlmann school became staunch

---

[83] "Muscular exertion" was an expression used by alpinist and poet Leslie Stephen, but the spirit of combining sport, spirituality, and scientific observation was promoted by British glaciologists such as James David Forbes and John Tyndall. See Bruce Hevly, "The Heroic Science of Glacier Motion," in Kuklick and Kohler, *Science in the Field* (cit. n. 2), 66–86, on 66, 84.

[84] Ahlmann to Eythórsson, 11 October 1936, JE Vatnajökull.

[85] See Maria Bohn in this volume.

skeptics of ideas of global warming, scrupulously refusing to leave the paradigm of
their founding father.[86] Inadvertently they illustrated the path dependence of field
practices and installations.

The understanding of climate change that Ahlmann undertook with his combination of elite and vernacular field science over five decades, spanning two world
wars and two polar regions, comes across today as an eminent example of a coproduction of knowledge in the field, a modernist paradigm of field and polar science,
and an admirable episode of North Atlantic mobilization of resources, private and
public, scientific and popular, local and metropolitan, Eastern and Western, military and peace-building. Still, through the usual irony of history, the knowledge and
the undisputable theoretical insights that were produced were soon superseded and
then largely forgotten. The reason we should remember Ahlmann's polar warming
is therefore not that it was a precursor to global warming, which it was not. It is precisely for the opposite reason: it cautions us to realize that the history of global warming was not a linear affair.

[86] Sörlin, "Narratives and Counter-narratives" (cit. n. 28), on 254–5.

Diagnosing the Dry:
Historical Case Notes from Southwest Western Australia, 1945–2007

by Ruth A. Morgan*

ABSTRACT

Long regarded for its reliable winter rainfall, the Southwest region of Western Australia was beset by unexpected dry conditions in the early 1970s whose persistence was baffling. The gradual growth of scientific interest in the region's rainfall, as this article contends, was strongly influenced by political, social, and economic concerns about the challenges posed by drought and climate change. The experience of rainfall decline coincided with international scientific and political interest in the global climate and the perception that it was deviating from its "normal" state. Indeed, this extended "dry" provided an Australian link to international concerns regarding anthropogenic global warming. This article argues that the historical, political, and economic importance of the Southwest's agricultural industries has led policy makers and researchers to perceive the region's changing climatic conditions as pathological and in need of diagnosis.

INTRODUCTION

Celebrated Australian poet Dorothea Mackellar deeply missed her homeland while visiting England in 1904. She longed for Australia, that land of "droughts and flooding rains," a brutal land of "flood and fire and famine," "a wilful, lavish land."[1] These lines from her poem "My Country" capture the distinctive variability that is characteristic of the Australian climate and environment. This variability has molded the face of the continent, leaving imprints on its landforms, soils, plant and animal life, and people. A particularly unusual characteristic of the Australian climate is its highly variable rainfall. Yet in a continent of extremes, there are also exceptions: for instance, the Southwest region of the state of Western Australia (WA). Its Mediterranean conditions of wet winters and dry summers were so favorable to European farming that geographer Griffith Taylor named the region "Westralia felix."[2] Indeed, the region has been progressively settled by Europeans since 1826 and is now home

---

* History M208, University of Western Australia, 35 Stirling Hwy., Crawley, Western Australia, 6009, Australia; ruthmorgan@mac.com.

The author is grateful to Jim Fleming and Vladimir Jankovic for organizing the conference from which this volume resulted; the participants at the conference for their feedback; and those who have given detailed and insightful comments on earlier drafts.

[1] Dorothea Mackellar, "My Country," in *The Witch-Maid and Other Verses* (London, 1914), 29–31.
[2] Henry A. Hunt, Griffith Taylor, and Edwin T. Quayle, *The Climate and Weather of Australia* (Melbourne, 1913), 37.

---

to over 80 percent of the state's population. Since the 1970s, however, persistent droughts and low rainfall have changed these perceptions of the region, particularly within scientific and policymaking circles. Scientific investigation into the causes and extent of the regional rainfall decline developed slowly. Only since the late 1990s has there been a close engagement between scientists and policy makers on the issue, arguably due to its association with anthropogenic global warming.

The story of Australia's political, social, and economic development is closely bound to the country's agricultural industries. Accordingly, the national scientific enterprise has been directed at enabling successive generations of farmers and their governments to transplant European farming practices successfully to the Australian continent. In the nineteenth and early twentieth centuries, according to Libby Robin, the applied sciences were vital to the fulfillment of the young nation's role in the imperial economy.[3] As a result, the relations between government and science are historically strong, founded on the notion that "scientific discovery could render valuable economic service to [the] developing country."[4] Boris Schedvin similarly describes the incorporation of applied sciences, such as biology and ecology, into the Australian "productive system."[5] In the face of economic and environmental volatility, farmers and elected officials saw in science a means to improve economic efficiency and protection from the elements.[6] It was in the postwar period, however, that Australians developed a renewed optimism and faith in regard to science and technology. As Roderick Home argues, "Australians looked to science to help them [overcome the environment], and Australian scientists willingly shouldered the responsibility."[7] Nowhere was this thinking more apparent than in the scientific effort to understand and predict the Australian weather and climate. The scientific research into the dry conditions of the Southwest of WA, therefore, is one of the more recent episodes in the historical relationship between the state and science in Australia.

The gradual growth of scientific interest in diagnosing the Southwest's "dry," as this article contends, was strongly influenced by political, social, and economic concerns about the challenges posed by drought and climate change. Although closely related, these concepts must be unraveled within the Australian context. Drought conditions are a familiar but feared enemy in Australia, with stories of dust, hunger, and hardship embedded in the national story since European settlement. In contrast, anthropogenic global climate change is a relatively recent specter for Australians.[8] It is a problem that is often depicted in the media and by policy makers as one that will be faced in the distant future, and indeed it appears that way, while drought is better understood as an immediate and very real experience. The dry spell in the Southwest of WA since the 1970s, however, conflates these concepts, as the Indian Ocean

[3] Robin, *How a Continent Created a Nation* (Sydney, 2007), 215.

[4] Libby Robin, "Ecology: A Science of Empire?" in *Ecology and Empire: Environmental History of Settler Societies*, eds. Tom Griffiths and Robin (Melbourne, 1997), 63–75, on 65.

[5] Schedvin, "Environment, Economy and Australian Biology, 1890–1939," *Historical Studies* 21 (1984): 11–28.

[6] Ibid., 15.

[7] Home, "Rainmaking in CSIRO: The Science and Politics of Climate Modification," in *A Change in the Weather: Climate and Culture in Australia*, eds. Tim Sherratt, Tom Griffiths, and Libby Robin (Canberra, 2005), 66–79, on 78.

[8] Many Australians were preoccupied with anthropogenic regional climate change from the nineteenth century. Some causes of such climatic changes were commonly thought to be agricultural development ("the rain follows the plow") and tree clearing. See, e.g., Neville Nicholls, "Climate and Culture Connections in Australia," *Australian Meteorological Magazine* 54 (2005): 309–19.

Climate Initiative (IOCI) has found that the enhanced greenhouse effect is partly responsible for the regional rainfall decline.[9] For several decades then, the Southwest has been experiencing those drier conditions that many scientists predict will afflict the southern region of Australia as carbon dioxide levels increase into the middle of the twenty-first century. Because of these relatively unusual circumstances, the Southwest has been likened to a "national canary"—a portent of Australia's climate future.[10]

Although the scientists involved in investigating this climatic exceptionalism of the Southwest rarely used such metaphors, their activities are analogous to a medical approach. In this scenario, an existing climate regime, as defined by scientific data and local experience, is seen as healthy or "normal" and any deviation as "patho-logical" or diseased. The affected region becomes the patient and the scientist its physician, who seeks the causes of the affliction and pursues remedies to ameliorate the patient's condition. James Lovelock, for instance, has adopted such language, describing himself as a "planetary physician" in his works on the Gaia hypothesis.[11] The implicit assumption in this medical scenario is that the patient can be cured and/or returned to its normal state, that its symptoms are merely temporary and can be overcome by the correct treatment. Attempts to overcome these symptoms are made difficult, however, by the relatively limited knowledge of regional climates.

This faith in science is also apparent in the political and cultural approaches to drought in Australia, particularly in rural settings. As farming methods and expecta-tions of agricultural productivity and profitability were developed in particular cli-matic conditions, it is important that these conditions remain constant. Yet as the national approach to drought has changed over time, so too have the physicians' di-agnoses of the Southwest's dry conditions. This article explores how the strands of drought and climate change have become intertwined with the development of scien-tific and political interest into their effects on the Southwest region of WA. Through the framework of the medical metaphor, it argues that political concern for the eco-nomic health of the Southwest's agricultural industries, combined with growing en-vironmental concerns and climate change discourses, has guided scientific research into the drying region.

DEFINING SOUTHWEST WESTERN AUSTRALIA

The Southwest of WA extends southwest of a line that connects the western coast-line south of Geraldton to the south coast near the town of Esperance (29°S to 35°S, 114°E to 120°E; see fig. 1).[12] This region enjoys annual rainfall ranging from about 400 millimeters in the east to about 1,400 millimeters on the south coast and cov-ers an area of approximately 314,000 square kilometers. It exhibits the character-istics of a Mediterranean climate of cool, wet winters and dry summers, with over

---

[9] Bryson C. Bates et al., "Key Findings from the Indian Ocean Climate Initiative and Their Impact on Policy Development in Australia," *Climatic Change* 89, nos. 3–4 (2008): 339–54.

[10] Brian Sadler, "Informed Adaptation to a Changed Climate State: Is Southwestern Australia a National Canary?" Indian Ocean Climate Initiative Web site, http://www.ioci.org.au/pdf/IOCI_Paper-Jan6.pdf (accessed 3 March 2008), 1.

[11] See James Lovelock, *Gaia: The Practical Science of Planetary Medicine* (Sydney, 1991); Lovelock, *Gaia: Medicine for an Ailing Planet* (London, 2005).

[12] Bates et al., "Key Findings from the Indian Ocean Climate Initiative" (cit. n. 9), 43.

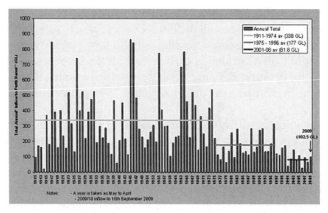

Figure 1. *Reduced inflows to Perth dams, 1911–2009. Image courtesy of the Water Corporation of Western Australia.*

80 percent of annual rain falling between April and October.[13] In the early 1970s, the region underwent a change in the state of its climate. Early winter (May, June, and July) rainfall underwent what water managers describe as a "step-decline," a sudden but persistent drop in rainfall, which particularly affected Perth's surface water supplies (fig. 1). Compared to the earlier half of the century, rainfall decreased by 10–15 percent in most parts of the region, and as much as 20 percent in some inland areas. In addition to this reduction in quantity, the spatial pattern of rainfall has altered, with the 170, 300, and 500 millimeter isohyets shifting toward the southwest coast.[14]

Postwar Economic Development

Since the European settlement of WA in 1826, agricultural development in the state's Southwest region has been an important priority of successive governments. In the early years, agriculture was necessary to feed the colony, with the long-term goal that farmers would be able to produce a surplus for export to other colonies and dominions in the British Empire. Yet the Southwest was not quite the "Westralia felix" that the colonists had hoped for. The continent's ancient but infertile soils raised barriers to the transplanting of European farming techniques. It was only through government and private investment in the soil and plant sciences that Western Australians were eventually able to overcome their environment, to tame the "wilful land" and pursue aggressive policies of agricultural development into the state's marginal lands.

These policies continued unabated into the post–World War II period. Although successive state governments focused on attracting industrial investment after the war, the traditional cornerstone of the WA economy, agriculture, was not overlooked. Indeed, an important feature of the government's development strategy was a com-

---

[13] Anne Brearley, *Ernest Hodgkin's Swanland: Estuaries and Coastal Lagoons of Southwestern Australia* (Crawley, 2005), 13–5.

[14] Don McFarlane, *Context Report on Southwest Water Resources for Expert Panel Examining Kimberley Water Supply Options* (Canberra, 2005), 19. Isohyets are lines representing areas with equal rainfall amounts.

mitment to "unbiased development," to both agricultural and industrial expansion.[15] A former director of the WA Department of Agriculture conveys this dominant approach to land use: "Our duty is to develop our virgin lands, to complete the development of the partially developed, and to increase the productivity of the developed lands until the maximum of which they are capable is reached."[16] This agenda of expansion was supported and sustained by technological and scientific endeavor, particularly in the areas of crop varieties, soils, farming machinery, and meteorology.

In 1943, the commonwealth government appointed the Rural Reconstruction Commission to establish a program of agricultural expansion that would ensure the nation's speedy recovery from wartime austerity. One of the tasks assigned to the commission was the settlement of returned servicemen and their families on the land. This led to the establishment of the War Service Land Settlement Scheme in 1945. In WA families of veterans were allotted properties throughout the Southwest, primarily in the south of the state near Albany and Esperance. By the late 1950s, this program had merged into the state government's New Farm Lands Scheme, which operated into the late 1960s. These programs involved the large-scale clearing of land, which the government boasted to have achieved at the rate of "a million acres a year." Land-clearing machinery had developed significantly since the Second World War to include surplus army tanks, bulldozers, crawler tractors, ship anchor chains, and heavy disc plows.[17] These methods were also used to clear the undeveloped land on established farms.[18] As a result of these destructive practices, the scale of land clearing was unprecedented in national history, with the area of land under cultivation in WA nearly doubling from 14 million to 25 million acres by the 1960s.[19]

The state and federal agricultural technocracy swelled to support this great national expansion on the land, and farmers were increasingly receptive to the latest scientific findings. The WA Department of Agriculture increased the number of its research stations and carried out field trials to assess fertilizers and trace-element needs for cereals and to improve the techniques of land preparation. Their findings were communicated to rural producers by an ever-growing network of advisory and extension services.[20] More scientific knowledge, more land under cultivation, the breaking of wartime shackles on overseas markets, and the increased production of Western Australian wheat and wool fostered optimism in the state's agricultural sector.

Farmers also benefited from the growing understandings of the Australian climate within the Commonwealth Bureau of Meteorology and the recently expanded Commonwealth Scientific and Industrial Research Organisation (CSIRO). Soon after the war, the bureau and CSIRO negotiated to divide the study of the country's weather. The bureau would collect data, monitor weather patterns, and issue forecasts, while

[15] Lenore Layman, "Development Ideology in Western Australia, 1933–1965," *Historical Studies* 20 (1982): 234–60.

[16] George L. Sutton, 1952, quoted in Barbara York Main, "Social History and Impact on Landscape," in *Reintegrating Fragmented Landscapes: Towards Sustainable Production and Nature Conservation*, eds. Richard J. Hobbs and Denis A. Saunders (New York, 1993), 23–64, on 50.

[17] George H. Burvill, "The Last Fifty Years, 1929–79," in *Agriculture in Western Australia: 150 Years of Development and Achievement, 1829–1979*, ed. George H. Burvill (Nedlands, 1979), 57–65.

[18] Main, "Social History and Impact on Landscape" (cit. n. 16), 50.

[19] Quentin Beresford, Hugo Bekle, Harry Phillips, and Jane Mulcock, *The Salinity Crisis: Landscapes, Communities and Politics* (Crawley, 2004), 61–3.

[20] These were similar in purpose and function to the experimental stations and extension services established in the United States from the 1880s. See Willard W. Cochrane, *The Development of American Agriculture: A Historical Analysis* (Minneapolis, 1993), 105–6.

CSIRO would undertake "fundamental research into the physics of the atmosphere."[21] The importance of science to national economic development had been recognized following World War I, and this role was consolidated by the contributions of these agencies to the Allied war effort in the Second World War. Schedvin convincingly argues that during the postwar period, scientists became "agents of economic development" at a "time of unprecedented confidence in scientific endeavor."[22] The outcomes of scientific research, applied science and technology, were the means by which the country's existing resources could be intensively exploited.[23] Indeed, policy makers focused especially on the agriculture sector to provide primary produce for export, which would help the nation avoid the balance-of-payment difficulties that it had faced in the interwar period.

Developments within the Bureau of Meteorology and CSIRO complemented this applied science and technology agenda. After the war, the meteorological community continued to support the fortunes of the agricultural sector by supplying farmers and the government with specialist knowledge of the WA climate. The bureau also established climatological sections in each state to provide information on the suitability of particular regions' climates for the economic development of agricultural and pastoral industries.[24] This research also served to define the limits of the normal climate conditions for the continent's farming areas. CSIRO also extended its research into the bureau's territory by creating a Division of Meteorological Physics. The mission of CSIRO was to "promote scientific research for the benefit of primary and secondary industries."[25] The new division's research would, therefore, improve the state of knowledge of the climate processes that had long affected the nation's primary producers.

Joy McCann argues that this postwar era is memorialized by the rural community as the "golden age" of the Australian wheat lands and, as a result, has "crystallized . . . as the period in which the normative state of Australian agriculture was achieved."[26] Indeed, it was during this period of favorable economic and weather conditions that expectations for the farming regions were set in terms of productivity, land clearing, profitability, and overall rural prosperity. These "healthy" and "normal" circumstances in the rural sector became the yardstick against which to measure future agricultural experience.

This prosperity, however, had its environmental costs, with soil salinity problems arising in some areas from the widespread and rapid land clearing. Although scientists were well aware of and concerned about such salinity problems, the WA government and farmers repeatedly ignored them in favor of continued development. Not only was addressing the problem too expensive and complex to deal with, but the problem

[21] Home, "Rainmaking in CSIRO" (cit. n. 7), 67.
[22] Boris Schedvin, *Shaping Science and Industry: A History of Australia's Council for Scientific and Industrial Research, 1926–49* (North Sydney, 1987), 346.
[23] Schedvin, "Environment, Economy and Australian Biology" (cit. n. 5), 11.
[24] Such arrangements were also present in the United States from at least the 1880s. See Christopher A. Fiebrich, "History of Surface Weather Observations in the United States," *Earth-Science Reviews* 93, nos. 3–4 (2009): 77–84.
[25] John R. Garratt, David Angus, and Paul Holper, *Winds of Change: Fifty Years of Achievements in the CSIRO Division of Atmospheric Research, 1946–1996* (Collingwood, 1998), 3.
[26] Joy McCann, "History and Memory in Australia's Wheatlands," in *Struggle Country: The Rural Ideal in Twentieth Century Australia*, eds. Graeme Davison and Marc Brodie (Melbourne, 2005), http://publications.epress.monash.edu/toc/sc/2005/1/1 (accessed 17 May 2009), chap. 3, 5.

itself revealed that agricultural expansion had environmental limits.[27] To accept the scientists' advice to limit land clearing would be to admit that nature could still, even in the face of postwar optimism, pose an impediment to economic development.

Although the agricultural industries prospered during this postwar period, regions in the east of the continent suffered periods of drought. The dry period lasted from about 1958 to 1970, affecting areas in Queensland, New South Wales, and Victoria with varying intensity. Public pressure, particularly from farming communities, mounted on the Bureau of Meteorology to modify the weather artificially through cloud seeding and to provide predictions of such dry periods. Droughts and flooding rains might have been expected as part of the continent's bodily rhythms, but it was economically important to know when the weather might depart from normal or expected conditions. Long-range weather forecasting would enable farmers to alter their seasonal farming and grazing practices to adapt to deviations in the weather. Although the media and farming communities seized eagerly upon forecasts made by colorful personalities such as Queensland's Inigo Jones and his assistant Lennox Walker, the ability to make long-range forecasts remained beyond the bureau's reach, while cloud-seeding experiments across Australia proved inconclusive.[28]

The End of the Boom

Having been spared most of the dry conditions of the past decade, Western Australian farmers had little warning before the arrival of the devastating drought in the winter of 1969. This drought was especially severe, and farmers required government assistance for fodder and grain, water, and the transport of stock to more favorable areas. As drought conditions persisted into the early 1970s, many farms in the Southwest region invested in private irrigation programs and undertook water conservation strategies. The plight of farmers was compounded by the introduction of wheat quotas in 1969. The quotas represented the Australian federal government's response to the oversupply of wheat stocks on the world market. They resulted in the reduction of the area sown with wheat by a third and compounded the effects of inflation on the farmers' costs of production.[29] The severity and impact of the drought led to research at the University of Western Australia's Department of Agronomy into how often such deficiencies in rainfall might occur and the effects of atmospheric circulations on the geographic pattern of rainfall.[30] Although these studies did not identify a pattern of recurrence, they documented a decrease in late-winter rainfall (August, September, October) since the early 1940s.[31] Furthermore, the exploration of the relationships between the region's climate and the atmosphere's general circulation indicated possibilities for making predictions of climatic variability, particularly in light of

[27] Geoffrey C. Bolton, *Spoils and Spoilers: Australians Make Their Environment, 1788–1980* (North Sydney, 1981), 138.

[28] Tim Sherratt, *Inigo Jones: The Weather Prophet* (Melbourne, 2007); Bryan F. Ryan and Brian S. Sadler, *Guidelines for the Utilisation of Cloud Seeding as a Tool for Water Management on Australia* (Canberra, 1995), http://www.cmar.csiro.au/e-print/open/cloud.htm (accessed 21 October 2009).

[29] David Black, "Liberals Triumphant: The Politics of Development, 1947–1980," in *A New History of Western Australia*, ed. Tom Stannage (Nedlands, 1981), 441–72.

[30] Eugene A. Fitzpatrick, *The Expectancy of Deficient Winter Rainfall and the Potential for Severe Drought in the Southwest of Western Australia* (Perth, 1970), 1, 3.

[31] Peter B. Wright, *Spatial and Temporal Variations in Seasonal Rainfall in Southwestern Australia* (Nedlands, 1971), 41–2.

CSIRO research on the Southern Oscillation and other large-scale circulation patterns.

Although most Australians were familiar with the vagaries of the weather, the events of the 1970s, both locally and overseas, raised concerns that the weather was changing in a dramatic fashion. Drought followed by heavy rains hit the eastern states of Australia between 1972 and 1974 as a result of the El Niño and La Niña phases of the El Niño-Southern Oscillation (ENSO). The rains caused major flooding but also led to above-average crops in well-drained lands. Although WA shared these conditions in those years, the rest of the decade was particularly dry for the state's agricultural areas. The state's farmers again urged the scientists of the Bureau of Meteorology and CSIRO to undertake cloud-seeding experiments over the drought-affected region. Although mild by comparison, these circumstances echoed the widely reported droughts in the Sahel (1972–3) and the Ukraine (1972) and the failure of the Indian Monsoon (1972). Adding to the gloom of widespread drought and famine was the 1972 publication of the Club of Rome's Malthusian *Limits to Growth*, which intensified global concerns that the world was rapidly depleting its limited resources.

Many scientists, meanwhile, were debating the causes and consequences of what appeared to be a global cooling trend. Some climatologists suggested that the world's climate was progressing toward another glacial phase, or a "little ice age."[32] With an ever-growing world population and most arable land already under cultivation, the scientific consensus was that any change in the climate, whether warming or cooling, would severely affect the world's food supply. Following the World Food Conference in November 1974, the federal Labor government requested that the Australian Academy of Sciences (AAS) investigate the possible effects of climatic change on world and Australian agricultural production.[33] While noting the dry period in the Southwest of WA, the AAS concluded that there was no evidence of significant climatic change, but cautioned that the effects of anthropogenic climate change might become apparent in the future.

It appeared that the era of postwar optimism about unlimited development was drawing to a close with insufficient food and energy resources to sustain the ever-growing global population. The severe weather conditions abroad resonated with the floods and droughts sweeping the Australian continent during the 1970s. Like its overseas counterparts, the Australian government was naturally concerned by these unusual circumstances, particularly their adverse impacts on agricultural production and rural livelihoods. The federal government turned, therefore, to the nation's physical scientists at the Bureau of Meteorology and CSIRO. Only through an enhanced knowledge of the intricacies of the Australian environment could the country's agricultural producers hope to better manage the risks faced by the continent.

COMORBID CONDITIONS

In the wake of the 1970s oil crises, researchers' and policy makers' attentions turned to the question of future levels of world energy use and alternative fuel sources. An outcome of this research was recognition of the relationship between the increasing use

[32] James R. Fleming, *Historical Perspectives on Climate Change* (New York, 1998), 132.
[33] Australian Academy of Science, *Report of a Committee on Climatic Change* (Canberra, 1976). Note that the AAS has no statutory obligations to government.

of fossil fuels and increases in the carbon dioxide content of the atmosphere.[34] There had been growing concerns regarding the effects of such increases on the world's climate since at least the 1940s.[35] Although some scientists supported a global cooling hypothesis, scientific research conducted in the United States in the late 1970s found that increasing levels of carbon dioxide created a "greenhouse effect," which induced a warming of surface air temperatures.[36] This warming would be "accompanied by shifts in the geographical distributions of the various climatic elements such as temperature, rainfall, evaporation and soil moisture."[37]

One of the recommendations of the World Meteorological Organization's Climate Conference in 1979 was the need for further research. The following year, scientists presented research relevant to the relationship between carbon dioxide and climate in the Australian context at an interdisciplinary symposium held by the AAS in Canberra.[38] Since the early 1970s, CSIRO scientists had been examining the carbon dioxide exchange between wheat plants and the atmosphere as part of a broader study on the effect of environmental factors on crop growth.[39] This research project expanded as international concerns grew about fossil fuel use and the rising concentrations of carbon dioxide in the atmosphere. Although scientific knowledge concerning climate processes in the Southern Hemisphere was relatively limited, the AAS symposium recognized the need to estimate the potential impacts of increased carbon dioxide on regional climates so as to enable the projection of the social and economic consequences of climatic change.[40] This task required the study of recorded precipitation patterns to assess the accuracy of the climate models used to predict climate conditions under increased levels of carbon dioxide.[41] One of the participating scientists, Barrie Pittock, who had been researching patterns of rainfall variation in the instrumental record since the early 1970s, suggested that regional climatic changes, such as those in the Southwest, were indicative of what might occur in Australia with further increases of carbon dioxide concentrations— that is, under enhanced greenhouse conditions.[42] In drawing this conclusion, Pittock interpreted the regional rainfall decline as a possible symptom of "the greenhouse syndrome."[43]

The onset of the severe 1982/3 El Niño, however, overshadowed the forecast of a greenhouse future, while ensuring that the climate and weather remained firmly in the minds of the public in Australia and around the world. In the eastern states of Australia, vast tracts of agricultural land were "drought declared" as farm dams were reduced to dusty basins and farmers were forced to hand-feed their stock. With little

[34] Graeme I. Pearman, "Preface," in *Carbon Dioxide and Climate: Australian Research*, ed. Pearman (Canberra, 1980), i–iv.

[35] Fleming, *Historical Perspectives* (cit. n. 32), 107–28; Fleming, *The Callendar Effect: The Life and Work of Guy Stewart Callendar (1898–1964), the Scientist Who Established the Carbon Dioxide Theory of Climate Change* (Boston, 2007), 65–87.

[36] Fleming, *Historical Perspectives* (cit. n. 32), 131.

[37] Barrie Pittock, "Towards a Warm Earth Scenario for Australia," in Pearman, *Carbon Dioxide and Climate* (cit. n. 34), 197–209.

[38] Pearman, *Carbon Dioxide and Climate* (cit. n. 34).

[39] Brad Collis, *Fields of Discovery: Australia's CSIRO* (Crows Nest, 2002), 353.

[40] Pittock, "Towards a Warm Earth Scenario" (cit. n. 37), 197.

[41] Stephen Schneider, "Climate Change and the World Predicament: A Case Study for Interdisciplinary Research," *Climatic Change* 1 (1977): 21–43.

[42] Pittock, "Recent Climatic Change in Australia: Implications for a CO_2-Warmed Earth," *Climatic Change* 5 (1983): 321–40.

[43] Jim Falk and Andrew Brownlow, *The Greenhouse Challenge: What's to Be Done* (Ringwood, 1989), 38.

vegetation to anchor them, the topsoils from these drought-affected areas eroded and were swept into dust storms that enveloped the region. These circumstances conspired to form the ideal conditions for the ferocious Ash Wednesday bushfires that swept through the states of South Australia and Victoria in February 1983.[44] Overseas, the El Niño event caused droughts across southern and southeast Asia and unseasonal typhoons in French Polynesia.[45] It was also associated with brush fires in the Côte d'Ivoire and Ghana and outbreaks of disease in the eastern United States. The global impacts of this El Niño event triggered a major research effort within the international scientific community toward understanding the ENSO phenomenon, particularly its origins in the interactions between the ocean and atmosphere.[46] Leading the endeavor in Australia was Neville Nicholls at the Bureau of Meteorology. As seasonal forecasting for primary producers was a desirable skill for the bureau, Nicholls thought that a better understanding of atmospheric processes, such as ENSO, might make the prediction of Australian droughts possible.[47] He would later attribute various features of the Australian climate, particularly its rainfall variability and the large spatial scales of its droughts and wet periods, to this elusive phenomenon.[48]

Following the severe Australian droughts of the early 1980s, calls for improved forecasting techniques continued throughout the decade. Australian drought-related research finally converged with the climate change agenda at the Greenhouse87 conference at Monash University in Victoria, Australia. This conference represented the culmination of Australian research efforts following two significant developments in the mid-1980s. In 1985, at the joint United Nations Environment Program, World Meteorological Organization, and International Council of Scientific Unions (ICSU) meeting in the Austrian town of Villach, participating scientists agreed that increasing concentrations of greenhouse gases would lead to an unprecedented rise in global mean temperature in the first half of the twenty-first century. More important for this argument, they also concluded that climate data from the past could no longer provide a reliable guide to future conditions.[49] The limits of "normal" required redefinition, and resource managers had to expect deviations from established norms. Any pathological conditions might therefore be related to increasing concentrations of greenhouse gases. The outcomes of the Villach meeting were reinforced by the publication of a major report by the Scientific Committee on Problems of the Environment, a committee of the ICSU.[50]

To inform policy makers and the general public of the implications of such scien-

[44] Peter Whetton, "Floods, Droughts and the Southern Oscillation Connection," in *Windows on Meteorology: Australian Perspective*, ed. Eric K. Webb (Collingwood, 1997), 180–99.

[45] Michael H. Glantz, *Currents of Change: Impacts of El Niño and La Niña on Climate and Society*, 2nd ed. (New York, 2001), 87–9.

[46] Michael H. Glantz, "Forecasting El Niño—Science's Gift to the 21st Century," in *El Niño: Overview and Bibliography*, ed. A. M. Babkina (New York, 2003), 29–40.

[47] See Neville Nicholls, "Towards the Prediction of Major Australian Droughts," *Australian Meteorological Magazine* 33 (1985): 161–6.

[48] See Neville Nicholls, "El Niño-Southern Oscillation and Rainfall Variability," *Journal of Climate* 1 (1988): 418–21.

[49] World Meteorological Organization, *Report of the International Conference on the Assessment of the Role of Carbon Dioxide and of Other Greenhouse Gases in Climate Variations and Associated Impacts* (Geneva, 1986), http://www.icsu-scope.org/downloadpubs/scope29/statement.html (accessed 8 April 2008).

[50] Bert Bolin, Bo R. Döös, Jill Jäger, and Richard A. Warrick, eds., *The Greenhouse Effect, Climatic Change, and Ecosystems* (New York, 1986). See also Matthias Dörries in this volume.

tific findings, the federal Labor government worked with CSIRO to convene the Greenhouse87 Conference. Although the Australian government had not been involved in the developing international dialogue on the greenhouse effect,[51] the growing public concern surrounding the issue resonated with Labor's environmental mandate. To demonstrate the Labor Party's green credentials, particularly to younger voters in an election year, the government provided strong support for national scientific research into the possible implications of the greenhouse effect. The basis of the 1987 conference discussions was Pittock's scenario of the possible effects of climate change on Australia by 2030.[52] The scenario depicted a significantly drier future for the Southwest of the continent, with reduced rainfall and stream flow in the region.[53] Participating hydrologists from the Water Authority of WA postulated that this expected shift in rainfall might have already commenced in about 1970.[54] The conference participants recommended greater understanding of factors affecting rainfall, such as ENSO, to improve seasonal rainfall prediction, and international cooperation on research projects, particularly within the South Pacific region and the Southern Hemisphere. The conference findings were disseminated to communities across the continent through the Greenhouse88 meetings of the following year. This was particularly timely, as the agricultural districts of Western Australia were experiencing drought conditions, which were affecting wheat returns and causing stock losses.

This emergent Australian greenhouse agenda mirrored international developments, which had accelerated during the 1980s. According to Daniel Bodansky's outline of the budding climate change regime, the Australian government and its overseas counterparts were engaged in the "agenda-setting phase . . . when climate change was transformed from a scientific into a policy issue."[55] Overseas, particularly in the United States, a small group of "environmentally concerned" scientists worked tirelessly to inform the Western public and its governments about the growing scientific knowledge of the greenhouse effect.[56] A series of well-publicized international meetings on the issue in the mid-1980s and the alarming testimonies of climate experts such as NASA's Jim Hansen before U.S. congressional committees resonated with the public's concern for other issues, such as the health of the ozone layer.[57] Their concerns were seemingly justified when the soaring temperatures of the North American summer of 1988 incited drought and extensive forest fires, echoing the Australian experience of the 1982/3 El Niño. These global anxieties over the possibility of a changing climate reinforced the view that deviations from normal climate conditions were pathological and, thus, cause for concern.

The seemingly increased volatility of the world's climate could not be easily dis-

[51] Robert J. Fowler, "Policy and Legal Implications of the Greenhouse Effect," in *Greenhouse: Planning for Climate Change*, ed. Graeme I. Pearman (East Melbourne, 1988), 694–707.

[52] Barrie Pittock et al., "Appendix: Climate Change in Australia to the Year 2030AD," in Pearman, *Greenhouse* (cit. n. 51), 737–40.

[53] Ibid.; Barrie Pittock, "Actual and Anticipated Changes in Australia's Climate," in Pearman, *Greenhouse* (cit. n. 51), 35–51.

[54] Brian Sadler, Bob Stokes, and Geoff Mauger, "The Water Resource Implications of a Drying Climate in Southwest Western Australia," in Pearman, *Greenhouse* (cit. n. 51), 296–311.

[55] Bodansky, "The History of the Global Climate Change Regime," in *International Relations and Global Climate Change*, eds. Urs Luterbacher and Detlef F. Sprinz (Cambridge, Mass., 2001), 23–40, on 23.

[56] Ibid., 27.

[57] Ann Henderson-Sellers, "Australian Public Perception of the Greenhouse Issue," *Climatic Change* 17 (1990): 69–96.

missed. Nor could the findings of various scientific meetings that increased carbon dioxide concentrations in the atmosphere would result in higher temperatures in the next century. For the Southwest, as scientists discovered, these were not separate problems: they merely represented two sides of the same climate coin.

CHECKING FOR SYMPTOMS

The concerns expressed by WA hydrologists at the Greenhouse87 conference led to closer scrutiny of the Southwest's rainfall decline under a new research agreement between CSIRO and the WA Environmental Protection Authority. Following a "decision in principle" at the Australian Environment Council meeting in July 1988, several of the state and territory governments had entered into research agreements with the CSIRO Division of Atmospheric Research to study the regional implications of the enhanced greenhouse effect.[58] These arrangements were prepared in the wake of the Toronto declaration earlier that year, which aimed to reduce carbon dioxide emissions to 80 percent of 1988 levels by 2005 in order to avoid or mitigate anthropogenic atmospheric change. Like its counterparts in New South Wales and Victoria, the WA government publicly adopted the Toronto target for "planning purposes."[59] The research agreement formed part of the broader CSIRO Climate Change Research Program that commenced in 1989 with funding from the federal government.

The Division of Atmospheric Research had gained increasing influence in the 1980s through its contributions to international research on the enhanced greenhouse effect and on the phenomenon of nuclear winter.[60] During the mid-1980s, the federal government had requested that the division investigate the potential climatic effects of nuclear war on Australia.[61] The government allocated extra funds for this research, which would be used to develop and apply computer climate models. The primary researcher on this project was Barrie Pittock, who continued to be closely involved in modeling the potential climate impacts of the greenhouse effect. Pittock understood the issues of nuclear winter and the greenhouse effect as pathological, as "major disturbances or changes in the climate system, taking it far from the present quasi-equilibrium situation."[62] He believed that further research was necessary to understand other possible outcomes under such scenarios.

By the end of the decade, environmental issues such as nuclear proliferation, air pollution, and the logging of old-growth forests were resonating with the Australian public. A growing proportion of voters considered the environment an important factor in how they would mark their ballot in the upcoming 1990 federal election.[63] The federal Labor government was perturbed by the related growth in support for the Australian Greens Party and set about proving itself as "green."

Although the environment may have been an issue of concern to most Australians,

[58] Barrie Pittock and Rob Allan, eds., *The Greenhouse Effect: Regional Implications for Western Australia, 1st Interim Report* (Perth, 1990), i, 1.
[59] Clive Hamilton, *Running from the Storm: The Development of Climate Change Policy in Australia* (Sydney, 2001), 31–2.
[60] See Matthias Dörries in this volume.
[61] Collis, *Fields of Discovery* (cit. n. 39), 353.
[62] Barrie Pittock, *Climatic Catastrophes: The International Implications of the Greenhouse Effect and Nuclear Winter* (Canberra, 1987), 19.
[63] Elim Papadakis, *Politics and the Environment: The Australian Experience* (St. Leonards, 1993), 142–4.

there were significant divisions within the electorate over environmental policies, particularly their potential economic impact. The nature and substance of the policies introduced by the federal Labor government in the late 1980s and early 1990s reveal its attempts to reconcile these divergent interests, which superficially dichotomize the environment and the economy and can be interpreted as responses to "brown" and "green" issues. The Australian public at large was primarily concerned with "brown" issues, ecological "risks" such as pollution, soil erosion, waste disposal, the greenhouse effect, and the ozone hole. Meanwhile, the audience for "green" issues, which was generally better educated, younger, urban, and left-leaning, was focused on conservation and the environmental effects of consumer capitalism.[64] As a mainstream party, Labor therefore had to accommodate with its environmental policies not only the desires of its economically concerned constituents but the differing interpretations of environmentalism expressed by its ecologically concerned constituents.

Although Labor had greatly benefited from the support of green groups in the 1983 election, it understood that the green stance could not be maintained in government. The conservationist green approach was perceived as too radical in demanding that Australians change their way of life to protect the environment. Concerned with the effects of environmental degradation on health and lifestyle, the brown position, however, presented a more palatable option for mainstream voters. This view promoted reformist and interventionist strategies, as opposed to disruption and dislocation, which could threaten economic prosperity.[65] With its emphasis on sustainable development, the 1987 Brundtland Report provided the federal Labor government an approach that could meet these demands on environmental policy.

The authors of the Brundtland Report argued that environmental protection and economic development were not mutually exclusive goals, and thus challenged governments to pursue ecologically sustainable development.[66] In 1989, the federal Labor government produced the first national Statement on the Environment and launched the "Decade of Landcare" (1990–2000). These strategies reflected the government's attempts to appeal to mainstream brown voters by addressing land degradation issues, particularly soil conservation.[67] Furthermore, the attention to rural sustainability was arguably a trade-off with primary producers for their support for the government's trade liberalization agenda.[68]

Buoyed by victory following the tightly contested 1990 federal election, the Labor government consolidated its social, economic, and sustainability policies through a range of schemes introduced in 1992, including the Ecologically Sustainable Development Strategy, the National Drought Policy, and the National Greenhouse Response Strategy. Although these policies superficially presented an environmentalist outlook, at their core was the pressing concern of protecting the national economy from the impacts of environmental change.

To ensure such protection, these strategies emphasized the need to understand and

[64] Jan Pakulski and Bruce Tranter, "Environmentalism and Social Differentiation: A Paper in Memory of Steve Crook," *Journal of Sociology* 40 (2004): 221–35.

[65] Ibid.

[66] Drew Hutton and Libby Connors, *A History of the Australian Environment Movement* (Oakleigh, 1999), 243.

[67] Diana G. Day, "Australia's First Environment Statement," *Environmentalist* 11 (1991): 9–17.

[68] Bill Pritchard, "Negotiating the Two-Edged Sword of Agricultural Trade Liberalisation: Trade Policy and Its Protectionist Discontents," in *Land of Discontent: The Dynamics of Change in Rural and Regional Australia*, eds. Pritchard and Phil McManus (Sydney, 2000), 90–104.

predict climate variability in order to ameliorate its impacts, particularly on agricultural activity. This need to improve the scientific understandings of the vagaries of the Australian climate reflected the Labor government's neoliberal view of drought, which aimed to promote "self-reliant approaches to managing for climate variability."[69] In this vein, drought and climatic variability were normalized so that they no longer constituted natural disasters but rather were part of rural life. They were to be expected and acceptable in a normal and healthy Australian climate.[70] To this end, improvements to the nation's forecasting capabilities were part of the government's rhetoric of demystifying drought to encourage farmers to prepare themselves for climate variability.[71] The significance of such forecasting was heightened by findings that the 1980s had been the hottest decade in more than a hundred years and that the Southern Hemisphere had experienced a general warming trend during the twentieth century.[72]

The focus of CSIRO research under the WA agreement turned to the possible impact of this warming trend on regional rainfall. As Pittock had posited in the early 1980s, the pattern of rainfall change during the twentieth century could reveal the possible implications of global warming for regional precipitation in the future. The researchers focused their analysis of Southwest rainfall records on the winter months of June, July, and August. These months are important for primary producers, following the seasonal "break" of autumn rains, which marks the beginning of the annual cropping cycle. Researchers again observed a downward trend since the 1940s, which became more marked from the mid-1960s.[73] Bureau of Meteorology researchers concurred with these findings and noted the significance of changes in regional rainfall for agricultural productivity.[74]

During the early 1990s, the federal and state governments directed the focus of national science toward regional climates in order to support the nation's international commitments to counter the enhanced greenhouse effect. For the federal Labor government, keenly aware of the need to navigate and manage electorally significant environmental issues, such greenhouse initiatives supported its claim of being the greenest of the major parties. Furthermore, the greenhouse issue complemented the government's other focuses on ecological sustainability and drought. These policies converged in the Southwest region of WA, where the prospect of an ongoing dry spell under enhanced greenhouse conditions focused research attentions on the impact of winter rainfall decline on agricultural productivity. The federal government's support of scientific investigation into the trio of greenhouse, drought, and sustainability therefore suggested its environmental credibility to both brown and green voters.

---

[69] Linda C. Botterill, "Government Responses to Drought in Australia," in *Beyond Drought in Australia: People, Policy and Perspectives*, eds. Botterill and Melanie Fisher (Collingwood, 2003), 49–56, on 52.

[70] Peter Hayman and Peter Cox, "Perceptions of Drought Risk: The Farmer, the Scientist and the Policy Economist," in Botterill and Fisher, *Beyond Drought in Australia* (cit. n. 69), 153–74.

[71] For the American context, see Ted Steinberg, "Introduction: Hometown Blues," in *Acts of God: The Unnatural History of Natural Disaster in America* (New York, 2000), xv–xxiii.

[72] Pittock and Allan, *Greenhouse Effect* (cit. n. 58), 15.

[73] Ibid., 41.

[74] Neville Nicholls and Beth Lavery, "Australian Rainfall Trends during the Twentieth Century," *International Journal of Climatology* 12 (1992): 153–63.

SYMPTOMS PERSIST

As the run of dry years continued into the mid-1990s, WA's water managers found it necessary to review their resource planning to ensure sufficient water supplies. They adjusted the expected water yield downward, particularly with regard to dams, based on the past forty years of observations. Demand for these limited resources, meanwhile, continued to increase. In spite of the revisions, the prevailing resource-management view of the water situation saw the rainfall conditions as a prolonged drought. With an overhaul of the management of the state's water resources came closer scrutiny of the lower levels of rainfall. In early 1996 the newly established WA Water and Rivers Commission (WRC) convened a seminar and workshop in Perth to address the recent low levels of rainfall and their effects on the hydrology of the Southwest region.[75] Scientists and water managers reviewed the existing state of knowledge about regional climate variability and the impacts of rainfall decline on water supply. The workshop represented a turning point for scientists and decision makers, as the rainfall conditions were no longer viewed as "an extreme run in a random process."[76] They now identified a "nonlinear jump" to a new regional climate equilibrium: a state of lower winter rainfall. This redefinition of the parameters of a normal climate to accommodate the pathological dry conditions mirrored earlier federal government efforts to normalize drought in rural communities. With this new perspective on the region's climate, the state's water managers further reduced the estimated long-term annual inflow to the Southwest water supply system. By this stage, the region's supply capacity was considerably exceeded by expected demand. In order to meet this demand and to avoid imposing water restrictions on consumers, the water authorities brought forward their plans to expand and develop the region's water supplies.[77]

Upon the recommendations of the WRC workshop, the WA coalition government established IOCI in 1998 as a five-year research project that consolidated the research conducted by the Bureau of Meteorology and CSIRO on the Southwest. This research body would provide state agencies with much-needed information on the climate variability affecting the region.[78] Its principal purpose was to analyze the possible relationships between sea temperatures in the Indian Ocean and the declining rainfall of the Southwest. IOCI was chaired by prominent local hydrologist Brian Sadler, who had presented the WA Water Authority's response to Pittock's climate scenario of lower Southwest rainfall and stream flow at the Greenhouse87 conference.

The establishment of IOCI represented more than a resource-management response to an increasingly pressing environmental problem. It reflected the uncertainty as to the possible causes of the regional rainfall decline. Air pollution, vegetation clearing, and the enhanced greenhouse effect were all possibilities. An interesting character

[75] John Ruprecht, Bryson Bates, and Bob Stokes, eds., *Climate Variability and Water Resources Workshop* (East Perth, 1996).

[76] Sadler, "Informed Adaptation to a Changed Climate State" (cit. n. 10), 3.

[77] Bryson C. Bates and Graeme Hughes, "Adaptation Measures for Metropolitan Water Supply for Perth, Western Australia," in *Climate Change Adaptation in the Water Sector*, eds. Fulco Ludwig, Pavel Kabat, Henk van Schaik, and Michael van der Valk (London, 2009), 187–204.

[78] Scott Power, Brian Sadler, and Neville Nicholls, "The Influence of Climate Science on Water Management in Western Australia: Lessons for Climate Scientists," *Bull. Amer. Meteor. Soc.* (June 2005): 839–44.

in the development of IOCI was the founding chairman of the state's Environmental Protection Authority (1971–7), Brian O'Brien. Because he was an established critic of climate change science and policymaking, O'Brien's support for the study of natural variability and forecasting suggests that he sought to see the Southwest situation as detached from the greenhouse discourse.[79] He believed that the "climate change projections were so uncertain that the information [on the future climate of the Southwest] should not be factored in at all."[80] In O'Brien's view, research priority should have been given to ENSO, as its impact on the continent's climatic variability was more pressing than the more abstract and nebulous concept of climate change.[81]

Support for IOCI also came from the state's farming community. As wool prices slumped in the 1990s, crop growing became a more financially appealing prospect for the Southwest's farmers. With lower levels of rainfall, crops took increasing precedence in the so-called wool belt areas of mixed farming, where the climate conditions had previously been too wet to make crops viable. Yet while some farmers adapted to these lower rainfall conditions, others became concerned about the changes in regional climate. As the Department of Agriculture observed, the dry run "complicat[ed] an already challenging list of management issues" facing Western Australian farmers, such as shifting product demand and salinity.[82] Representing these rural interests to the government was the state's deputy premier and leader of the WA National Party, Hendy Cowan. His earlier support for a government-funded dry-land salinity program to remedy the degraded landscape of the wheat belt region had been thwarted because of its rural focus. Research into climate variability in the Southwest region, however, could benefit both urban and rural Western Australians. Hence, it was Cowan who in 1997 presented the proposal to the coalition cabinet to establish IOCI.[83]

By the end of the decade, the issue of regional rainfall decline was absorbed into competing but not exclusive scientific and policymaking positions on climate change, drought, and climatic variability. The water managers of the Southwest did not have the luxury of waiting for more definite conclusions on the possible impacts of climate change. In their view, the climate in the Southwest had changed, and it was adversely affecting the region's water supplies. Although the collaborative research effort embodied by IOCI reflected a range of concerns for the region, these concerns were all directed at understanding the climate processes that affected it. As a response to agricultural and environmental concerns about the causes and implications of the pathological conditions of the Southwest climate, IOCI represented the latest chapter in a long history of state-sponsored scientific investigation into Australian climate variability.

IOCI research to date has been undertaken in successive stages through collaboration between WA resource managers and researchers from the Bureau of Meteorology and CSIRO. It has confirmed the findings of earlier studies that the Southwest

[79] Sadler, "Informed Adaptation to a Changed Climate State" (cit. n. 10), 4.

[80] Power et al., "Influence of Climate Science on Water Management" (cit. n. 78), 840.

[81] Brian J. O'Brien, "Greenhouse Impacts on the Southwest of Western Australia," *Mining Review* 17 (1991): 19–20.

[82] Luke Morgan et al., *Climate Change, Vulnerability and Adaptation for Southwest Western Australia, 1970–2006* (South Perth, 2008), 54.

[83] IOCI Panel, *Report on Second Year—the Indian Ocean Climate Research Initiative—for Western and Southern Australia* (East Perth, 2000), 2.

region is experiencing substantially lower rainfall now than during the mid-twentieth century.[84] IOCI was also complemented by research arising from the Intergovernmental Panel on Climate Change's *Second Assessment Report* (1995), which sought more data on climatic extremes and variability to make an evaluation of recent global changes.[85] Research from the Bureau of Meteorology and Monash University concurred with IOCI findings that extreme rainfall events in the Southwest during the twentieth century and early twenty-first century have declined in the early winter months of May, June, and July.[86]

The publication of the first stage of IOCI research in 2002 coincided with one of the Southwest's driest years and the lowest stream flows on record. Many farmers faced the worst drought in more than a century and applied for financial assistance from the federal government. Although the affected area included the southeast parts of the Southwest of WA, other areas experienced good harvests because the dry conditions overcame their usual problems of waterlogging.[87] IOCI forecasts of even drier conditions in the future captured the headlines, conjuring a vision of the Southwest deserted for want of water.[88] Ambitious politicians even dusted off and revived plans to pipe water to the Southwest from the Kimberley region in the far north of the state. Most important, this dry spell reinforced the significance and relevance of the IOCI mission: not only to predict a distant future but also to provide farmers with insights into more immediate climate conditions.

A CONTAGIOUS CLIMATE

As the nation heralded the beginning of a new century, its celebrations were muted by the spread of the dry conditions experienced in the Southwest region of WA to the eastern states. This so-called millennium drought affected areas of southwestern Queensland, western New South Wales, eastern South Australia, northwestern Victoria, and the Gascoyne region of WA. These circumstances were exacerbated by the 2002/3 El Niño event, which triggered soaring temperatures across the country. Emotive media coverage of the drought and campaigns to raise funds for struggling rural families led to the revival of public debate on issues such as "drought proofing" the continent.[89] With the entire country seemingly in the grip of its old foe, Australians and their governments again turned to their scientists for answers.

In WA, the newly elected state Labor government continued to support the IOCI arrangement established by its coalition predecessor. The 2001 state election campaign had been largely fought over environmental issues, particularly the logging of old-growth forests in the Southwest region. Promising to protect the remaining

---

[84] David Day, *The Weather Watchers: 100 Years of the Bureau of Meteorology* (Carlton, 2007), 465.

[85] Neville Nicholls and Lisa V. Alexander, "Has the Climate Become More Variable or Extreme? Progress 1992–2006," *Progress in Physical Geography* 31 (2007): 77–87.

[86] See Lisa V. Alexander et al., "Trends in Australia's Climate Means and Extremes: A Global Context," *Australian Meteorological Magazine* 56 (2007): 1–18.

[87] Morgan et al., *Climate Change, Vulnerability and Adaptation* (cit. n. 82), 56.

[88] Day, *Weather Watchers* (cit. n. 84), 465.

[89] Åsa Wahlquist, "Media Representations and Public Perceptions of Drought," in *Beyond Drought in Australia*, 67–86 (cit. n. 69). The idea of drought proofing the continent has captivated the imaginations of many Australians since the late nineteenth century. See Pandora K. Hope, Neville Nicholls, and John L. McGregor, "The Rainfall Response to Permanent Inland Water in Australia," *Australian Meteorological Magazine* 53 (2004): 251–62.

old-growth forests, Labor achieved a landslide victory against the coalition and quickly carried out its election promises to cease logging in the Southwest forests. Attuned, therefore, to its predominantly urban constituency's environmental concerns, which had turned to climate change, Labor developed a new state greenhouse strategy. This plan "raised the priority of greenhouse issues and committed departments . . . to a range of actions aimed at reducing emissions and adapting to climate change."[90] With its successful record of providing research support for the land and water management and agricultural sectors, IOCI was identified as a cornerstone of the government's new policy.[91]

The incorporation of IOCI into the state Labor government's stance on climate change was in stark contrast to its cautious origins under the previous government. The state government also differed from the position of the federal coalition government on the issues of drought and climate change. From its election in 1996, the coalition (of the Liberal and National parties) had worked tirelessly to suppress and undermine greenhouse debate within Australia and to stymie attempts to negotiate an international climate change regime. The government argued that any global agreement on reducing greenhouse emissions would be detrimental to the resource-dependent Australian economy. This position reflected the coalition's traditional support base, the business, mining, and agriculture sectors. As drought conditions persisted in the eastern states, however, this antigreenhouse stance was proving increasingly difficult for the government to rationalize to the community. Recognizing the community's fears of worsening drought under greenhouse conditions, the coalition government reoriented its position on the environment to introduce water reform, thus turning "the climate problem into a water problem."[92]

This reform focused particularly on the nation's "food bowl," the drought-stricken Murray-Darling Basin, which dominates the surface water resources of southeast Queensland, New South Wales, Victoria, and eastern South Australia. The Bureau of Meteorology observed that southeastern Australia was now exhibiting the same symptoms of rainfall decline that had presented in the Southwest of the continent several decades earlier.[93] During the first decade of the twenty-first century, inflow into this river system was just a third of the long-term average.[94] The persistence and spread of drought conditions during the late 1990s and into the twenty-first century reinforced and extended state and federal governments' commitments to support scientific research into the Australian climate. To facilitate further research into the climate variability of the Southeast region, several state and commonwealth agencies established the South East Australian Climate Initiative (SEACI) in 2006. Modeled closely on IOCI, this undertaking across the continent recognized the ongoing success of the Southwest research, particularly its provision of information relevant to state agencies.[95] The research undertaken on behalf of each initiative paralleled emerging scientific understandings of the

[90] Morgan et al., *Climate Change, Vulnerability and Adaptation* (cit. n. 82), 56.

[91] Bates et al., "Key Findings from the Indian Ocean Climate Initiative" (cit. n. 9), 351.

[92] Sarah Bell, "Concerned Scientists, Pragmatic Politics and Australia's Green Drought," *Sci. Pub. Pol.* 33 (2006): 561–70, on 563.

[93] Barrie Pittock, *Climate Change: An Australian Guide to the Science and Potential Impacts* (Canberra, 2003), 51.

[94] Åsa Wahlquist, *Thirsty Country: Options for Australia* (Crows Nest, 2008), 164.

[95] Bates et al., "Key Findings from the Indian Ocean Climate Initiative" (cit. n. 9), 352.

climate processes that influence the conditions experienced in the southern areas of the continent.

As Sheldon Ungar argues, "Environmental claims are most often honored when they can piggyback on dramatic real-world events."[96] For the outcomes of scientific research, it mattered little whether the research brief was informed by the politically sensitive issue of climate change or by the more populist issue of water resources. In the Southwest at least, the issues of drought and climate change had long been inseparable. The scientific research into the climate processes affecting the region provided insights into both problems, connecting the dry conditions of the present to the potentially drier conditions predicted for the future.

CONCLUSION

Agricultural production has contributed significantly to the economic development of Australia, a settler society. The nation's scientists have played an ongoing role in supporting this agricultural enterprise. The establishment and expansion of the Western Australian agricultural sector was founded upon the faith of the state's farmers in applied science and technology. In this environment of extremes, the scientific endeavor to glean the secrets of the continent's weather and climate has been no less important for the nation's farmers than research into soils, fertilizers, and crops. The research of the Bureau of Meteorology, CSIRO, and various universities contributed to a broader national and historic project of "science for development," underwritten by those governments that have viewed the weather as significant to their country's fortunes.[97]

Long regarded for its reliable winter rainfall, the Southwest region of Western Australia was beset by unexpected dry conditions in the early 1970s whose persistence was baffling. The regional circumstances first attracted the attentions of the state's university researchers. Government investment and scientific interest in the nation's weather and climate for agricultural production grew as the volatile climate conditions of the 1970s at home and abroad combined with fears that the world was fast reaching its limits to growth. The widespread impact of the 1982/3 El Niño event and the emergent international greenhouse discourse consolidated the relationship between science and government on climate matters. In the late 1980s, the federal government renewed its commitment to science to investigate the twin threats of drought and climate change, to ensure the protection of Australia's economic prosperity, and to prove the Labor government's green credentials to an environmentally concerned public. For the Southwest, the resulting research predicted that the region's dry conditions would persist into the twenty-first century.

By the early 1990s, therefore, the Southwest of WA had become a climate patient and the nation's scientists and resource managers its physicians. As the decade progressed, WA water managers became increasingly concerned that the dry conditions had continued unabated. In response, the state government, its resource managers, and the nation's scientists collaborated in IOCI to improve the understanding of the region's climate in the hope that this would shed light on the causes and extent of the

[96] Ungar, "The Rise and (Relative) Decline of Global Warming as a Social Problem," *Sociol. Quart.* 33 (1992): 483–501, on 483.
[97] See Adrian Howkins in this volume.

rainfall decline. The onset of widespread drought conditions across the continent in the twenty-first century reinforced a growing concern that the Australian climate was shifting away from those climate parameters best suited to the nation's agricultural industries. Yet the idea of anthropogenic climate change was unpalatable to the conservative federal coalition government. Instead, the government reconceptualized the climate issue as a water problem, requiring water reform across the country.

The understanding of the climate of the Southwest of WA has thus been shaped and molded since the 1960s by events and debates regarding weather, agriculture, and environmental concerns at the local, national, and international levels. The region's experience of rainfall decline coincided with a renewed scientific and political interest in the global climate and the perception that it was deviating from its normal state. In this context, Australian fears of drought converged with the new threat of anthropogenic climate change to focus attentions on regional climate variability and vulnerability, as observed in the Southwest. These attentions brought together state and federal agencies and funding on a familiar project to understand the Australian climate and its voting public. Diagnosing the dry conditions of the Southwest of WA since the 1970s, therefore, has a long history founded in Australia's colonial past, a history that ensures that Australian science remains "science for development" in the twenty-first century.[98]

[98] Robin, "Ecology: A Science of Empire?" (cit. n. 4), 65.

SOCIAL CONTEXTS

The Letter from Dublin:

Climate Change, Colonialism, and the Royal Society in the Seventeenth Century

by Brant Vogel[*]

ABSTRACT

This article discusses an anonymous letter published in the *Philosophical Transactions* in 1676 that reports the theories of American colonists about the cause of their warming climate (cultivation and deforestation), and offers Ireland's colonial experience as a counterexample: Ireland was a colony with decreased cultivation, but the same perceived warming. That such an objection seemed necessary to the author shows that anthropogenic climate change could be a subject of debate and that the concept of climate was tied into theories of land use and to the colonial enterprise. Since he was liminal to both the Royal Society of London and the intellectual circles of Dublin, his skepticism, contextualized here, questions both the elite discourse and the discourse at the colonial periphery.

INTRODUCTION

In July of 1676, Henry Oldenburg saw fit to lead off number 127 of the *Philosophical Transactions of the Royal Society of London* with an anonymous letter from Dublin, written May 10 of the same year. Besides an interesting zoological observation about the smell of a muskrat's testicles when kept in the pocket for three months, as practiced by sailors; an analysis of a warm winter in Ireland, with a summary of the author's weather register; a description of an insect found in Virginia; the smell of water during an ocean voyage; and a description of a novel hygrometer of the author's own invention, the letter contained a narrative related by a "Master of several ships" about a magnetic marvel. The ship master, "a man of good credit," knew a captain who had been traveling out of New England to the Barbados, when, near Bermuda, his ship was struck by lightning. The ship's compass reversed polarity as a result of the strike. These observations alone made the letter worthy of the *Philosophical Transactions*. But perhaps the most striking inclusion in the letter is what appears to be a refutation of claims about the role of cultivation in climate change in the North American

[*] 135 North 6th Street, Brooklyn, N.Y., 11211; brantvogel@earthlink.net. Previous versions of this article were presented at the "Three Societies Meeting" (Oxford, 2008) and "Climate and Cultural Anxiety" (Colby College, Me., 2009). I would like to thank James Fleming and Vladimir Jankovic, the anonymous *Osiris* reviewers, and Don Garden and the Colby conference participants for their invaluable suggestions and critiques. Support for the final revisions of this article was provided by the 2010 Irish-American Research Travel Fellowship from the American Society for Eighteenth-Century Studies and by Michael Volchok.

colonies. The argument implies that the colonists' observations of and opinions about such climate change, conceived of as a local phenomenon in the seventeenth century, were contradicted by comparative observation on a trans-Atlantic scale.[1]

This part of the letter is intriguing not only as an early refutation of North America's climate narrative (amelioration through stewardship), but also because of its political and trans-Atlantic vectors. In brief, the anonymous correspondent, having heard that many of the North American colonists claimed that their weather had become more temperate because of the introduction of European agriculture, deforestation, and increased population, found their reasoning specious, in that Ireland's weather had also become warmer, despite a decrease in population and a subsequent decrease in agricultural production and land clearing. The change in weather must, he argued, have come from some other cause.

"Climate" was not his term. He spoke of "temperature." But this use of "temperature," meaning the "temper of the air," besides having medical nuances, is situated within two distinct contemporary understandings of climate. The term "climate" itself had taken on several metaphoric meanings by the seventeenth century, but its natural philosophical meaning usually followed one of two concepts inherited from antiquity. One notion of "climate," deriving from the Hippocratic works, conveyed the sense of the qualities of a specific place. The other (and this is a gross generalization) conceived of "climate" as the differences in temperature between different parts of the globe according to latitude.[2] The seventeenth-century writer would see this latter concept as Aristotelian.[3] Both concepts had medical and (human) geographical implications, as well as economic and "natural" ones. The letter from Dublin mixes these two concepts, refuting the local arguments of North American observers and expressing angst over a general temperature rise in the northern temperate zone.

Recent literature discusses an eighteenth- and early nineteenth-century debate over the perceived climate change in North America and the role of human activity in producing it. This discussion grew out of both Enlightenment thought and local promotion.[4] It is true that the bulk of the early scientific discussions of North American climate change by European *philosophes* and their American counterparts was published in the eighteenth century, especially the latter half. But a skeptical opinion published during the Stuart Restoration, not to mention a discussion of climate change at all, is worthy of closer examination. Published philosophical discussions of this instance of climate change were sparse in the Anglophone world before 1676. To whom was the Dublin author responding?

This article discusses a rejoinder to the theories of American colonists about the cause of their warming climate and offers Ireland's colonial experience as a counterexample. Ireland was a colony with decreasing cultivation, but the same perceived warming. That such an observation seemed necessary to the author shows that anthropogenic climate change was a subject of debate and that the concept of climate was

[1] "An Extract of a Letter &c. from Dublin May the 10th, 1676," *Phil. Trans. Royal Soc. London* 11 (1676): 647–53.
[2] Clarence J. Glacken, *Traces on the Rhodian Shore: Nature and Culture in Western Thought from Ancient Times to the End of the Eighteenth Century* (Berkeley, Calif., 1967).
[3] Craig Martin, "Experience of the New World and Aristotelian Revisions of the Earth's Climates during the Renaissance," *History of Meteorology* 3 (2006): 1–15.
[4] James Rodger Fleming, *Historical Perspectives on Climate Change* (New York, 1998), 11–32; Jan Golinski, "American Climate and the Civilization of Nature," in *Science and Empire in the Atlantic World*, eds. James Delbourgo and Nicholas Dew (New York, 2008), 153–74.

tied into theories of land "improvement" and to the colonial enterprise as a whole. To fully explicate this letter, we need to examine its seventeenth-century context in terms of natural philosophy, colonial politics, and political economy, as well as the author's social and institutional position.

Historiography

The history of this idea of climate requires a somewhat interdisciplinary approach, particularly because it intertwines with the history of European expansion. The disciplines of historical geography, the history of science, and environmental history are prerequisites to understanding the interaction of climate and colonialism. As Europe expanded to other parts of the globe, it brought with it certain assumptions about questions relating to climate, such as whether Europeans could even live in other parts of the world, whether climate determined human character, or human character and activity determined climate, whether "untouched" nature and "uncivilized" humans were good or evil, and whether a world would be found that would be more profitable than Europe or could be remade into a new Europe.[5] The exegesis of this particular text also requires an understanding of Atlantic studies issues and of Ireland's role as an English colony.[6]

The notion of climate in the early modern period (after the critique of the more global Aristotelian model) had its roots in ancient medicine, starting with the Hippocratic writings. Climate was associated at that time with place, whereas now we understand it as *global* as well as *local*. Climate was seen as affecting the health in such works as *Airs, Waters, Places*,[7] but also, perhaps as an extension, as affecting character. Medicine, geography, morality, and politics intersected and, through much of the ancient tradition, were attached to particular localities. This concept saw expression in the writings of Jean Bodin (1530–96), who thought that national character derived from climate. Having become a subject of political philosophy, the discussion of the effect of climate on civilizations would be carried on by Jean-Baptiste Du Bos, Montesquieu, and David Hume, who, in the mid-eighteenth century, sparked the debate about the reality and benefits of climate change. Cultivation and civilization, it was assumed, brought with them a more temperate climate. Climate affected the character of people, and people changed the climate by their improvement of the land. And this changed the history of Europe. It was in these terms that the discussion of climate change in North America became controversial. It was on this locus that, in the eighteenth century, European philosophers and gentleman farmers on the other

[5] Glacken, *Traces* (cit. n. 2); Fleming, *Historical Perspectives* (cit. n. 4); William Cronon, *Changes in the Land: Indians, Colonists, and the Ecology of New England* (New York, 1983); Alfred W. Crosby, *Ecological Imperialism: The Biological Expansion of Europe, 900–1900* (New York, 1986); Richard Grove, *Green Imperialism: Colonial Expansion, Tropical Island Edens, and the Origins of Environmentalism, 1600–1860* (New York, 1995); Grove and Vinita Damodaran, "Imperialism, Intellectual Networks, and Environmental Change: Origins and Evolution of Global Environmental History, 1676–2000," *Economic and Political Weekly*, November 14, 2006, 4345–54, and November 21, 2006, 4497–505; Carolyn Merchant, *The Death of Nature: Women, Ecology, and the Scientific Revolution* (San Francisco, 1980).

[6] For the colonial story of Ireland (and America), see J. G. A. Pocock, "British History: A Plea for a New Subject," *J. Mod. Hist.* 47 (1975): 601–21. See also a review article by Jane Ohlmeyer, "Seventeenth-Century Ireland and the New British and Atlantic Histories," *Amer. Hist. Rev.* 104 (1999): 446–62.

[7] Francis Adams, trans., *Hippocratic Writings* (Chicago, 1952).

side of the Atlantic, such as Thomas Jefferson, became involved in the debate as to whether cultivation of the land improved the climate of the place, and why.[8]

This debate had a perceived change in climate in the American colonies as an antecedent. Actual textual evidence of a pre-eighteenth-century debate is sparse. Other colonizing societies, namely, those from the Iberian Peninsula, had talked for two centuries about the effects of deforestation on local climates.[9] But the Royal Society of London for the Improvement of Natural Knowledge, on the other hand, discussed this change rarely, which suggests that these were relatively new ideas in Anglophone natural philosophy. If the discussion existed in seventeenth-century London, it existed in coffeehouses and taverns, in the anecdotes of colonial settlers and returning sailors, and in the rhetoric of promoters. People's memories tend to exaggerate the harshness of past weather. But within the stories told by the early settlers, weather had changed on the local level, and there is reasonable evidence that agriculture did affect local climate.[10] However, reliable quantitative data for the period are also sparse, so claims about climatic changes are inherently anecdotal.

Climate was an acknowledged problem in the colonization of North America, and the contemporary literature mentioned it abundantly. This has been analyzed thoroughly in the secondary literature, from Karen Ordahl Kupperman's 1982 article to recent works by Jan Golinski, for example.[11] Kupperman noted the tenacity of Old World climatic expectations even as early settlers encountered the harsh conditions of the mini–ice age. North America should have been like Europe. In some sense, anthropogenic climate change through agriculture extended a hope of realizing such expectations; thus, recommendations for *ways* of improving conditions were offered. Only later in the seventeenth century, as "evidence" in the form of colonial reports accumulated, did this climate change become a commonplace and the foundation for a sort of climate-based nationalism in America.[12]

Discussion of colonial climates in early Restoration London was a rarity, and discussions of change even more so. The letter from Dublin, in presenting this reportage, was a novelty. That the letter was critical of the reports was equally a novelty, as was the accompanying account of a warming trend in Ireland that seemed, to the writer, to be somehow cut loose from the comforting idea of improvement.

The Argument

The anonymous author was a peripheral figure to the network of natural philosophy he sought to join, with only one letter, which was read twice in London, published once, and cited once in Dublin. Aside from his letter's being a historical curiosity, it is of interest because of its context, drawing together popular perceptions with commercial interests and the elite discourse, as well as connecting peripheral localities with the philosophic center. The author is interesting in light of whom he

---

[8] See Fleming, *Historical Perspectives* (cit. n. 4), 3–54.
   [9] Starting with Columbus, they believed that trees drew rain, and deforestation changed rain patterns. Ibid., 27.
   [10] Cronon, *Changes* (cit. n. 5), 109–26.
   [11] Kupperman, "The Puzzle of the American Climate in the Early Colonial Period," *Amer. Hist. Rev.* 87 (1982): 1262–89; Golinski, *British Weather and the Climate of Enlightenment* (Chicago, 2007); Golinski, "American Climate" (cit. n. 4).
   [12] Fleming, *Historical Perspectives*; Golinski, "American Climate" (Both cit. n. 4).

was responding to—American colonists—and how their experience of the weather clashed with his own. As to the texts that might represent this context, I will look at (1) colonial chronicles, accounts, and journals (eyewitness accounts of change were available and could have come to the author's attention); (2) promotional literature (ranging from ephemera to substantial volumes, meant to drum up support and recruit settlers for colonial ventures); (3) travelers' and explorers' answers to queries (examples of accounts by reliable returning travelers, and those of actual settlers, of the climatic lore of faraway localities in response to natural curiosity from the cultural center); and (4) elite works of natural philosophy by members of the Royal Society and their milieu.

This last link turns out to be a dead end. The letter from Dublin predates the natural philosophical discussion in the author's network. John Evelyn's *Sylva* has been cited as an early Anglophone discussion of the relation of deforestation to local climate change.[13] Evelyn, a well-placed founding fellow of the Royal Society, whose often-reprinted work concerned the perennial English lumber crisis, discussed the relationship between deforestation and the tempering of the climate and articulated it best in the 1706 edition, although he had been developing that edition since the 1680s.[14] The earliest commentator on this climate change was St George Ashe, whose "Discourse of the Air" treats of the same concerns, but was only presented to the Philosophical Society of Dublin in 1684. The Dublin letter was an antecedent to Evelyn and Ashe and had evidently been read as well as cited by Ashe. In a very limited way, it may have played a role in opening a new avenue of discussion and inquiry to the members of the two philosophical societies, and it should be classed as both a travel account and an account of natural philosophical activity.

THE LETTER

The 1676 letter from Dublin situates the contrast of local and transoceanic climate change within a discourse of history, politics, and economics. First, reporting the American claim, the correspondent to the Royal Society observes:

> That in *America* (at least as far as the English plantations are extended) there is an extraordinary alteration, as to temperature, since the *Europeans* began to Plant there first, is the Ioynt assertion of them all; neither had it near so many admirers, as witnesses: in regard that this change of temperature, is, and not without some reason, generally attributed to the cutting down of vast woods, together with clearing and cultivating of the Country.

This is the emerging North American climate trope, which only came under notable critique over a century later[15]—except in this case. He continues,

---

[13] Peter Eisenstadt, "The Weather and Weather Forecasting in Colonial America" (PhD diss., New York Univ., 1990), 230–1; Fleming, *Historical Perspectives* (cit. n. 4), 27.

[14] John Evelyn, *Sylva*, 4th ed. (London, 1706), chap. 1, pt. 6. Note that the fourth edition of *Sylva* was a substantial reworking of Evelyn's 1664 monograph, which did not discuss climate change. The fourth edition was at least sixteen years in the making; see "Diary and Correspondence of John Evelyn," http://www.archive.org/stream/diarycorresponde03eveluoft/diarycorresponde03eveluoft_djvu .txt (accessed 24 January 2011).

[15] Fleming, *Historical Perspectives* (cit. n. 4), 31–2, 45–54.

but that *Ireland* should also considerably alter without any such manifest cause, doth certainly, either invalidate the reason generally admitted for the alteration of *America* newly mentioned, or els evince, that quite different causes may produce the same effect.

This is the challenge, based on personal observation and parsimony. It is accompanied by a historical account of the political and economic contrasts between the colonies:

> For if it be true, as some compute, that this Kingdom was better inhabited and husbanded before the late bloody war, than at present, it should, according to the reasons alledged for the change of temperature in America, be rather grown more intemperate, *viz*: for want of cultivation: But contrary is observable here, and every one almost begins to take notice, that this country becomes every year more and more temperate. Now whether there were more inhabitants in *Ireland* before the late war than at present, I shall not here insist upon, neither do I think it an easy matter to determine, yet sure I am, that there hath been no such increase of people here within these 16 or 20 years, nor such improvements as to be accountable for the great change in temperature that is late observed.[16]

The author, who tells the reader he has lived in several places in Ireland and has been to the West Indies, found that the theories (or perceptions) of the colonial witnesses and the testimonies of the promotional admirers did not follow logically, because he himself was a witness to a warming trend in Ireland that lacked the same causal antecedents. One colony was growing in population and cultivated acreage, the other colony lying fallow after the Cromwellian Wars and the difficulties of the Stuart Restoration, yet both were becoming more temperate. His subtext is a perceived neglect of the Irish Plantations and the boosterism driving colonization abroad. The climate-agriculture bond is not only philosophical, but political and economic. With that in mind, we must ask, To what, or whom, was he responding? We can see that notions of property and its improvement, as well as the expansion of colonialism, were involved.

EARLY COLONIAL CHRONICLES

From the beginning of England's colonial adventure, reports came back of either commodious or incommodious weather. And reports came back of climate change. The term "commodious weather" as used in the seventeenth century implied weather that was both "healthful" and "productive of commodities," and the effects of "improved" weather, or its opposite, on both health and wealth were not lost on seventeenth-century gentlemen. Such conjunctions of health, economics, politics, and climate would become further articulated in the eighteenth century. The colonies' weather and climate and their similarity or dissimilarity to the English clime were central to this story.

Some of the earliest concerns of the North American settlement were cold winters and "unhealthful" air. As for the pliability of the climate, several early writers gave advice on how to moderate extreme weather problems on what we would call

---

[16] "Letter" (cit. n. 1), 648–9 (in all quotations in this article, emphasis is the same as in the original text). The "late war" refers to the English Civil War and Oliver Cromwell's reconquest of Ireland (events beginning around 1641 and concluding in 1651 and later); the "16 or 20 years," the resolution of these events and the restoration of Charles II in 1660.

a micrometeorological scale—planting trees as windbreaks, cutting trees to warm the soil, clearing marshes by one's house—often in reference to Newfoundland.[17] Some, such as Richard Eburne (1624), would extend the benefits of improvement to entire plantations, declaring that the "Healthinesse of any Countrie by plantation and inhabitation must needs be much increased," necessitating deforestation and wetland clearing with the usual result of "clensing the aire" and "clearing the fogges."[18] And Richard Whitbourne (1620) offered land clearing as a way of moderating Newfoundland's cold. Exposure of the soil to the rays of the sun would "make the Countrey much the hotter Winter and Summer."[19] However, these are recommendations for settlers, not observations that improvement had actually worked.

One early American colonist, William Wood, claimed a change in rain and climate in New England as early as 1634: "Of late the seasons be much altered, the rains coming oftener but more moderately."[20] The weather was becoming more moderate. Deforestation and then full-on European-style cultivation were thought of more and more as being factors in the moderation of the climate. In the sixteenth century, Richard Hakluyt expressed concerns about the possible harm to the economic value of the Americas from human activity—that humans could change a local environment was not in question.[21] After settlement, those in the colonies came to perceive, rightly or wrongly, that they were having some positive local impact. Admittedly, Wood was acting as a promoter, countering classical climate concepts by offering up examples of how both American climate and English bodies were adaptable, and in some sense more vigorous than their European counterparts, but the most salient argument is that things were improving.[22]

The early seventeenth-century climates can only be seen through anecdote—the freezing of rivers, major gales, remarkably bad winters. There was very little of what we call data, yet those like Wood nevertheless thought that something was going on. No doubt they were trying to compensate for the acknowledged widespread death (because of ill planning and misfortune) in early settlements, but also to apologize for the deaths due to the "seasoning" of the earlier settlers. "Seasoning," which later became a term of the slave trade, initially included the conditioning of all settlers as they acclimated to America. The term could be taken in the sense of the hardening of the constitution of an individual, or the culling of the weak from a population, insofar as seasoning came to be used by promoters as a rationalization for high mortality rates. Throughout, reports back home were concerned with whether the land was healthy and profitable for Englishmen. Opinion was not united in the English colonies. From Hakluyt's fear of damage to Wood's positive assessment, coming a generation after the brutal seasonings that characterized colonization under the Tudors and early Stuarts, we see discussion of climate, but only in a scattered way. The discussion seems to have intensified as the colonial adventure began to be profitable.

[17] Kupperman, "Puzzle" (cit. n. 11), 1287–8.
[18] Eburne, *A plaine path-vvay to plantations* (London, 1624), 22.
[19] Whitbourne, *A discourse and discouery of Nevv-found-land* (London, 1620), 57.
[20] Fleming, *Historical Perspectives* (cit. n. 4), 27.
[21] Timothy Sweet, "Economy, Ecology, and Utopia in Early Colonial Promotional Literature," *American Literature* 71 (1999): 399–427.
[22] See Timothy Sweet, *American Georgics: Economy and Environment in Early American Literature* (Philadelphia, 2002), 50–73.

PROMOTIONAL LITERATURE

In contrast to the settlers' testimony, the testimonials of promotional literature for the most part emerged from the world of colportage and only thinly disguised their economic motivation. The promotional literature of the colonial enterprise in the Restoration period, at the service of investors such as the Virginia Company, the Massachusetts Bay Company, the Somer Isles Company, and the Proprietors of Carolina, as well as any number of smaller adventures, had a vested interest in describing the climate as healthful and commodious. And after a certain point, particularly after the expanding public awareness of the early hardships in Virginia, the stories of winter starvation, and other mishaps, it needed to say that things were getting better. William Wood welcomed the improvement of climate in 1634; the pamphleteers of the Restoration, "admirers" of America, promoted new adventures such as Carolina and the tropics, lauded climate change, and welcomed stories from veteran colonials that the weather was becoming more temperate, particularly in colonies that needed more free settlers, as opposed to non-free or indigenous workers. One suspects these tracts were not entirely expositions of natural history (which is not to say that the works of Robert Plot were not advantageous to the enclosure movement).[23]

Early seventeenth-century accounts often discussed the climate of a place and whether that climate was agreeable. But the concept of climate change was introduced slowly over the course of the century. When J. Day's account of the Guiana plantation (1632) told about climate, it discussed conditions as found and why they might be healthy for English settlers.[24] That the climate was survivable and profitable seemed the most important thing. How would the crops be, and how would the seasoning be? When Wood wrote about North America in 1634, he wrote about a place with harsh weather.

Moving into the decade when the letter from Dublin was written, we see, at first, concern for the local climate as is, as for instance in Samuel Clarke's 1670 account of North American colonies:

> *Virginia* is a Country in *America*. . . . The Plantation which was begun in the year 1606. was under the Degree of 37. 38. and 39. where the tempreture of the air, after they were well seasoned, agreed well with the constitutions of the *English.* They found the *Summer* as hot as in *Spain:* the *Winter* as cold as in *France* or *England:* The heat of *Summer* is in *June, July,* and *August,* but commonly a cool Briess asswages the vehemency of the heat: The chiefest *Winter* is in half-*December, January, February,* and half *March.*[25]

The seasons are compared to those in European places of which a reader might have some anecdotal weather knowledge, adhering to the notion of the similarity of climate in the same latitudes.[26] Improvement of the weather in the colonies was not noted.

However, in the ill-reputed cartographer Richard Blome's 1672 description of

---

[23] E.g., Plot, *Natural History of Oxford-shire* (Oxford, 1677).

[24] Day, *A publication of Guiana's plantation* (London, 1632).

[25] Clarke, *A true and faithful account of the four chiefest plantations* (London, 1670), 5.

[26] One of the most pervasive ideas about climate inherited from the classical period was that similar latitudes had similar temperatures, from the "torrid," to the "temperate," to the "frigid." See Fleming, *Historical Perspectives* (cit. n. 4), 22.

Virginia,[27] we see North America's climate changed by forest clearing, directly softening the seasoning. Under the heading "Its Air and Temperature," he reports, "This *Countrey* is blest with a sweet a[n]d wholesome *Air,* and the *Clime* of late very agreeable to the *English,* since the clearing o[f] *Woods;* so that now few dyeth of the *Countreys* disease, called the *Seasoning.*"[28] Blome and his "experienced persons" do not mention such change in other colonies, just Virginia, with an eye to its difficult past. Over the course of the decade, the Virginia story picked up details as it became much more the standard account in the literature of colportage, as we see in the description given by the compiler of the 1676 epitome of *Speed's Theatre of the Empire of Great Britain* (for which the source is quite probably Blome as well):[29] "The Air of *Virginia* is accounted of a temperature very wholsome and agreeable to English constitutions, (especially since by the cutting down of the Woods, and the regulation of diet, the seasonings have been abated) only within the present limits of *Virginia* it is somewhat hotter in Summer than that part called *Mary-Land;* and the seasoning was formerly more violent and dangerous here to the English at their first landing."[30] This brings us to the year the Dublin letter was written, just before Bacon's Rebellion would change the colony in another way (making a slave economy suddenly more attractive than one based on free settlers, and thus ending that phase of Virginia's promotion).[31] The center of the discourse is the improvement of Virginia. Both the promotional writers (or "admirers") and the letter writer claim to be relying on experienced witnesses, which adds credibility to their accounts, but the Dubliner casts doubt on both witnesses and admirers. We will likewise see that the genre of "answers" published by the Royal Society not only relies on local witnesses, but uses these accounts to add credibility.

Promotional literature of the 1680s would come to make even greater claims, when the Proprietors of Carolina made increased efforts to sell America. Carolina continued to be salubrious and profitable, but was also subject to improvement. A favorite example from 1682 speaks volumes in its title, but little in its content. The title is *CAROLINA; OR A DESCRIPTION Of the Present State of that COUNTRY, AND The Natural Excellencies thereof, viz. The Healthfulness of the Air, Pleasantness of the Place, Advantage and Usefulness of those Rich Commodities there plentifully abounding, which much encrease and flourish by the Industry of the Planters that daily enlarge that Colony.*[32] The body of this tract, written by "T. A., a

---

[27] *Oxford Dictionary of National Biography* (hereafter cited as *ODNB*), s.v. "Blome, Richard (*bap.* 1635?, *d.* 1705)," by S. Mendyk, http://www.oxforddnb.com/view/article/2662 (accessed 24 January 2011).

[28] Richard Blome, *A description of the island of Jamaica; with the other isles and territories in America, to which the English are related, viz. Barbadoes, St. Christophers, Nievis or Mevis, Antego, St. Vincent, Dominica, Montserrat, Anguilla, Barbada, Bermudes, Carolina, Virginia, Maryland, New-York, New England, New-Found-Land. /Taken from the notes of Sr. Thomas Linch, knight, governour of Jamaica; and other experienced persons in the said places* (London, 1672), 141–2.

[29] Blome published epitomes of Speed (e.g., *Speed's maps epitomiz'd: or the maps of the counties of England alphabetically placed* [London, 1681]), in addition to plagiarizing Speed. The 1676 epitome is either his work or that of a competitor or collaborator.

[30] *An epitome of Mr. John Speed's theatre of the empire of Great Britain And of his prospect of the most famous parts of the world. In this new edition are added, the descriptions of His Majesties dominions abroad, viz. New England, New York, Carolina, Florida, Virginia, Maryland, Jamaica, Barbados, . . .* (London, 1676), 209.

[31] See Edmund Sears Morgan, "Rebellion," in *American Slavery, American Freedom: The Ordeal of Colonial Virginia* (New York, 1975), 250–70.

[32] Published in London, 1682.

Gentleman," discusses the flora and fauna of the Carolinas, as well as the behavior of the natives, their mode of hunting, strange agriculture, and lack of dress, and, indeed, anything but manmade climate change. But the publishers and author knew the value of a good title. The pamphlet was "to be Sold by Mrs. *Grover* in *Pelican Court* in *Little Britain*," that is, by the sort of bookseller who was a class below ordinary publishers, working in a neighborhood known for purveyors of the literature of colportage.[33] The title page mentions that it was printed for a W. C., who was probably William, the Earl of Craven, one of the eight lords proprietors of the Carolina grant.[34] Proprietor Anthony Ashley Cooper, first Earl of Shaftesbury, and the other proprietors, in their general attempts to restart the cultivation of the Carolinas along the extractive model, had with the rest of the grantees commissioned a number of tracts, often emphasizing the salubrious climate.[35] John Locke, FRS, had been Shaftesbury's secretary and had drafted the constitution of the colony in 1670,[36] to be included in the Dublin edition of *Carolina* in 1684.[37] The usual argument for the advantage of settling the area was based on the concept of like latitudes. The promotional tracts, usually printed in London or Dublin, the two main sources of immigration, often emphasized the more temperate latitude of Carolina, as one where such things as grapes and olives could grow easily once the land was improved (echoing the dreams of the sixteenth-century settlers). During his European exile in the late 1670s, Locke had taken notes on Mediterranean agricultural products, which he sent to Shaftesbury in 1680 as suggestions for Carolina.[38] These were passed down to the pamphleteers. The classical ideas of climate in the large sense (latitude) and in the smaller sense (place and improvement) inform the language of promoters.

However, these admirers were not universally believed. In 1682, John Dryden, a political enemy to Shaftesbury's Whigs, offered this satire of promotional hyperbole and its emphasis on airs:

> Since faction ebbs, and rogues grow out of fashion,
> Their penny-scribes take care t'inform the nation
> How well men thrive in this or that plantation;
>
> How Pennsylvania's air agrees with Quakers,
> And Carolina's with Associators:
> Both e'en too good for madmen and for traitors.[39]

---

[33] Elizabeth Lane Furdell, *Publishing and Medicine in Early Modern England* (Rochester, N.Y., 2002).

[34] *ODNB*, s.v. "Craven, William, Earl of Craven (*bap.* 1608, *d.* 1697)," by R. Malcolm Smuts, http://www.oxforddnb.com/view/article/6636 (accessed 24 January 2011).

[35] David Armitage, "John Locke, Carolina and the *Two Treatises of Government*," *Political Theory* 32 (2004): 602–27; *ODNB*, s.v. "Cooper, Anthony Ashley, First Earl of Shaftesbury (1621–1683)," by Tim Harris, http://www.oxforddnb.com/view/article/6208 (accessed 24 January 2011).

[36] Locke, *The fundamental constitutions of Carolina* (London, 1670).

[37] *Carolina described more fully then* [sic] *heretofore being an impartial collection made from the several relations of that place in print, since its first planting (by the English): from diverse letters from those that have transported themselves (from this kingdom of Ireland), and the relations of those that have been in that country several years together: whereunto is added the charter with the fundamental constitutions of that province . . .* (Dublin, 1684).

[38] Armitage, "John Locke" (cit. n. 35), 611.

[39] Quoted ibid., 613.

Discussion of climate in the English colonial context was already seen as worse than naive. Positive assessments of climate change, such as appear in the Virginia literature, should be viewed in this light.

INQUIRIES AND ANSWERS

The Dublin correspondent could have been reacting to settlers and the hacks of the popular press—"witnesses" and "admirers." However, these same concerns about climate and health, and about deforestation and cultivation, did on occasion surface in the new "scientific" press.

Members of the Royal Society were in the habit of giving lists of inquiries to travelers, some of which were published in its *Philosophical Transactions*, but others not, either because they predated the journal or because they were only of particular interest to the individuals composing the questions. To quote the preamble to the first published list by Lawrence Rooke in volume 1, "They have from time to time given order to several of their Members to draw up both *Inquiries* of things Observable in forrain Countries, and *Directions* for the Particulars, they desire chiefly to be informed about. . . . Considering with themselves, how much they may increase their *Philosophical* stock by the advantage, which England injoyes of making Voyages into all parts of the World, they . . . set down some *Directions* for *Sea-men* going into the *East & West-Indies*."[40]

Dr. Rooke's inquiries, touching upon cartography, the declination of the needle, meteorology, the drawing of prospects, and other matters, were quite mundane. Robert Boyle's early inquiries (also published in volume 1), besides a bit of astrology, stick to the basics, which include the meteors and the "salubrity or insalubrity of the Air."[41] Sir Robert Moray, Robert Hooke, and numerous others composed questions, some universal, some very particular. In volume 2, there are some inquiries for the East Indies about the verity of folkloric items related by sailors and by sources like John Mandeville's travels, such as "whether . . . there be a Plant, . . . whence fall little Leaves, which turned into Butter-flies?"[42]

I have yet to find a list of inquiries from Royal Society fellows that mentions anthropogenic climate change, or even an affinity of rain for trees, but I have found answers regarding such questions, indicating that lists of them were transmitted in private. Query lists that were published were examples; the published answers, more concrete. This points to a broader culture outside the pages of *Philosophical Transactions* or the Royal Society's meeting rooms.

The first answer of interest, published anonymously in *Philosophical Transactions* in 1666, predates the Dublin letter and concerns observations in the West Indies of the association of trees with rain, "there being certain Trees which attract the Rain, though Observations have not been made of the *kinds*; so as that if you destroy the woods, you abate or destroy the *Rains*. So *Barbados* hath not now half the

---

[40] "Directions for Sea-Men, Bound for Far Voyages," *Phil. Trans. Royal Soc. London* 1 (1665–6): 140–3.

[41] Boyle, "General Heads for a Natural History of a Countrey," *Phil. Trans. Royal Soc. London* 1 (1665–6): 186–7.

[42] "Inquiries for Suratte, and Other Parts of the East-Indies," *Phil. Trans. Royal Soc. London* 2 (1666–7): 417.

Rains, it had, when wooded. In *Jamaica* likewise at *Guanaboa* they have diminisht the Rains as they extended their Plantations." The report out of which this observation came was certainly written in answer to queries, but in 1666, with fewer precedents, the "trees which attract rain" may or may not have been mentioned in answer to a specific question. This observation by local informants, who were either early settlers or natives, is followed up by the observer's empirical corroboration and his apology for a lack of full identification of the tree in question.[43] This answer is also similar enough to the earlier non-English accounts to be said to fall within that established tradition. And the observation seems to have been a novelty to the Royal Society.

Two decades later, in 1688, reports from Cape Corse in West Africa responded more directly to inquiries. A Mr. J. Hillier responded to a question about the benefits of deforestation: "I remember for the Unhealthfulness of the Place, you proposed to inquire if it were Woody, and if any good might be done to it by cutting down the Woods, as has happen'd in many American Plantations."[44] Here we see the colonial story that the Dubliner had questioned being put to the test—Could Africa also benefit from the same treatment? By this time, twelve years after the Dublin letter, the Europeans had naturalized the American experience. The benefit of deforestation, rather than a decrease in rain, fog, or cold winters, would be an increase in "healthfulness" through the softening of tropical conditions.

The cutting of trees and shrubs would make a cold climate warmer (Virginia and New England), a damp climate drier (the West Indies), and an unhealthful (i.e., hot) climate more healthful (Africa). The notion that civilization follows the tempering of a climate conforms with the classical climate model and its revival in the eighteenth century. And as the English explorers, observers, and settlers represented the civilized, the reproduction of English land management in other parts of the globe increased industry, health, and civility.

In his own report, our author, who had traveled to Virginia, had more in common with these traveling correspondents to the *Philosophical Transactions* than with the other classes of writers discussed above. These correspondents were all marginal to the Royal Society but reported back to it in the discourse of the study of nature, not selling anything but their own credibility as reporters, and occasionally making some attempt to offer philosophical analyses of their observations. They could be critical of other eyewitnesses and native informants who were further from the intellectual center, and they could be contradictory to the notions in the literature of colportage. This is not to say that the interests of the Royal Society were pure; rather, they were always intertwined in economics and the good fortunes of the membership.

And we do know that the Dublin correspondent read these answers to inquiries. Later in his letter to Oldenburg, he cited the anonymous 1666 correspondent quoted above, affirming that writer's account of the quality of shipboard water drawn from the Thames with the observation that New England water also became potable after a noisome interlude.[45] The Dublin author answered the water quality question, but did not tie it to the climate question. He was not writing about Jamaica, but about the Vir-

[43] "Observations Made by a Curious and Learned Person, Sailing from England, to the Caribe-Islands," *Phil. Trans. Royal Soc. London* 2 (1666–7): 497.

[44] Hillier, "Part of Two Letters from Mr. J. Hillier, Dated Cape Corse, Jan. 3. 1687/8 and Apr. 25. 1688.," *Phil. Trans. Royal Soc. London* 19 (1695–7): 690.

[45] "Letter" (cit. n. 1), 652–3; "Observations" (cit. n. 43), 495.

ginia observations, which were more pertinent to the Irish condition. But given this context, why did he express skepticism when none was sought?

MR. HENRY NICHOLSON AND HIS WORLD

The writer from Dublin had a name: Mr. Henry Nicholson. The letter that formed the basis for the publication in *Philosophical Transactions* got a hearing at two successive meetings of the Royal Society, on June 1 and June 8, 1676, according to the minutes transcribed in Thomas Birch's *History of the Royal Society*. At the first reading, much notice was taken of the incident of the reversed compass:

> Mr. Oldenburg read a letter written to him from Dublin, dated 10th May, 1676, by Mr. Henry Nicholson, relating a strange effect of thunder upon a magnetic sea-card, whose north and south points had changed positions in such a manner, that though the master of the ship had with his finger brought the flower de lys to point directly north, it would as soon as at liberty return to the new unusual posture. Besides, upon examination it was found, that every compass in the ship was of the same humour. Capt. Grofton of New England was said to be the master of the ship, to which this accident happened; and Mr. Howard, master of several ships, and a man of good credit, was the relater of this accident.

The attention of the amanuensis, and presumably of the meeting as a whole, was solely on the magnetic marvel and on the credibility of Nicholson, an Irish newcomer, and his American witnesses: "It was ordered, that these persons be inquired after and examined concerning the truth of this relation." There was no notice yet of the weather, nor of the long list of other observations and experiments.[46]

The minutes for the June 8 meeting mention Nicholson's weather concerns, experiments, and analyses, as well as the muskrat: "A letter from Mr. Henry Nicholson, dated at Dublin, 20th [*sic*] May, 1676, relating a strange effect of thunder upon a magnetic sea-card, its north and south points having changed positions irrecoverably; and containing likewise some observations about the alteration of the temperature of Ireland and other countries; as also the contrivance of an hygroscope; together with an experiment proposed for discovering the use of respiration; and some observations concerning the American flying hart, and the strong musky scent of the animal called musk-quash."[47] The language "Ireland and other countries" should be noted. In this meeting that discussed Nicholson's observations with greater credence than that of the week before, the minutes set up an equivalency between Ireland and "countries" like Virginia. Nicholson was given a hearing. However, this letter and its anonymous publication in the *Philosophical Transactions* are the last heard of this Henry Nicholson during this phase of the Royal Society.

The name Henry Nicholson was not uncommon, but there is one candidate who fits the chronology. There was a Henry Nicholson in the register for Trinity College Dublin, on whom T. Percy C. Kirkpatrick, registrar of the Royal College of Physicians, offered a brief biography in his *History of the Medical Teaching in Trinity*: "Henry Nicholson, also a Sizar, entered Trinity College at the age of 17 on December 3, 1667,

---

[46] Thomas Birch, *The history of the Royal Society of London for improving of natural knowledge*, vol. 3 (London, 1757), 317–8.

[47] Ibid., 318.

and proceeded to the degree of M.B. on July 7, 1674, being . . . one of the first who is recorded as having taken this degree." If this was our Henry, he would have been a man in his midtwenties when he wrote his letter, which was not unusual for a correspondent of the Royal Society.[48] This Henry, who had received only his "Batchelor of Physick in 1674" after protracted study as a charity student,[49] had settled into a living situation more friendly to his natural studies and observations after travels and other "less settled" circumstances (described below) in 1674, and nothing was heard from him since. Full-time natural philosophers had to be financially secure. Henry the sizar was not. If he was the author of our letter, we have a marginal figure in natural philosophy who nevertheless had educational ties to the one institution, Trinity, that allowed Dublin to play a significant role in the new philosophy. Trinity taught nearly half of the members of the Dublin Society and hosted the first meetings,[50] as well as educating the key Irish fellows of the Royal Society. The Anglican Trinity circle often traveled between Dublin and London. The notes of the two societies, along with those of the society at Oxford, were read aloud at Dublin Society meetings. More important, this Henry received the same degree as two well-placed members of the Dublin Society, future Irish College of Physicians president John Madden, who received his MB the same day as Nicholson, and Allen Mullin (MB, 1676), a physician who was very active in natural philosophy.[51] It is only through such connections that an outsider could even dream of getting a letter to Oldenburg. Nicholson was perhaps just a figure who was not rich enough, talented enough, or leisured enough to become a fixture in these circles. He shows up in the papers of the Royal Society in the 1670s, but never appears in the papers of the Dublin Society in the 1680s and '90s, except as a footnote in a paper by Ashe. But rather than speculate, we can better learn how Nicholson's ideas fit into this milieu by returning to the text.

First, he writes of his long wanderings: "For my own part, I was never furnished with leisure nor conveniences before this year, to make observations in particular of this kind; my occasions being such as required a removal from place to place, and for some time to the *West-Indies*."[52] He does not specify where in the West Indies he went, but does later describe a specific journey from one part of North America to another. His observations of the qualities of shipboard water indicate the extent of his travels, but not his age: "I can affirm upon my own knowledge, that Water taken aboard at *New London* in *New England*, though in eight days time it stunk intolerably, yet when we came to *Virginia,* it recovered . . . "[53] The travels might explain the length of his school career.

As for his personal experience of the conditions in Ireland, after writing about what he saw as the lack of agricultural and demographic progress since the wars "16 or 20 years ago," he laments that no one took an "account of the weather a dozen to

---

[48] Kirkpatrick, *History of the Medical Teaching in Trinity College Dublin and of the School of Physic in Ireland* (Dublin, 1912), 87. Kirkpatrick went on to identify this student with Dr. Henry Nicholson, who was appointed the first botany professor at Trinity. Although this is chronologically possible, Kirkpatrick would later express doubt about the identification in his notes held at the Royal Society of Physicians of Ireland.

[49] Ibid., 56.

[50] K. Theodore Hoppen, *Papers of the Dublin Philosophical Society, 1683–1709* (Dublin, 2008), 1:xxi–xxii.

[51] Ibid., 2:953, 959–60; Kirkpatrick, *Medical Teaching in Trinity* (cit. n. 48), 56–7.

[52] "Letter" (cit. n. 1), 649.

[53] Ibid., 652.

fourteen years past."[54] He remembers that period, which may have included his child-hood, as being frosty, in contrast to the mild present.

The new natural philosophy was for the most part in Ireland the game of younger men, and of the educated. Nicholson's weather diary, which showed a good familiarity with current instrumental meteorology, his quotations of *Philosophical Transactions*, and his natural historical documentation as presented show us a well-educated individual trying his best to ingratiate himself to the Royal Society, to show his worth and credibility. He was an outsider showing everything he knew in one outpouring.

His doubts about the causes of climate change, tinged by concern about the state of the Irish plantations—a worry that would affect the Whig Anglicans at both Trinity and London—place him in the larger intellectual milieu of the leading Dublin natural philosophers and their closest friends in England. These include Sir William Petty, surveyor of Ireland to the Cromwells and Charles II, who was involved in the colonial adventure for many years but allied himself with the new philosophy, playing roles in the beginnings of both the Royal Society and the Dublin Society. Petty's ideas about population growth and economics in England, Ireland, and America in his *Political Arithmetick* of 1683 share Nicholson's concerns for the management of the Irish plantations.[55] Politics and economics are inseparable. Petty's younger friend William Molyneux, FRS, a natural philosopher and political writer, was both scientifically adept and concerned with the lot of the Anglo-Irish overlords of the Irish colony.[56] Molyneux, one of John Locke's closest friends, was at school at Trinity at the same time as Nicholson. He would go on to form (with Petty) the Dublin Society—his network, and its relationship to London and Oxford, defines what it meant to be a philosophic insider in Dublin, and this was the network to which Nicholson aspired.

Another friend of Molyneux, St George Ashe, a Trinity scholar, became provost of Trinity and a Church of Ireland bishop. He was active in natural philosophy and secretary of the Dublin Society.[57] He was the only natural philosopher of the London-Oxford-Dublin network to quote Nicholson. Ashe presented a paper to the Dublin Society in 1684 on the nature of air. His discourse was in keeping with the current work in London by Boyle, Hooke, and others at the Royal Society and touched upon classical climatology and medical meteorology as well as instrumental meteorology. And his judgment of the effect of cultivation is that it was a good thing, falling within the Whiggish ideal of improvement:

> The general causes of insalubrity in the air proceeds [*sic*] from excess of heat and moisture, sudden changes of winds, many standing waters, lowness of shores, and inconstancy of the weather. And the most universal remedy against all these is well peopling of a country, which constrains the inhabitants to cultivate the land, drain stagnant waters, cut down woods, and the like. This is manifest in Holland where all the forementioned causes would certainly concur to make the air most unwholesome, were they not in a great measure prevented by the industry of the people in draining and clearing the country. In America the alteration of the temperature of the air for the better is yet more

[54] Ibid., 649.

[55] The concern for Ireland is much expanded in the posthumous edition of *Political Arithmetick* (London, 1690); *ODNB*, s.v. "Petty, Sir William (1623–1687)," by Toby Barnard, http://www.oxforddnb.com/view/article/22069 (accessed 24 January 2011).

[56] Molyneux, *The case of Ireland's being bound by acts of Parliament in England* (Dublin, 1698).

[57] *ODNB*, s.v. "Ashe, St George (1658–1718)," by Hermann J. Real, http://www.oxforddnb.com/view/article/750 (accessed 24 January 2011).

remarkable since the Europeans began to plant there first, and the Roman Campagne (which once, if we believe Cicero and other old writers), abounded with wholesome and pleasant valleys, being wasted and dispeopled by the incursions of barbarians, has now a most unhealthful infected air. But we need not seek instances so far from home to confirm this truth. Our own country, which is infamous in old historians for its bogs, rains, and mists, is so considerably mended by no other method than culture and good husbandry, that it may even compare with the best of our neighbouring nations.

Ashe was working within the same vocabulary as Nicholson, but found climate change good, something that makes a colony as respectable as the mother country.[58] His classical and Continental reading is clear. He goes on to describe a joint venture in comparing Dublin and Oxford weather diaries, at the other end of which was either Locke or Ashmolean curator Robert Plot.[59] He writes that but for some mountains in Wicklow, Dublin's weather would be just as good as Oxford's.

One would hope Nicholson would have been cited in this context, but instead he was cited anonymously for his hygroscope, which was listed along with Hooke's and Molyneux's designs, among others.[60] When Ashe's paper was read at the Dublin Society, on October 25, 1686, the minutes only discussed what causes the weather in Dublin to be unwholesome.[61]

Nicholson once again proved to be a marginal figure. The reading of his climate questions, along with what had become by 1686 a climate trope in the popular press, were commonplaces not worthy of comment. That a natural philosopher and cleric found it a subject worth taking up shows us that the idea of climate change through land improvement was becoming naturalized even in this intellectual network. By the time John Evelyn was revising his *Sylva*, it seemed as though enough credible witnesses had testified to make it a matter of fact. And in the European and colonial contexts, it became a matter of land management and improvement. This would bolster the political philosophy and actions of the empire builders.

CONCLUSION

John Locke's activities in the 1670s included working for Shaftesbury in devising the extractive and cruel Carolina colony, going into exile for other Whig work with the earl, writing and revising the *Constitutions of Carolina*, working on his *Two Treatises on Government*, and writing on scripture and religious toleration, but neglecting the weather diaries that he had started in the 1660s and resumed in the 1680s (after being in exile), when he was still revising the Carolina work and conceiving the *Essay Concerning Human Understanding*.[62] That is, the letter from Dublin was published in a period when we also see in the person of Locke the nascence of Whig politics, the expansion of British colonialism, the growth of capitalism, the codification of empirical epistemology, and the establishment and spread of instru-

[58] Hoppen, *Papers* (cit. n. 50), 153.
[59] Plot and Molyneux had been making such exchanges with each other in the mid-1680s; R. T. Gunther, *Early Science in Oxford*, vol. 12 (Oxford, 1923–67), 138–9.
[60] Hoppen, *Papers* (cit. n. 50), 151.
[61] At the same meeting, Petty's 1687 edition of *Political Arithmetick* was presented to the society, with Molyneux acting as respondent. Ibid., 93–4.
[62] Locke, *The fundamental constitutions of Carolina* (London, 1670); *Two treatises of government* (London, 1690); *An essay concerning humane understanding* (London, 1690).

mental science. While much of the *Two Treatises* discusses social contract and establishing modern liberal government within European nations, Locke's agrarian capitalism can be found in chapter 5 of the second treatise,[63] which gives his theory of property. In brief, as opposed to the noble but simple habitation of land by the natives in America and other such savages, European-style property rights derive from the improvement of the land, that is, from establishing boundaries and changing the land so as to allow for surplus production. He who has done so has earned certain rights to the land. This theory can be seen as justifying the Caribbean and Carolina land thefts, and likewise those in Ireland, as well as legitimating enclosure back in England. Any such lands that have not been taken under control, deforested, planted, and brought to profit he calls "vacant" places.[64] One should consider the resemblance of these vacant wastes to the tabula rasa of the young mind as seen in the *Essay* and later works, also vacant until improved and cultivated, and even to the blank pages of Locke's weather diary, to be filled with data, or the blank chapters of Boyle's *General History of Air*, waiting to be improved by his followers in natural philosophy.[65] But how does this help us understand the letter from Dublin? Likewise, we can look at Petty's economics, which examines the relationships between population and land use. Does it justify sending the Irish to the Americas, and the English and Scots to Ireland, as had happened throughout the seventeenth century? Or does it explain the celebration by the fellows of the Royal Society, Ashe, and later Evelyn of the benefits of "improvement"?

These questions contextualize Nicholson's moment of doubt about the human causation of climate change in the North American case. If this climate improvement were truly due to cultivation, he would not fear a greater change in temperature. But given Ireland's lack of cultivation, he worried about why that country was getting warmer too.

To understand Nicholson's denial of North American claims about climate, perhaps it is best to see it partially in terms of claims about Irish climate, which involve the supposed improvement wrought by English settlement in very specific areas. The primary concern of writers on Irish climate was *wetness*, in the form of rain, mists, and bogs. And colonial concerns superseded those of political party: Gerard Boate, in his 1652 *Natural History of Ireland*, dedicated by Samuel Hartlib to Cromwell, celebrated the elimination of bogs and reduction of rains. This same text, essentially a manual for colonial management, would be reprinted with additions from Sir Thomas Molyneux in 1726.[66] William Molyneux and Nicholson were both cited by Ashe for improving Robert Hooke's hygrometer.[67] Ashe's "Discourse" was also focused on humidity and precipitation. William King's paper on bogs blamed the laziness of the natives for these geographical features of Ireland.[68]

The English "cult of improvement" had already made climate changeability a

[63] Neal Wood, *John Locke and Agrarian Capitalism* (Berkeley and Los Angeles, 1984).

[64] Locke, *Two Treatises* (cit. n. 62), book 2, chap. 5, sec. 36.

[65] Robert Boyle, *The general history of the air* (London, 1692).

[66] Gerard Boate, *Irelands naturall history* (London, 1652), 168–9; Boate, *A natural history of Ireland, in Three parts* (Dublin, 1726). See also S. Mendyk, "Gerard Boate and 'Irelands Naturall History,'" *J. Roy. Soc. Antiquar. Ireland* 115 (1985): 5–12

[67] See also Molyneux, "A Letter from William Molyneux Esq; Sec. of the Dublin Society; To One of the S. of the R. S. concerning a New Hygroscope, Invented by Him," *Phil. Trans. Royal Soc. London* 15 (1685): 1032–5.

[68] Hoppen, *Papers* (cit. n. 50), 217–26.

commonplace notion in lands close to home.[69] Nicholson's dilemma was that the requisite economic and demographic development seemed lacking in Ireland. And local knowledge involved moisture, not temperature. If the warm winter just experienced was not explained by the comforting ideology of "improvement," it left Nicholson open to apocalyptic anxiety.

Nicholson saw that weather in America and Ireland should be compared, but was this evidence of a more general conception of climate? His conception was at least latitudinal, and his perception that both places grew warmer, yet agriculture did not appear to be the logical reason, led him to look for a warming pattern that could be demonstrated by instrumental records and comparison (see Hooke's scheme in the 1660s).[70] And he seemed to fear a general, progressive warming:

> Neither can I impute this extraordinary alteration to any fortuitous concourse of ordinary circumstances requisit to the production of fair weather; because it is manifest, that it hath proceeded gradually, every year becomeing more temperate than the year preceding. If any in this city or country hath kept an exact account of the weather for at least a dozen or fourteen years past, I doubt but their Iournalls will verify, what I have only general observed.[71]

Coincidentally, according to Gordon Manley's twentieth-century reconstruction, there was a peculiarly warm winter in 1675/6.[72] This spurred Nicholson's memories of the winters when rivers froze over. It would seem he had some sort of misgivings about a change in the climate with an unknown cause. His response was recourse to instruments and organized weather observations.

> This last winter newly ended, I have Kept an exact account of wind and weather (as I intend to doe, God willing, for the future) being well provided with a *Barometer, sealed Thermometers, Hygroscopes*, and all things requisit to the performance of so nice and necessary a Task.[73]

Quantitative measurement would legitimize his subjective impressions. Had the North American theory concerning improvement applied to Ireland, the warming weather would have been less unsettling to him.

To the Royal Society, Nicholson was a presenter of novelties, magnetic oddities, and instrumental contrivances. His contributions were interesting enough to print, but, whether because of his lack of social attainment, insufficient evidence for his claims, lack of patronage, or the peculiarity of his observations, he did not earn public credit, at least in print. As for his opinions and fears about climate, even in his doubt he was well within the discourse of his milieus, both popular and philosophical. That doubt is interesting not in itself, but in how it exposes the political currents beneath claims about climate in the North Atlantic world of the seventeenth century.

---

[69] Richard Drayton, *Nature's Government: Science, Imperial Britain, and the "Improvement" of the World* (New Haven, Conn., 2000), 52–4.

[70] Thomas Sprat, *The history of the Royal-Society* (London, 1667), 173–9. See also Andrea Rusnock, "Correspondence Networks and the Royal Society," *Brit. J. Hist. Sci.* 32 (1999):155–69.

[71] "Letter" (cit. n. 1), 649.

[72] Manley, "Central England Temperatures: Monthly Means, 1659 to 1973," *Quarterly Journal of the Royal Meteorological Society* 100 (1974): 389–405. See also Georgina Endfield in this volume.

[73] "Letter," 649–50.

Inventing Caribbean Climates:
How Science, Medicine, and Tourism Changed Tropical Weather from Deadly to Healthy

*by Mark Carey**

ABSTRACT

This article examines how four major historical factors—geographical features, social conditions, medicine, and tourism—affected European and North American views of the tropical Caribbean climate from approximately 1750 to 1950. It focuses on the British West Indies, a region barely examined in the historiography of climate, and examines the views of physicians, residents, government officials, travelers, and missionaries. International perceptions of the tropical Caribbean climate shifted markedly over time, from the deadly, disease-ridden environment of colonial depictions in the eighteenth century to one of the world's most iconic climatic paradises, where tourists sought sun-drenched beaches and healing breezes, in the twentieth. This analysis of how environmental conditions, knowledge systems, social relations, politics, and economics shaped scientific and popular understandings of climate contributes to recent studies on the cultural construction of climate. The approach also offers important lessons for present-day discussions of climate change, which often depict climate too narrowly as simply temperature.

Tropical climate provoked anxieties among Europeans and North Americans for centuries because many believed tropical regions were disease ridden, morally degrading, and even deadly.[1] In the nineteenth century, however, many Europeans and North Americans were emphasizing more positive aspects of the tropical climate from India and Madagascar to the West Indies.[2] By the twentieth century, Europeans

---

* Robert D. Clark Honors College, University of Oregon, Eugene, OR 97403; carey@uoregon.edu.

[1] Richard Ligon, *A True and Exact History of the Island of Barbadoes* (1657; repr., London, 1673), 27. See also Karen Ordahl Kupperman, "Fear of Hot Climates in the Anglo-American Colonial Experience," *William Mary Quart.* 41 (1984): 213–40; Kupperman, "The Puzzle of the American Climate in the Early Colonial Period," *Amer. Hist. Rev.* 87 (1982): 1262–89; Mart A. Stewart, "'Let Us Begin with the Weather?' Climate, Race, and Cultural Distinctiveness in the American South," in *Nature and Society in Historical Context*, eds. Mikuláš Teich, Roy Porter, and Bo Gustafsson (New York, 1997), 240–56.

[2] Eric T. Jennings, *Curing the Colonizers: Hydrotherapy, Climatology, and French Colonial Spas* (Durham, N.C., 2006); Georgina H. Endfield and David J. Nash, "'Happy Is the Bride the Rain Falls On': Climate, Health and 'the Woman Question' in Nineteenth-Century Missionary Documentation," *Trans. Inst. Brit. Geogr.* 30 (2005): 368–86; Mark Harrison, *Climates and Constitutions: Health, Race, Environment and British Imperialism in India, 1600–1850* (New York, 1999); Dane Kennedy, "The Perils of the Midday Sun: Climatic Anxieties in the Colonial Tropics," in *Imperialism and the Natural World*, ed. John M. MacKenzie (New York, 1990), 118–40; David N. Livingstone, "Tropical Climate and Moral Hygiene: The Anatomy of a Victorian Debate," *Brit. J. Hist. Sci.* 32 (1999): 93–110.

---

and North Americans were flocking to the hot tropical climate in areas like the Caribbean to bask in the sun, sand, and sea.[3] The Caribbean had become a climatic paradise, a refuge from icy northern winters. These changing views of tropical climate in some ways parallel the transformation of perceptions of wilderness from dangerous wasteland to enticing recreation destination.[4] Closer inspection of published and unpublished sources that described the Caribbean climate from approximately 1750 to 1950, however, reveals the complexity of this transformation—as well as the multiple ways in which climate is historically and culturally constructed.

European and North American understandings of tropical climate have varied widely over time because most observers conceptualized it through the lenses of colonialism, race relations, economics, and medicine.[5] They also conceived contradictory characterizations of the tropical climate—the lushly productive Eden and the deadly, morally depraved Hell.[6] Even when tourism boosters were marketing the Caribbean climate as refreshing and healthful during the twentieth century, many continued to wonder if tourists could survive the lurking dangers of unruly natives, political instability, tropical diseases, and exotic creatures such as parasites, snakes, and insects. Theories of climatic determinism—the notion that climate determined human capacity and thus defined societies' and races' level of civilization—also continued well into the twentieth century and led to stereotypically negative depictions of tropical inhabitants.[7] Tropical climate is clearly a rich topic for historical analysis because of the multiple meanings embedded in it.

Relatively few studies have examined climate history in the Caribbean, which lies almost entirely in the tropical belt: the Tropic of Cancer runs along Cuba's northern coast.[8] Research on the region's climate has focused on climatic reconstructions and hurricanes rather than cultural histories of climate.[9] The lack of research on the his-

[3] Ian Littlewood, *Sultry Climates: Travel and Sex* (Cambridge, Mass., 2001); Mimi Sheller, *Consuming the Caribbean: From Arawaks to Zombies* (New York, 2003); Krista Thompson, *An Eye for the Tropics: Tourism, Photography, and Framing the Caribbean Picturesque* (Durham, N.C., 2007); Ian G. Strachan, *Paradise and Plantation: Tourism and Culture in the Anglophone Caribbean* (Charlottesville, Va., 2002).

[4] Roderick Nash, *Wilderness and the American Mind*, 3rd ed. (New Haven, Conn., 1982).

[5] E.g., Philip D. Curtin, *Death by Migration: Europe's Encounter with the Tropical World in the Nineteenth Century* (New York, 1989); Georgina H. Endfield and David J. Nash, "'A Good Site for Health': Missionaries and the Pathological Geography of Central Southern Africa," *Singapore J. Trop. Geogr.* 28 (2007): 142–57; Kenneth F. Kiple and Kriemhild Conee Ornelas, "Race, War and Tropical Medicine in the Eighteenth-Century Caribbean," in *Warm Climates and Western Medicine: The Emergence of Tropical Medicine, 1500–1900*, ed. David Arnold (Amsterdam, 1996), 65–79; David N. Livingstone, "The Moral Discourse of Climate: Historical Considerations on Race, Place and Virtue," *J. Hist. Geogr.* 17 (1991): 413–34.

[6] Nicolás Wey Gómez, *The Tropics of Empire: Why Columbus Sailed South to the Indies* (Cambridge, Mass., 2008).

[7] Ellsworth Huntington, *Civilization and Climate*, 3rd ed. (1915; repr., New Haven, Conn., 1924); Livingstone, "Tropical Climate and Moral Hygiene" (cit. n. 2).

[8] Important exceptions include Mark Harrison, "'The Tender Frame of Man': Disease, Climate, and Racial Difference in India and the West Indies, 1760–1860," *Bull. Hist. Med.* 70 (1996): 68–93; Richard Grove, *Green Imperialism: Colonial Expansion, Tropical Island Edens and the Origins of Environmentalism, 1600–1860* (New York, 1995).

[9] Michael Chenoweth, *The 18th Century Climate of Jamaica Derived from the Journals of Thomas Thistlewood, 1750–1786* (Philadelphia, 2003); Louis A. Pérez, Jr., *Winds of Change: Hurricanes and the Transformation of Nineteenth-Century Cuba* (Chapel Hill, N.C., 2001); Matthew Mulcahy, *Hurricanes and Society in the British Greater Caribbean, 1624–1783* (Baltimore, 2006).

torical construction of Caribbean climate is notable because the region has become one of the world's most iconic destinations for climate-related tourism.[10] This article analyzes European and North American understandings of the Caribbean climate during the period of significant perceptual and scientific changes from the mid-eighteenth to the mid-twentieth century. It examines the observations of physicians, scientists, residents, government officials, travelers, missionaries, and tourism boosters. While focusing primarily on the British West Indies, it points to broader trends in the Caribbean and offers insights on the cultural construction of tropical regions more generally.

Throughout this history, scientific and popular understandings of climate were shaped by many environmental, social, cultural, political, intellectual, and economic factors. Most Europeans and North Americans recognized that the tropical or Caribbean climate was not singular or monolithic. They distinguished various climatic characteristics even on small islands. With such specific, complex understandings of the region's climate, they also identified certain areas and climate types as healthy, while they classified others as dangerous. Over time, new forces affected views of tropical climate, especially medical breakthroughs and the tourism economy after the late nineteenth century. Climate constructions thus changed over time and were dependent upon where observers went, why they went there, which geographical features they saw, who they encountered, how landscapes looked to them, and what scientific knowledge they had. In short, four principal historical factors influenced European and North American understandings of tropical Caribbean climate over time: geographical features, social conditions, medicine, and tourism. Perceptions of climate, as scholars are increasingly showing, had as much to do with these factors as with scientific knowledge and atmospheric conditions.[11] Environmental knowledge was produced over time through the dynamic interplay among these various forces.[12]

---

[10] For other health- and climate-driven travel in the nineteenth century, see John Beckerson and John K. Walton, "Selling Air: Marketing the Intangible at British Resorts," in *Histories of Tourism: Representation, Identity and Conflict*, ed. John K. Walton (Buffalo, N.Y., 2005), 55–68; Harriet Deacon, "The Politics of Medical Topography: Seeking Healthiness at the Cape during the Nineteenth Century," in *Pathologies of Travel*, eds. Richard Wrigley and George Revill (Amsterdam, 2000); Vladimir Jankovic, "The Last Resort: A British Perspective on the Medical South, 1815–1870," *Journal of Intercultural Studies* 27 (2006): 271–98; Judith T. Kenny, "Climate, Race, and Imperial Authority: The Symbolic Landscape of the British Hill Station in India," *Ann. Assoc. Amer. Geogr.* 85 (1995): 694–714; Simon M. Kevan, "Quests for Cures: A History of Tourism for Climate and Health," *International Journal of Biometeorology* 37 (1993): 113–24; Helen Tiffin, "Colonies, Consumption and Climate," in *Crabtracks: Progress and Process in Teaching the New Literatures in English*, eds. Gordon Collier and Frank Schulze-Engler (Amsterdam, 2002), 267–82.

[11] Wolfgang Behringer, *A Cultural History of Climate* (Malden, Mass., 2010); Vladimir Jankovic, *Reading the Skies: A Cultural History of English Weather, 1650–1820* (Chicago, 2001); James Rodger Fleming, Vladimir Jankovic, and Deborah R. Coen, eds., *Intimate Universality: Local and Global Themes in the History of Weather and Climate* (Sagamore Beach, Mass., 2006); Christian Pfister, "Climatic Extremes, Recurrent Crises and Witch Hunts: Strategies of European Societies in Coping with Exogenous Shocks in the Late Sixteenth and Early Seventeenth Centuries," *Medieval History Journal* 10, nos. 1–2 (2007): 33–73.

[12] Paul S. Sutter, "Nature's Agents or Agents of Empire? Entomological Workers and Environmental Change during the Construction of the Panama Canal," *Isis* 98 (2007): 724–54; Mark Carey, "Latin American Environmental History: Current Trends, Interdisciplinary Insights, and Future Directions," *Environ. Hist.* 14 (2009): 221–52.

GEOGRAPHIES OF CARIBBEAN CLIMATE

Many factors determined how Europeans and North Americans historically understood and classified the Caribbean climate. On the one hand, doctors, scientists, residents, and travelers occasionally measured weather systematically, noting precise
temperatures, dates of hurricanes, precipitation amounts, wind velocity, and other
characteristics.[13] This tendency corresponded with a broader scientization of meteorology among Western societies.[14] On the other hand, those who wrote about the
Caribbean generally described many other diverse climatic characteristics. They analyzed how geographical conditions, wind, and topography all varied throughout the
region and shaped perceptions of climate in distinct ways. These factors also changed
over time and, in many cases, led chroniclers to identify good and bad climates on
the basis of their assemblage of characteristics. Climate was never conceptualized
in simplistic or monolithic terms, either in 1700 or in 1900. James Johnson, one of
the most famous English physicians of the early nineteenth century, captured this
broader tendency to understand climate through what he called "the influence of locality." Johnson was a member of the Royal College of Physicians of London, and
the numerous editions of his book *The Influence of Tropical Climates on European
Constitutions*, first published in 1813, circulated widely among physicians, military
leaders, and government officials traveling to tropical regions, including the Caribbean.[15] His concept of locality can be seen in the writings of many others who defined
climate on the basis of local geographical features.

Some environments generally inspired fear about a region's climate, such as
swamps and marshes, which supposedly emitted bad air. As a 1775 publication on
animal husbandry explained, "A dry and pure air, such as is found on hilly or mountainous sea coasts, free from marshes or swamps, is always healthy ... but when
a meridian sun unites with a marshy rotten soil, in which the heavy rains stagnate,
then it is impossible for a country to be tolerably healthy."[16] Marshes were linked
to climate because of the understanding of miasmas, the pre–germ theory idea that
poisonous vapors or bad air caused diseases. Many prominent physicians of the late
eighteenth and early nineteenth centuries noted similar negative health consequences
resulting from contact with marshy air. Benjamin Moseley, a Chelsea Hospital physician and member of the Royal College of Physicians, explained in 1792 that swamps
and stagnant water caused fevers and liver disease.[17] Colin Chisholm, a distinguished
physician in the United States and Europe and inspector general of hospitals in the
West Indies, concurred. In his 1822 book submitted to the commander in chief of the
British army, Chisholm even complained that marshes made people unhappy.[18] This

[13] E.g., Chenoweth, *18th Century Climate of Jamaica* (cit. n. 9); Thomas W. Jackson, *Tropical Medicine, with Special Reference to the West Indies, Central America, Hawaii, and the Philippines, Including a General Consideration of Tropical Hygiene* (Philadelphia, 1907), 34–5.
[14] James Rodger Fleming, *Historical Perspectives on Climate Change* (New York, 1998); Katharine
Anderson, *Predicting the Weather: Victorians and the Science of Meteorology* (Chicago, 2005).
[15] James Johnson, *The Influence of Tropical Climates on European Constitutions*, 4th ed. (London,
1827), 465; also see Harrison, "Tender Frame of Man" (cit. n. 8).
[16] Harry J. Carman, ed., *American Husbandry* (1775; repr., New York, 1939), 408–9.
[17] Moseley, *A Treatise on Tropical Diseases; on Military Operations; and on the Climate of the West
Indies* (London, 1792), 51.
[18] Colin Chisholm, *Manual of the Climate and Diseases of Tropical Countries; in Which a Practical
View of the Statistical Pathology, and of the History and Treatment of the Diseases of Those Countries,
Is Attempted to Be Given* (London, 1822), 1–3.

relationship between marshes, climate, and health persisted through the nineteenth century as well.[19] It led physicians and other visitors to classify parts of the Caribbean as healthier than others according to local climatic conditions. Barbados, for example, with its dry, rocky ground and few marshes, was consistently portrayed as one of the healthiest Caribbean islands.

Commentators also referred to mountains as important factors influencing tropical climate, noting in particular that higher elevations had more salubrious climates than lowlands.[20] Most agreed throughout the nineteenth century that the climate of Jamaica's mountainous areas was healthier than that of coastal Kingston.[21] Henry Walter Bates, the famous English naturalist well known for his Amazon studies with Alfred Russel Wallace, commented in the 1870s that hilly parts of the larger Antilles (Cuba, Jamaica, Hispaniola, and Puerto Rico) were "healthy and enjoyable," with a mild climate, in contrast to the "decidedly unhealthy" lowland areas where the climate produced epidemics such as yellow fever.[22] Many British colonial officials maintained that troops should be stationed in mountainous terrain rather than along coasts. The late nineteenth-century British surgeon D. H. Cullimore, who was stationed in India for a period and who wrote about the relationship between climate and health, believed yellow fever never reached the elevation of Camp Jacob (1,600 ft.) in Martinique or the hill stations in Jamaica. He even classified Jamaica's Newcastle hill station as a potential health resort for U.S. residents.[23]

Wind was the most common signal of a healthy climate, especially the trade winds that blew across islands such as Barbados. Wind has had various meanings over time, often negative.[24] And though British perceptions of wind became more embedded in purely scientific and chemical descriptions during the Victorian era, views of Caribbean breezes remained relatively stable through time because, as almost every commentator pointed out, the winds provided a refreshing respite from tropical heat. Ever since the initial colonization of Barbados in the seventeenth century, Europeans frequently described it as the healthiest of all Caribbean islands because of its cool winds.[25] There were similar views of other islands' breezes and in subsequent periods as well. Noting the positive influence of wind on the island of Santa Cruz near Puerto Rico, the English banker and Quaker minister Joseph John Gurney pointed

[19] E.g., *Statistical Report on the Sickness, Mortality, and Invaliding among the Troops in the West Indies, Prepared from the Records of the Army Medical Department and War-Office Returns* (London, 1838), 32, 102; S. Edwin Solly, *A Handbook of Medical Climatology: Embodying Its Principles and Therapeutic Application with Scientific Data of the Chief Health Resorts of the World* (Philadelphia, 1897), 435.

[20] E.g., Chisholm, *Manual of the Climate and Diseases of Tropical Countries* (cit. n. 18), 1.

[21] Anthony Trollope, *The West Indies and the Spanish Main* (Leipzig, 1860), 39–40; see also Charles Denison, *Rocky Mountain Health Resorts: An Analytical Study of High Altitudes in Relation to the Chronic Pulmonary Disease*, 2nd ed. (Boston, 1880), 10.

[22] Bates, *Central America, the West Indies, and South America*, 3rd ed. (London, 1885), 144.

[23] D. H. Cullimore, *The Book of Climates; Acclimatization; Climatic Diseases; Health Resorts and Mineral Springs; Sea Sickness; Sea Voyages; and Sea Bathing*, 2nd ed. (London, 1891), 237; see also Ira Nelson Morris, *With the Trade-Winds: A Jaunt in Venezuela and the West Indies* (New York, 1897), 47.

[24] Vladimir Jankovic, "Gruff Boreas, Deadly Calms: A Medical Perspective on Winds and the Victorians," *Journal of the Royal Anthropological Institute* 13 (2007): S147–S164.

[25] Ligon, *True and Exact History* (cit. n. 1), 27; Edmund Hickeringill, *Jamaica Viewed: With all the Ports, Harbours, and their Several Soundings, Towns, and Settlements Thereunto Belonging* (London, 1705), 4–5; Larry Dale Gragg, *Englishmen Transplanted: The English Colonization of Barbados, 1627–1660* (New York, 2003), 15–7.

out that the delightful easterly breeze saved him from the hot, stagnant air common in other areas.[26] Many others echoed Gurney by suggesting that the local climatic conditions that allowed refreshing winds to blow in certain regions did more than cool the weather. For many, these winds had the potential to purify the spirit, rejuvenate the body, and prevent illness—as if the wind blew away diseases before they could infect a person.[27]

SOCIAL CONDITIONS AND CLIMATE

Climate was always about much more than environmental conditions, even if marshes, mountains, and winds all shaped views of air quality and weather. European and North American perceptions of the landscape—itself a human cultural construction—also influenced understandings of climate.[28] The extent of agriculture on cleared lands and fertile soil figured strongly into the assessment of whether a climate was beneficial. The belief that land clearing, deforestation, and cultivation affected climate extends back to ancient times, and Europeans brought the same notion to their American colonies in the West Indies and North America.[29] In short, many Europeans believed that harsh climates could be improved by draining marshes, clearing woodlands, and farming. Social conditions therefore affected understandings of climate. George Pinckard of the Royal College of Physicians exemplified this view when he reported in 1806 that a place "can only be made healthy by the unceasing toil of man." Land clearing and productive agriculture improved both the land and climate, he concluded.[30] But commentators were careful to indicate that too much removal of trees and brush would make the country drier and less fertile, reducing its capacity for sugar production.[31]

Related to the degree of land clearing for agriculture, a place's perceived level of economic development and civilization were additional aspects of the human landscape that could shape perceptions of climate. This link between climate and social conditions was not unique to the West Indies or even the American tropics. British patients and physicians turned the island of Madeira on the Atlantic coast of North Africa into a health resort in the nineteenth century, though it later declined from that position both because there was no scientific proof of healing and because the place did not offer ideal European amenities.[32] Thus, the degree to which a place resembled Europe mattered in how Europeans perceived its climate. While European cultivation and so-called civilization supposedly improved climates, social upheaval and delinquency made them more dangerous. In Haiti, for example, the revolution that put former slaves in charge of the island had, to a Polish soldier visiting in 1803,

[26] Gurney, *A Winter in the West Indies, Described in Familiar Letters to Henry Clay, of Kentucky* (London, 1840), 10.

[27] Robert T. Hill, *Cuba and Porto Rico with the Other Islands of the West Indies: Their Topography, Climate, etc.* (London, 1898), 374.

[28] On cultural constructions of landscape, see Simon Schama, *Landscape and Memory* (New York, 1995).

[29] Fleming, *Historical Perspectives on Climate Change* (cit. n. 14), chap. 2.

[30] George Pinckard, *Notes on the West Indies: Written during the Expedition under the Command of the Late General Sir Ralph Abercromby* (London, 1806), 386–7.

[31] R. Montgomery Martin, *History of the British Colonies* (London, 1835), 4:6, 124; Charles Kingsley, *Westward Ho! Or the Voyages and Adventures of Sir Amyas Leigh* (Cambridge, 1855), 229.

[32] Jankovic, "Last Resort" (cit. n. 10).

completely transformed the island's climate: "The air here is most unhealthy, especially since the time of the black revolt twelve years ago."[33] Haiti's social revolution
altered European views of its climate, demonstrating how social and class relations
affected classifications of climate. Poor climates were also sometimes linked to a
lack of hygiene or poor sanitation measures in certain regions. The solution to such
unhealthy climates was to institute not only public health campaigns, but also more
social control and police patrols.[34]

Other social conditions also affected perceptions of climate, such as proximity to
population centers. Nineteenth-century physicians often equated urban poverty with
unhealthy Caribbean climates. A crowded city would lead observers to describe the
climate as hotter, less windy, more stifling, and less healthy.[35] Less densely populated
areas, on the other hand, tended to inspire comments about more favorable climates
with healing winds. The U.S. traveler Ira Nelson Morris captured these sentiments in
the late nineteenth century, pointing out that foreign residents "have their homes on the
outskirts of the town [in St. Thomas], or back among the hills, where they enjoy
the breeze of the trade-winds and a cleanliness not to be found in the towns themselves."[36] Similar sentiments determined the fate of a health resort in Cape Colony,
South Africa, from the 1830s to 1860s. As the surrounding area became more urban
and poorer, Europeans changed their view of the climate's healthfulness and moved
their resort inland to suburbs.[37] Such preferences for the climate of rural, less crowded
areas corresponded with nineteenth-century romanticism, which tended to privilege
pastoral, rural, and even sublime landscapes over those of increasingly industrialized
and polluted urban areas.

Diverse views of the Caribbean climate reveal how numerous historical forces affected scientists', physicians', residents', and travelers' understanding of the region's
environment. Just as there was no singular West Indian or Caribbean climate, nor was
there any universal consensus that tropical climate was unhealthy, even during the
eighteenth century. Further, the human landscape and social conditions influenced
how Europeans and North Americans perceived the region's climate, suggesting
that climatic determinism may not have been as strong or ubiquitous in this period
as scholars sometimes contend. In fact, it was not only native residents' behavior
that shaped climatic understandings, but also Europeans' activities once they arrived
in the West Indies.[38] Moderate exercise, temperance in eating and drinking, sleep,
cleanliness, and general avoidance of vice were common behaviors mentioned as
being required to thrive in hot climates.[39] Such sentiments about the importance of
human behavior over deterministic climatic forces demonstrate both that complex

---

[33] Quoted in Laurent Dubois, *Avengers of the New World: The Story of the Haitian Revolution* (Cambridge, Mass., 2004), 230.

[34] A Resident, *Sketches and Recollections of the West Indies* (London, 1828), 216.

[35] E.g., Jackson, *Tropical Medicine* (cit. n. 13), 66.

[36] Morris, *With the Trade-Winds* (cit. n. 23), 13. Other examples include Richard Henry Dana, Jr.,
To Cuba and Back: A Vacation Voyage (Boston, 1859), 261–2; and Hill, *Cuba and Porto Rico* (cit.
n. 27), 58–9.

[37] Deacon, "Politics of Medical Topography" (cit. n. 10).

[38] Livingstone, "Tropical Climate and Moral Hygiene" (cit. n. 2), 104.

[39] Moseley, *Treatise on Tropical Diseases* (cit. n. 17), 2; Johnson, *Influence of Tropical Climates
on European Constitutions* (cit. n. 15), 5; Chisholm, *Manual of the Climate and Diseases of Tropical
Countries* (cit. n. 18), iii, 6; James Cecil Phillippo, *The Climate of Jamaica* (London, 1876), 34; for
counterexamples, see *Four Year's Residence in the West Indies, During the Years 1826, 7, 8, and 9: By
the Son of a Military Officer* (London, 1833), 56, 494, 584.

factors shaped climate perceptions and that the West Indian climate was not nearly as inherently unhealthy as some in Europe depicted it to be.

MEDICINAL CLIMATES

Travel for health dates back millennia, but it became more systematic after the eighteenth century.[40] By then, some health seekers were going regularly to the British seacoast for a change of air.[41] Others turned to hydrotherapy, soaking in French medicinal spas, particularly between 1815 and 1850.[42] Those suffering from tuberculosis and asthma visited mountainous areas including the Alps, the Rocky Mountains, and elsewhere after the mid-nineteenth century.[43] Some Europeans went farther abroad, to the Mediterranean or to areas colonized by their nations, such as the hills of India, the outskirts of Cape Town, and the temperate regions of Australia, New Zealand, and Canada.[44] The Caribbean, however, has been largely overlooked in this literature, even though French colonists began visiting Guadeloupe for its healthful springs and fresh mountain breezes in the 1830s.[45]

Positive views of the Caribbean climate and its healing influences date back much earlier than scholars often assume. Even in the seventeenth century there was some praise for the healing capacity of the Caribbean climate. In 1679, Thomas Trapham completed his study of health in Jamaica and suggested that the island was healthier than England.[46] One of the most famous health seekers, George Washington, went to Barbados in 1751 with his brother Lawrence, who was suffering from pulmonary disease.[47] Like many other North Americans, Washington was escaping a northern winter and hoping the island's warm tropical air would help heal him. He was also lured to Barbados because of the high regard for Bridgetown physician Dr. William Hillary. Hillary subsequently wrote on the effects of rainfall and temperature on diseases, explicitly following the theories of Hippocrates and referring to the ancient thinker as "that wise father and prince of physicians."[48]

Europeans and North Americans identified selected places in the Caribbean as healthy more frequently during the nineteenth century. In Jamaica, for example, a region once known as Hellshire Hills became Healthshire Hills.[49] The English banker

[40] Kevan, "Quests for Cures" (cit. n. 10).

[41] John K. Walton, *The English Seaside Resort: A Social History, 1750–1914* (New York, 1983).

[42] Douglas Peter Mackaman, *Leisure Settings: Bourgeois Culture, Medicine, and the Spa in Modern France* (Chicago, 1998).

[43] Gregg Mitman, "Geographies of Hope: Mining the Frontiers of Health in Denver and Beyond," *Osiris* 19 (2004): 93–111; Mitman, *Breathing Space: How Allergies Shape Our Lives and Landscapes* (New Haven, Conn., 2007).

[44] John Pemble, *The Mediterranean Passion: Victorians and Edwardians in the South* (New York, 1987); Dane Kennedy, *The Magic Mountains: Hill Stations and the British Raj* (Berkeley and Los Angeles, 1996); Deacon, "Politics of Medical Topography"; Kenny, "Climate, Race, and Imperial Authority"; Tiffin, "Colonies, Consumption and Climate"; Jankovic, "Last Resort" (All cit. n. 10).

[45] Jennings, *Curing the Colonizers* (cit. n. 2), chap. 3.

[46] Cited in Kiple and Ornelas, "Race, War and Tropical Medicine" (cit. n. 5), 66.

[47] George Washington, *The Daily Journal of Major George Washington, in 1751–2, Kept While on a Tour from Virginia to the Island of Barbadoes, with His Invalid Brother, Maj. Lawrence Washington*, ed. J. M. Toner (Albany, N.Y., 1892), 4.

[48] William Hillary, *Observations on the Changes of the Air and the Concomitant Epidemical Diseases, in the Island of Barbados* (London, 1759), preface; see also Jan Golinski, *British Weather and the Climate of Enlightenment* (Chicago, 2007), 144–6.

[49] Frank Fonda Taylor, *To Hell with Paradise: A History of the Jamaican Tourist Industry* (Pittsburgh, Pa., 2003), 13.

and minister Gurney noted in 1840 that fellow passengers on a ship headed from New York to the West Indies were going "mostly in search of a warmer climate and better health."[50] Robert Schomburgk's 1848 history of Barbados recognized the West Indian climate as a pleasurable escape from icy winters. Born in Germany, Schomburgk was a surveyor for England's Royal Geographical Society. Queen Victoria knighted him for his extensive Caribbean and South American surveys, and in the 1840s he served as a British diplomat in Barbados. Like others writing about aspects of the West Indian climate, Schomburgk wondered why European physicians familiar with the consistent daily and annual temperatures of Barbados did not recommend it more often as a healing place for invalids, particularly those suffering from pulmonary diseases.[51] James Phillippo, a British medical doctor and member of the Medico-Chirurgical Society of Edinburgh, echoed these sentiments a few years later. He believed that Jamaica's climate had long been misrepresented as "deadly." He instead saw the island as a "health resort for the invalid."[52] Many others also tried to change what they considered an unfair prejudice against Caribbean climate, noting that those who actually visited parts of the Caribbean recognized the climate as healthful rather than deleterious.[53]

Medical breakthroughs regarding the role of mosquitoes in disease transmission—as well as improved understanding of germ theory, microbiology, and parasitology—also transformed perceptions of tropical climate by removing the false belief that tropical diseases came from hot climates. Physicians could blame insects and germs, rather than just climate, for the illnesses that Europeans and North Americans contracted in the tropics. Much of this research on so-called tropical medicine was motivated by colonial expansion and the European and North American quest to make tropical areas in Africa, Asia, and the Americas safer for colonizing officials, colonial troops, and laborers.[54] Medical experts thus not only identified mosquitoes as the source of malaria, yellow fever, and other diseases, but also waged campaigns against the insects to reduce disease outbreaks. Engineering projects and sanitation campaigns—that is to say, greater environmental management—helped make the tropics safer for foreigners, which led some to question theories of environmental determinism.[55] Of course, climatic determinism did not vanish altogether, as colonialist and racist notions of the "white man's burden" persisted at the turn of the twentieth century.[56] In the greater Caribbean region, diseases such as malaria and yellow fever had crippled U.S. troops in the Spanish-American War and workers building the Panama Canal. In response, sanitation engineers removed yellow fever from Havana

[50] Gurney, *Winter in the West Indies* (cit. n. 26), 2.

[51] Robert H. Schomburgk, *The History of Barbados, Comprising a Geographical and Statistical Description of the Island* (London, 1848), 78.

[52] Phillippo, *Climate of Jamaica* (cit. n. 39), 2–3.

[53] E.g., J. H. Sutton Moxly, *An Account of a West Indian Sanatorium and a Guide to Barbados* (London, 1886), v.

[54] Steven Palmer, "Migrant Clinics and Hookworm Science: Peripheral Origins of International Health," *Bull. Hist. Med.* 83 (2009): 676–709; Julyan G. Peard, *Race, Place, and Medicine: The Idea of the Tropics in Nineteenth-Century Brazilian Medicine* (Durham, N.C., 1999).

[55] Jackson, *Tropical Medicine* (cit. n. 13), v.

[56] Warwick Anderson, "Disease, Race, and Empire," *Bull. Hist. Med.* 70 (1996): 62–7; Kennedy, "Perils of the Midday Sun"; Livingstone, "Tropical Climate and Moral Hygiene" (Both cit. n. 2); Michael Worboys, "Manson, Ross and Colonial Medical Policy: Tropical Medicine in London and Liverpool, 1899–1914," in *Disease, Medicine, and Empire: Perspectives on Western Medicine and the Experience of European Expansion*, eds. Roy MacLeod and Milton Lewis (New York, 1989), 21–37.

by 1902. In the Panama Canal Zone, disease-prevention programs caused the death rate to fall from three times that of the United States in 1905 to half the U.S. rate in 1914.[57] Science and medicine thus helped make the tropics safer for Europeans and North Americans—whether colonial administrators or sun-seeking tourists.[58]

Nineteenth- and early twentieth-century references to the healthful Caribbean climate reveal four important issues in the social and geographical construction of climate. First, many physicians and other writers who categorized the climate as healthy were explicitly hoping to overturn popular (mis)conceptions of the region's climate as deadly. The sources analyzed in this article generally do not discuss the climate as deleterious, thus indicating either that there was a gap between popular understandings in Europe and firsthand experiences in the Caribbean or that the nineteenth century was a transition period when previous notions of the deadly climate still lingered despite the increasing number of positive depictions by visitors and physicians. Second, the growing interest in escaping the cool northern winters of North America indicates a view of climate among many in the northern United States divergent from that of southerners. Gurney's 1840 observation that fellow passengers on the ship from New York to the West Indies were mostly traveling in search of a warmer winter came at a time when southern slave owners were using climatological theories as a pretext for maintaining slavery. Africans could more effectively labor under the hot sun, they argued.[59] Tropical climate conjured something quite different for the southern plantation owners who saw it as linked to inferior African slaves than it did for the northerners to whom it offered a respite from winter snow.

Third, nineteenth-century Europeans and North Americans tended to identify salubrious climates in the same places that previous observers had classified as healthy on the basis of geographical features and social conditions. Physicians' assurances of a healthy climate in the nineteenth century thus corresponded with these other socioenvironmental factors that travelers, residents, and officials had already noted. Fourth, science and medicine altered public perceptions of the climate because the new knowledge—and the public health campaigns that followed—showed Europeans and North Americans that they could live more safely under certain climatic and environmental conditions. Medicine thus not only influenced the social and cultural construction of climate, but also affected people's interaction with tropical regions.

TOURISM AND CLIMATE AS A COMMODITY

International tourism helped turn the Caribbean climate into a commodity after the mid-nineteenth century. Several forces explain how Caribbean tourism rose to prominence between roughly 1850 and 1950. Steamship travel facilitated access. Tourists ventured beyond their own countries because of growing interest in exotic destinations, new romantic notions of the "tropical picturesque," a quest to escape

---

[57] J. R. McNeill, *Mosquito Empires: Ecology and War in the Greater Caribbean, 1620–1914* (New York, 2010), 308–12; see also Paul Sutter, "Tropical Conquest and the Rise of the Environmental Management State: The Case of the U.S. Sanitary Efforts in Panama," in *Colonial Crucible: Empire in the Making of the Modern American State*, eds. Alfred W. McCoy and Francisco A. Scarano (Madison, Wis., 2009), 317–26.

[58] G. M. Giles, *The Outlines of Tropical Climatology* (London, 1904), 88.

[59] Stewart, "Let Us Begin with the Weather?" (cit. n. 1).

unhealthy industrialized cities, and an increasing desire to be outdoors.[60] In Europe, many people began to seek out the sun after the late nineteenth century, when architects and planners increasingly designed open, sunlit areas; by the 1920s, the suntan had become a fad.[61] Further, the marketing of verdant tropical landscapes and exotic populations lured foreigners who sought not only relaxation and romance but also power over much poorer resident populations. Entrepreneurs and colonial officials within the Caribbean also played a role distributing tourist propaganda to transform the sun, sand, and sea into commodities. After the nineteenth century, a range of historical actors marketed the Caribbean climate for foreign consumption. Instead of inventing a brand-new climate, however, tourism boosters built on previous classifications of healthful climates in specific geographical places with particular social conditions.

Church officials and military personnel were among the first to explicitly market the Caribbean climate in order to attract new missionaries. In 1865, the Church of England reported that many British colonial possessions were on sunny islands in tropical and southern regions, such as the Bahamas. Living on these beautiful islands, the report continued, could hardly be construed as "ministerial self-sacrifice" because the climate and fresh air were so enjoyable.[62] The report reads like tourism propaganda rather than an examination of missionary activities. Two decades later, Sutton Moxly, the English chaplain for British troops, became yet another advocate for changing European opinions about Caribbean climate. He believed medical knowledge exposed the folly of theories about climatic determinism because good hygiene could protect troops from tropical diseases.[63] He was promoting climatic conditions in the British Empire to attract soldiers and allay recruits' fears.

Residents and businesses within the West Indies also became tourism boosters by marketing the healthful, positive aspects of the Caribbean climate. Many boosters pointed to the same places that earlier commentators had classified as healthy—areas with consistent winds, productive agriculture, more developed economies, uncrowded settlements, and appealing landscapes. Barbados, a center of Caribbean tourism, was one of Britain's wealthiest West Indian colonies, as well as a place with dry ground and pleasant breezes from the trade winds. A 1905 hotel brochure from Hastings, Barbados, bragged that its location and tourist attractions made it the most "ideal winter resort of the tropics for tourists, invalids and those seeking a genial clime."[64] The brochure made a point of explaining that the island looked like "carefully cultivated English hills," thereby noting the landscape and social conditions as well as topographical and climate features that all combined to make the island appealing for tourists.

Other tourism propaganda highlighted the healthful and relaxing aspects of the Caribbean climate, which some argued was superior to that of Florida, Italy, or southern

[60] Sheller, *Consuming the Caribbean* (cit. n. 3); Walton, *English Seaside Resort* (cit. n. 41).

[61] Ken Worpole, *Here Comes the Sun: Architecture and Public Space in Twentieth-Century European Culture* (London, 2000); Louis Turner and John Ash, *The Golden Hordes: International Tourism and the Pleasure Periphery* (New York, 1976), 78–9.

[62] Church of England, *Work in the Colonies: Some Account of the Missionary Operations of the Church of England in Connexion with the Society for the Propagation of the Gospel in Foreign Parts* (London, 1865), 91, 99–100.

[63] Moxly, *Account of a West Indian Sanatorium* (cit. n. 53), v, 3–7.

[64] Marine Hotel Brochure, Hastings, Barbados, 1905, National Archives of Barbados, Bridgetown (hereafter cited as NAB), file no. S. Ref Pam X 85, 1.

France.[65] A 1921 government brochure even boasted that Barbados was *"the healthiest spot on the globe"* and could thus serve as a *"sanatorium of the West Indies."* It continued, "But the charm of the climate of Barbados lies in the fact that, being comparatively flat, it is swept by the sea breezes and health bearing trade winds."[66] These were the same environmental conditions in the Caribbean that observers had identified in the eighteenth century as being more healthful than marshy, lowland areas. Barbadians began marketing their climate even more vigorously and systematically after the 1930s.[67] Others in the growing tourism industry did as well. The SS *Berlin*, for example, distributed a brochure in 1937 calling for "sunshine seekers" interested in cruises from New York to Jamaica and Cuba.[68] International visitors also helped change perceptions of the West Indian climate. In 1898, United States Geological Survey scientist Robert Hill reported that tourism had become the second most lucrative industry for the Barbados economy after sugar. Tourism had accelerated, he explained, because Barbados was a desirable winter destination for Europeans and North Americans, because the English went to "see the colonies," because of its prominent place in international shipping, and because of its fame as a health resort for the English of Trinidad and Guiana.[69] In 1937, Raymond Savage wrote nostalgically about the "sunny days and clear blue skies" of the West Indies. He urged readers to take additional time to "laze on the verandah inhaling the marvelous air."[70]

The marketing of the region's enticing climate was cemented into policy during the post–World War II era as colonial and then national policy makers put tourism at the center of their economic development plans. The Anglo-American Caribbean Commission, formed secretly in 1942 as a joint U.K.-U.S. endeavor to plan the postwar Caribbean political and economic landscape, took a particular interest in developing tourism as a way to diversify and expand the region's economy.[71] Of course the dual legacies of tourism—its profound importance for the region's economy and its embeddedness in historical inequality and exploitation—did not disappear.[72] The commission recognized that marketing campaigns would have to be launched, infrastructure built, accessibility improved, and diseases eliminated. But it maintained consistently that "the Caribbean area as a whole is healthy and remarkably free from diseases. . . . The health-giving properties of the sunshine and of the sea air of the Caribbean are recognized as important assets to the area in respect to its develop-

[65] Algernon E. Aspinall, *The Pocket Guide to the West Indies* (London, 1907), v, 59–60; James H. Stark, *Stark's History and Guide to Barbados and the Caribbee Islands* (London, 1903), 110; DaCosta and Company, *Barbados Illustrated: Historical, Descriptive, and Commercial* (Barbados, 1911), NAB, file no. S Ref F 2041 D35, 18.

[66] "British West Indies as a Health and Tourist Resort" (Barbados, 1921), NAB, file no. Pam A 1248, 2–3 (emphasis in the original).

[67] Barbados Publicity Committee, "Souvenir of Ever British Barbados" (Glasgow, 1936), NAB, file no. Pam A 831; Jean Holder, "A Brief History of Travel from Ancient to Modern Times and Its Relationship to 20th Century Tourism" (Barbados, 1988), NAB, file no. Pam C 778.

[68] National Archives II, College Park, Md., record group 43, Records of International Commissions, lot 53D466, Anglo-American Caribbean Commission (hereafter cited as NA II), box 44, file E13-1.

[69] Hill, *Cuba and Porto Rico* (cit. n. 27), 376; also see Morris, *With the Trade-Winds* (cit. n. 23), 53.

[70] Raymond Savage, *Barbados: The Enchanting Isle* (Philadelphia, 1937), 13, 68–70; see also E. A. Hastings Jay, *A Glimpse of the Tropics, or, Four Months Cruising in the West Indies* (London, 1900), 38–40.

[71] For commission documents related to tourism, see NA II, boxes 44–8.

[72] Vincent Vanderpool-Wallace, "Foreword," in *New Perspectives in Caribbean Tourism*, eds. Marcella Daye, Donna Chambers, and Sherma Roberts (New York, 2008), ix–x; Strachan, *Paradise and Plantation* (cit. n. 3), chaps. 1–2.

ment as a major field for vacation travel."[73] For the commission, the only trouble-some aspect of the Caribbean climate was hurricanes. Otherwise, it adopted the same framework for understanding and promoting climate that commentators had used since the eighteenth century. Breezes, mountains, trees, and temperature were the main criteria used to define salubrious versus unhealthy climates. Mountainous areas allowed the tourist to move up the slopes to cooler weather, but anyone blocked from the trade winds behind mountains would swelter. Anglo-American Caribbean Commission reports consistently praised Barbados for having a good climate because it lacked mountains, and thus the breezes could blow freely across the island. They also suggested that trees and forests "improved" the climate; the commission generally expressed preferences for rural rather than urban climates.

The long-standing tendency to associate social conditions with regional climate also appeared in tourism propaganda that depicted the Caribbean as a place with both warm sun and civilized society. Referring to Barbados as "Little England"—an old nickname but one that increasingly appeared in tourism promotions—underscored how socioeconomic characteristics merged with climatic and environmental conditions. Tourism pamphlets often pictured a landscape that would appeal to many Europeans and North Americans, with a caption that focused on the region's healthy climate. These brochures from the 1940s to the 1960s had slogans such as "Barbados: Land of Abiding Sunshine" (1946); "Barbados: The Riviera of the Caribbean . . . Where It Is Always Summer" (1953); and "Barbados, The West Indies: Island in the Sun" (1962).[74] The warm tropical climate, far from a deadly deterrent, was now the Caribbean's best economic asset.

*   *   *

By the mid-twentieth century, most Europeans and North Americans found sunny tropical beaches quite alluring, as is still the case today. But this was not always so. International perceptions of the tropical climate changed over time, most broadly from seeing it as dangerous in the seventeenth century to considering it healthy and relaxing in the twentieth. The transformation began as early as the late eighteenth century, when physicians, scientists, government officials, travelers, residents, and missionaries began identifying specific parts of the Caribbean climate as healthy, but the transition was never complete; all of these groups of observers continued to see other areas of the Caribbean as unhealthy. Their views were shaped by four principal factors beyond meteorological characteristics: geographical features, social conditions, medical knowledge, and the tourism industry. The ways in which these forces influenced people's understanding of climate helps to reorient present-day discussions about climate change, which often focus solely on temperature or the impact of global warming.[75] In tropical regions and elsewhere, societies have always understood climate through a range of lenses, and science has been only one among many equally important others. Thus, in the past and the present, the science of climate must be put in dialogue with culture, ideology, economics, politics, and social relations.

---

[73] Anglo-American Caribbean Commission, *Caribbean Tourist Trade: A Regional Approach* (Washington, D.C., 1945), NAB, file no. Pam. A 152.

[74] Barbados Tourism Advertisements, NAB, file no. Srl 468; NA II, box 48, file E13-6.

[75] Diana M. Liverman, "Conventions of Climate Change: Constructions of Danger and the Dispossession of the Atmosphere," *J. Hist. Geogr.* 35 (2009): 279–96.

Reculturing and Particularizing Climate Discourses:

Weather, Identity, and the Work of Gordon Manley

*by Georgina Endfield**

ABSTRACT

During the last three decades, popular discourses of climate have been replaced by a global and scientific metanarrative focusing on climate change and global warming. Recent climate scholarship has thus highlighted the need for a reexamination of the idea of climate and its culturally and spatially variable dimensions, though anxieties over the reemergence of climatic determinism in a number of high-profile publications have compounded a reluctance to engage in scholarship exploring the influence of climate on society. This article focuses on the work of British meteorologist and geographer Gordon Manley, whose publications, particularly those produced for popular audiences in the second half of the twentieth century, focus on the relationships between climate, place, and identity in postwar Britain. I argue that his approach may have renewed significance given recent efforts to reparticularize and reculture climate change discourses.

INTRODUCTION

Throughout the nineteenth and early twentieth centuries, climate represented an "exploitable hermeneutic resource" that gave meaning to cultural and environmental distinctiveness. Classical theories of climate were reworked to explain racial, pathological, economic, and moral characteristics and distinctions between different parts of the world.[1] Such ideologies gave rise to an environmental determinism that provided an ethno-climatological influence on geographical debates into the twentieth century and have been particularly associated with the work of Ellsworth Huntington (1876–1947), a U.S.-based "unrestrained and undisciplined geographic determinist,

* School of Geography, University of Nottingham, University Park, Nottingham NG7 2RD, U.K.; Georgina.Endfield@nottingham.ac.uk.

[1] David N. Livingstone, "Tropical Hermeneutics: Fragments for a Historical Narrative; An Afterword," *Singapore J. Trop. Geogr.* 21 (2000): 92–8, on 93; Livingstone, "Climate's Moral Economy: Science, Race, and Place in Post-Darwinian British and American Geography," in *Geography and Empire*, eds. Neil Smith and Anne Godlewska (Oxford, 1995), 137; Vladimir Jankovic, "Intimate Climates: From Skins to Street, Soirees to Societies," in *Intimate Universality: Local and Global Themes in the History of Weather and the Climate*, eds. James Rodger Fleming, Jankovic, and Deborah R. Coen (Sagamore Beach, Mass., 2006), 1–34.

eugenicist, and popular writer."[2] Huntington was broadly concerned with the "response of living things . . . to their environment," with the impacts of climate and climate change on the development and location of civilization, and with the influence of weather on human efficiency, reflecting the intellectual milieu of the period.[3] The misappropriation of environmental deterministic ideas in the early decades of the twentieth century to support racist eugenic theories, coupled with growing criticism that such arguments diminished the importance of human agency in environmental change and were based on flawed techniques, meant that work by Huntington and others adopting deterministic perspectives was increasingly discredited.[4] The geographical community in particular strove for a shift away from what were increasingly regarded as socially dysfunctional climatic explanations of history.[5] There was a "flight from determinism" toward cultural explanations of social change whereby nature was relegated to setting the limits of choice and adaptation for human societies.[6] More recently, scholarship on social vulnerability to environmental change has highlighted the need for investigations that consider interactions between social and ecological systems,[7] while climate history research has advanced debates about the role of climate in human affairs and is offering important insights into how different societies have responded and adapted to climatic variability, and weather events, in the past.[8]

Yet academic anxieties over the emphasis placed on the role of climate (and the environment) in shaping human histories have recently resurfaced in response to a number of high-profile publications. Jared Diamond's *Guns, Germs and Steel* (1997) and his sequel, *Collapse: How Societies Choose to Fail or Succeed* (2005), it has been argued, mark a worrying new trend toward Eurocentric, environmentally deterministic, and reductionist perspectives on human history and fail "to take into account many of the advances in human-environmental thought since the early twentieth

[2] James Rodger Fleming, *Historical Perspectives on Climate Change* (Oxford, 1998), 95.

[3] Gordon Manley, "The Revival of Climatic Determinism," *Geogr. Rev.* 48 (1958): 98–105, on 98; Kent McGregor, "Huntington and Lovelock: Climatic Determinism in the 20th Century," *Physical Geography* 25 (2004): 237–50; Fleming, *Historical Perspectives* (cit. n. 2), 95.

[4] Richard Peet, "The Social Origins of Environmental Determinism," *Ann. Assoc. Amer. Geogr.* 75 (1985): 309–33, on 327; McGregor, "Huntington and Lovelock" (cit. n. 3), 243.

[5] Gabriel Judkins, Marissa Smith, and Eric Keys, "Determinism within Human-Environment Research and the Rediscovery of Environmental Causation," *Geogr. J.* 174 (2008):17–29; James Blaut, "Environmentalism and Eurocentrism," *Geogr. Rev.* 89 (1999): 391–408.

[6] See Mike Hulme in this volume; Gordon R. Lewthwaite, "Environmentalism and Determinism: A Search for Clarification," *Ann. Assoc. Amer. Geogr.* 56 (1966): 1–23, on 12; Peet, "Social Origins of Environmental Determinism" (cit. n. 4); Carl Sauer, "The Agency of Man on Earth," in *Man's Role in Changing the Face of the Earth*, ed. W. L. Thomas (Chicago, 1956); Sauer, *Land and Life: A Selection from the Writings of Carl Ortwin Sauer*, ed. J. Leighly (Berkeley, 1963).

[7] P. Billie Lee Turner et al., "Illustrating the Coupled Human-Environment System for Vulnerability Analysis: Three Case Studies," *Proceedings of the National Academy of Sciences* 100 (2003): 8080–5.

[8] E.g., Emmanuel Le Roy Ladurie, *Histoire du climat depuis l'an mil* (Paris, 1983); Robert I. Rotberg and Theodore K. Rabb, eds., *Climate and History* (Princeton, N.J., 1981); Christian Pfister, "Climate and Economy in Eighteenth Century Switzerland," *J. Interdis. Hist.* 9 (1978): 223–43; Hubert H. Lamb, *Climate: Past, Present and Future* (London, 1977); Lamb, *Climate, History and the Modern World* (London, 1982); T. M. L. Wigley, M. J. Ingram, and G. Farmer, eds., *Climate and History: Studies in Past Climates and Their Impact on Man* (Cambridge, 1981), 479–513; Christian Pfister and Rudolf Brazdil, "Climatic Variability in Sixteenth Century Europe and Its Social Dimension: A Synthesis," *Climatic Change* 43 (1999): 5–53; Mariano Barriendos, "Climatic Variations in the Iberian Peninsula during the Late Maunder Minimum (AD 1675–1715): An Analysis of Data from Rogation Ceremonies," *Holocene* 7 (1997): 105–11; P. D. Jones, A. E. J. Ogilvie, T. D. Davies, and K. R. Briffa, eds., *History and Climate: Memories of the Future?* (London, 2001).

century," particularly concerning human agency in environmental modification.[9] Citing scholarship produced by the paleoenvironmental community, geographers Paul Coombes and Keith Barber have expressed concern over a similar "revival of environmental determinism," manifest in a number of studies in which abrupt climate change is invoked to explain catastrophe and civilization collapse.[10] Although this scholarship emphasizes the significance of climatic factors in an explanatory reductionist sense, the branding of such work as generically "deterministic" reveals a pervasive fear within an academic community sensitized to the potential pitfalls of engaging in research that explores climate's role in shaping societies.

There are fears that the histrionics associated with climatic determinism, and its apparent revival, might lead to a reluctance to engage in investigations of climate and society at a time when the need to (re)connect climate and human histories is arguably more important than ever.[11] Indeed, over the last three decades, climate change has become a dominant environmental narrative of the twenty-first century.[12] Yet while there is a consensus that the global climate is changing, that human activities are exacerbating natural climatic variability, and that climate change will pose new and significant challenges for global society at large, the precise impacts for different social, economic, and ecological systems are less clearly understood.[13] It is essential to try to obtain a better understanding of how different groups of people in different spatial contexts conceptualize climate and its fluctuations.[14] Moreover, in the process of climate change's propulsion into the global arena, notwithstanding developments in vulnerability and climate history scholarship, there has been something of a deculturalization of the idea of climate.[15] Contemporary debates over the imminent climate change threat, coupled with the fixation on the apparent acceleration in anthropogenic global warming and the dominance of climatic modeling in climate change studies, have obscured the distinctive meaning that climate holds, and has held in the past, for different people and places. There have, therefore, been calls for a "reexamination of climate change" that challenges global perspectives on climate and incorporates "contributions from the interpretive humanities and social sciences," such that climate's culturally and spatially variable dimensions and meanings can be

[9] Blaut, "Environmentalism and Eurocentricism"; Judkins et al., "Determinism within Human-Environmental Research," 17 (Both cit. n. 5).

[10] Coombes and Barber, "Environmental Determinism in Holocene Research: Causality or Coincidence?" *Area* 37 (2005): 303–11, on 303.

[11] John McNeill, "Diamond in the Rough: Is There a Genuine Environmental Threat to Security? A Review Essay," *International Security* 30 (2005): 178–95, on 179.

[12] Mike Hulme and John Turnpenny, "Understanding and Managing Climate Change: The UK Experience," *Geogr. J.* 170 (2004): 105–15.

[13] Neil W. Adger, "Social Capital, Collective Action and Adaptation to Climate Change," *Economic Geography* 79 (2003): 387–404, on 387; Stephen H. Schneider, "What Is Dangerous Climate Change?" *Nature* 411 (2001): 17–9.

[14] Fekri Hassan, "Environmental Perception and Human Responses in History and Prehistory," in *The Way the Wind Blows: Climate, History and Human Action*, eds. R. J. McIntosh, J. A. Tainter, and S. K. McIntosh (New York, 2000), 121–40; Adger, "Social Capital, Collective Action and Adaptation" (cit. n. 13); Adger, N. W. Arness, and E. L. Tompkins, "Successful Adaptation to Climate Change across Scales," *Global Environmental Change* 15 (2005): 77–86; Christian Pfister, "Learning from Nature-Induced Disasters: Theoretical Considerations and Case Studies from Western Europe," in *Natural Disasters, Cultural Responses: Case Studies towards Global Environmental History*, eds. Christof Mauch and Pfister (New York, 2009), 17–40.

[15] Mike Hulme, "Governing and Adapting to Climate: A Response to Ian Bailey's Commentary on 'Geographical Work at the Boundaries of Climate Change,'" *Trans. Inst. Brit. Geogr.* 33 (2008): 424–7, on 424.

reintroduced into contemporary climate debates. More specifically, there is a need to "reinvent" global climate change discourses "as discourses about the relationship between local weather and physical objects and about the relationship between local weather and cultural practices."[16]

Various scholars have highlighted the need to explore how "ordinary people" understand and talk or write about the weather, to investigate the ways in which people engage with and ascribe meanings to the weather and make sense of it.[17] A number of publications have focused on cultural histories of attitudes toward the weather, the meteorological tradition of Enlightenment Britain, Victorian weather forecasting, and eyewitness narratives of historical weather events.[18] Recent volumes have also explored the myriad ways in which humans have understood the idea of climate and the history of debates over climate change.[19] Building on earlier work on cloud art and landscape meteorology, John Thornes has illustrated the significance of visual culture and the representation of the sky, atmosphere, and weather in major historical artworks, paralleling "a new awareness of cultural physical geography."[20] Such cultural climatology approaches are affording a more nuanced, spatially, temporally, and culturally specific understanding of climate and, it follows, of a changing climate.

Yet there are earlier examples of this genre of scholarship. The remainder of this article focuses on the culural climatology of British meteorologist and geographer Gordon Manley (1902–80). While some of Manley's perspectives might be seen to resonate with climatic deterministic arguments, he explores the powerful relationship between climate, weather, and culture in postwar Britain, and his approach may have renewed significance given recent calls to "reculture" or reparticularize climate change discourses by exploring "the relationship between local weather and cultural practices."[21]

THE POPULAR SCHOLARSHIP OF GORDON MANLEY:
INFLUENCES, CONTEXT, AND CONTRIBUTION

Having studied engineering in Manchester, followed by geography at Cambridge in the 1920s, Manley joined the Meteorological Office in 1925, before accepting

---

[16] Mike Hulme, "Geographical Work at the Boundaries of Climate Change," *Trans. Inst. Brit. Geogr.* 33 (2008): 5–11, on 6.

[17] Sarah Strauss and Benjamin S. Orlove, "Up in the Air: The Anthropology of Weather and Climate," in *Weather, Climate, Culture*, eds. Strauss and Orlove (Oxford, 2003), 3–14, on 7; Vladimir Jankovic and Christina Barboza, eds., *Weather, Local Knowledge and Everyday Life* (Rio de Janeiro, 2009); Jan Golinski, *British Weather and the Climate of Enlightenment* (Chicago, 2007); Fleming et al., *Intimate Universality* (cit. n. 1).

[18] Golinski, *British Weather* (cit. n. 17); Lucian Boia, *The Weather in the Imagination* (London, 2005); Vladimir Jankovic, *Reading the Skies: A Cultural History of the English Weather, 1650–1820* (Manchester, 2001); Katherine Anderson, *Predicting the Weather: Victorians and the Science of Meteorology* (Chicago, 2005); Richard Hamblyn, "Introduction," in *The Storm*, by Daniel Defoe (London, 2003), x–xl.

[19] Fleming et al., *Intimate Universality* (cit. n. 1); Fleming, *Historical Perspectives* (cit. n. 2).

[20] C. J. P. Cave, *Clouds and Weather Phenomena for Artists and Other Lovers of Nature* (Cambridge, 1926); Leo C. W. Bonacina, "Landscape Meteorology and Its Reflection in Art and Literature," *Quarterly Journal of the Royal Meteorological Society* 65 (1939): 485–97; Thornes, "Cultural Climatology and the Representation of Sky, Atmosphere, Weather and Climate in Selected Artworks of Constable, Monet and Eliasson," *Geoforum* 39 (2008): 570–80, on 571; Ken J. Gregory, *The Changing Nature of Physical Geography* (London, 2000), cited in Thornes, 571.

[21] Hulme, "Geographical Work at the Boundaries" (cit. n. 16), 6.

lecturing posts first at Birmingham University and latterly at Durham. In 1939, he returned to Cambridge and thence to Bedford College, London, where he was professor of geography from 1948 to 1964. He later went on to establish a new Department of Environmental Studies at Lancaster University before retiring in Cambridge in the late 1960s. While studying at Cambridge, Manley developed an interest in polar and high-altitude environments, taking inspiration from Frank Debenham, professor of geography from 1930 to1949 and geologist on the British Antarctic Expedition of 1910–3 under Captain Robert F. Scott.[22] It was at Durham, however, that Manley began to focus specifically on the British climate and on the collection and analysis of long-term early instrumental weather records for the United Kingdom, as well as the people who created them.[23] It is from such sources that he assembled the Central England temperature series of monthly mean temperatures stretching back to 1659. His record was compiled from overlapping series of observations at different sites in a triangular area enclosed by Bristol, Manchester, and London, took over three decades to produce, and constituted the longest historical temperature series based directly on thermometer readings (taken monthly from 1659, and daily from 1772).[24] This is still treated as "the firmest piece of our knowledge of climatic events since the middle of the 17th century" and stands as a test for all the proxy records of climatic change for this period.[25]

Manley worked within the tradition of a field scientist and natural historian. Echoing early nineteenth-century meteorologists who "turned toward investigating the atmosphere as a laboratory,"[26] he regarded the country as his "laboratory." He kept a detailed diary of the British weather and maintained his own field station at Morehouse, on the peak of Great Dunn Fell, one of the bleakest summits of the northern Pennines in England.[27] He published widely in geography and meteorology and on a range of other subjects, including antiquarian maps and atlases, energy and fuel, transport problems, and the geography of Durham, producing over 182 papers. He was a long-serving member of the Royal Geographical Society and was also a leading member, and one-term president, of the Royal Meteorological Society.[28]

Keen to promote the study of climate and the weather, Manley was particularly interested in the interrelationship between climate, culture, and place.[29] His now-classic text *Climate and the British Scene*, first published in 1952, demonstrated the role that climate has played in shaping British landscapes, national identity, character, and "the many other elements that go to make up the British scene."[30] He fo-

[22] Michael J. Tooley and Gillian M. Sheail, "The Life and Work of Gordon Manley," in *The Climatic Scene*, eds. Tooley and Sheail (London, 1983), 1–16, on 2–3.

[23] Manley, "The Weather and Diseases: Some Eighteenth Century Contributions to Observational Meteorology," *Notes and Records of the Royal Meteorological Society* 9 (1952): 300–7.

[24] Manley, "Central England Temperatures: Monthly Means, 1659–1973," *Quarterly Journal of the Royal Meteorological Society* 100 (1974): 389–405; D. E. Parker, T. P. Legg, and C. K. Folland, "A New Daily Central England Temperature Series, 1772–1991," *International Journal of Climatology* 12 (1992): 317–42; Tooley and Sheail, "Life and Work" (cit. n. 22), 7.

[25] Hubert H. Lamb, "The Little Ice Age Period and the Great Storms within It," in Tooley and Sheail, *Climatic Scene*, 104–31, on 104; see also Tooley and Sheail, "Life and Work," 7 (Both cit. n. 22).

[26] Jankovic, *Reading the Skies* (cit. n. 18), 165; Tooley and Sheail, "Life and Work" (cit. n. 22), 5.

[27] Kenneth Hare, "Review of *Climate and the British Scene*," *Geogr. Rev.* 44 (1954): 323–4; Peter Marren, *The New Naturalists* (London, 1995), 168.

[28] Tooley and Sheail, "Life and Work" (cit. n. 22), 1.

[29] Manley, "The Geographer's Contribution to Meteorology," *Quarterly Journal of the Royal Meteorological Society* 73, nos. 315–6 (1947): 1–10.

[30] Manley, *Climate and the British Scene*, New Naturalist, no. 22 (1952; repr., London, 1962), 4.

cused on "the flow of the air, the changeable weather and the rubato embroidery upon the fundamental rhythm of the seasons" as key influences in "moulding the British mind,"[31] and he highlighted the connections between climate and the peculiarities of people and places across the United Kingdom by studying the British climate as "an expression of our integrated experiences of 'weather.'"[32] *Climate and the British Scene* represented the twenty-second book in the New Naturalist series, the most famous series on British natural history, and reflected its aim to promote new ways of observing and understanding nature and to provide "a new survey of Britain's natural history . . . popular in price, presentation and appeal."[33] Manley thus deliberately set out to engage the general public, and English climatologist Hubert Lamb described the book as "a feast" of Manley's "most attractive writing."[34]

The temporal context within which Manley was working was also an important influence on his approach. The New Naturalist series coincided with the coming-of-age of the "new" science of ecology in the second half of the twentieth century and the institutionalization of field study and nature conservation in a postwar society looking toward a more peaceful future.[35] Manley's publication also reflects a more general impulse in the United Kingdom, beginning in the years following World War I, toward regional survey, the capturing and sharing of local knowledge, and the promotion of citizenship. As David Matless has illlustrated, this was partly inspired by Patrick Geddes (1854–1932), the Scottish biologist-sociologist-geographer and educationalist. Geddes had a particular influence on British geography after 1918 and believed that people and environment were dialectically interdependent, and that local nature observation, in its broadest sense, would foster good citizenship and create a sense of belonging and identity.[36] There are earlier precedents elsewhere drawing such links between the environment and national identity. The Abbé du Bos, a French diplomat, historian, and critic, argued that changes in climate could explain "the rise and decline of creative spirit in particular nations."[37] Alan Bewell has also demonstrated how a comprehensive understanding of meteorology was deemed by Thomas Jefferson to be fundamental to the construction of the American republic, its people, and its future in the eighteenth century. Efforts to assimilate regional variations in meteorogical conditions within the new republic were also linked to a belief that deliberate anthropogenic environmental changes might nurture a temperate climate that would serve as a "nursery of genius, learning, industry and the liberal arts."[38] In France, Germany, and England around the same time, climate similarly became central to the description and

[31] Manley, cited in Allen Perry, "Classics in Physical Geography Revisited," *Progress in Physical Geography* 25 (2001): 541–3, on 542.

[32] Manley, *Climate and the British Scene* (cit. n. 30), 1.

[33] Billy Collins, founder of the New Naturalist series, quoted in Marren, *New Naturalists* (cit. n. 27), 15.

[34] Lamb, "The Life and Work of Professor Gordon Manley (1902–1980)," *Weather* 36 (1981): 220–31.

[35] Marren, *New Naturalists* (cit. n. 27), 13; see also Sharon Kingsland, *Modelling Nature: Episodes in the History of Population Ecology* (Chicago, 1988).

[36] David Matless, "Regional Surveys and Local Knowledges: The Geographical Imagination in Britain, 1918–1939," *Trans. Inst. Brit. Geogr.* 17 (1992): 464–80.

[37] Fleming, *Historical Perspectives* (cit. n. 2), 12.

[38] Karen Ordahl Kupperman, "The Puzzle of the American Climate in the Early Colonial Period," *Amer. Hist. Rev.* 87 (1982): 1262–89; Hugh Williamson, *Observations on the Climate of Different Parts of America* (New York, 1811), cited in Fleming, *Historical Perspectives* (cit. n. 2), 25.

understanding of nations and people through contemporary political, literary, scientific, medical, and colonial writing.[39]

Manley, however, was working at a time when there was something of a post–Second World War revival in public engagement with the natural environment and identity in Britain,[40] and his publication, like the others in the New Naturalist series, can be interpreted in this context. He was keen to illustrate the distinctive "personalities of regions" and to demonstrate how "regions acquired an identity through their weather and local weather acquired the identity of its regions."[41] In this respect, his writing resonates with an earlier chorographic tradition.[42] He drew out similarities and differences between places, delving into local histories, customs, and folklore and the "character" of local geology, soils, and the landscape as well as the weather, often demonstrating distinctions between regions. This focus on identity politics was mirrored in other highly influential texts of the 1950s. William George Hoskins's *Making of the English Landscape*, published only three years after *Climate and the British Scene*, for example, illustrated the ways in which the English people came to understand their local landscapes and histories. Hoskins's extraordinary detailed study offered a historical and geographically informed perspective on the English landscape and, like Manley's work, was a "pursuit of the unique."[43]

Manley's interest also extended beyond the British Isles. He regularly compared the "erratic" nature of the British climate with climates overseas, contrasting, for example, the rainfall of London with altogether wetter Moscow, Chicago, and Rome,[44] or highlighting the "sombre monotony of the brassy Mediterranean."[45] He explored the similarities and differences between the climates of temperate Atlantic regions and monitored conditions elsewhere, though often with a view to predicting their implications for weather conditions over Britain.[46] His record of publication might suggest that he had less concern for the climate characteristics of British colonies relative to those of the British Isles, a function perhaps of the period in which he was writing, when many former colonies were gaining independence or had already done so and were thus subject to less scrutiny by U.K.-based scholars. Manley does, however, appear to have been influenced by Huntington in his consideration of the implications of climates in different global regions. He undertook a comparative study of altitudinal change in the Pennines of northern England and the mountains of New England, Huntington having previously extolled the "stimulating virtues" of the latter's climate.[47] Moreover, some of Manley's academic papers demonstrate a sympa-

[39] Alan Bewell, "Jefferson's Thermometer: Colonial Biogeographical Constructions of the Climate of America," in *Romantic Science: The Literary Forms of Natural History*, ed. Noah Heringham (New York, 2003), 111–38.

[40] Matless, "Regional Surveys and Local Knowledges" (cit. n. 36).

[41] E. W. Gilbert, "The Idea of the Region," *Geography* 45 (1960): 157–75, on 158, cited in Vladimir Jankovic, "The Place of Nature and the Nature of Place: The Chorographic Challenge to the History of British Provincial Science," *Hist. Sci.* 38 (2000): 79–113.

[42] Jankovic, "Place of Nature" (cit. n. 41), 81.

[43] Charles Phythian-Adams, "Hoskins's England: A Local Historian of Genius and the Realisation of This Theme," *Transactions of the Leicestershire Archaeological and Historical Society* 66 (1992): 143–59.

[44] Manley, "This Year's Dry Spring: Prospects of Drought," *Manchester Guardian*, May 1, 1956.

[45] Manley, "The Summer of 1954: Not Quite the Worst," *Manchester Guardian*, August 26, 1954.

[46] Manley, "The Effective Rate of Altitudinal Change in Temperate Atlantic Climates," *Geogr. Rev.* 35 (1945): 408–17; Manley, "Depressions Off Their Courses Upset November's Weather," *Manchester Guardian*, December 1, 1954.

[47] Manley, "Effective Rate of Altitudinal Change" (cit. n. 46), 408.

thy with Huntington's "forthright promulgation of his doctrine of the climate—and therefore of climatic variations—on the welfare of peoples" in different parts of the world.[48]

While Manley acknowledged that doubt had been cast on Huntington's "somewhat crude basis of assessment," he seems to have supported his general perspective and advocated the adoption of "some of those oft despised geographical virtues" in studies of climate.[49] Huntington had attempted to draw associations between geography and human history, mapping "climatic energy zones" in order to explain the rise and fall of great civilizations.[50] In a set of lecture notes entitled "Geography of British History," written from Bedford College in the 1960s, Manley reflected on these links: "Have you ever wondered why we in this country have been so active? Have you ever looked at a world population map and wondered why the great 'blobs' of world population are where they are? Have you ever wondered why the active subscribers of world civilisations are primarily in Europe?" Sometimes, he argued, "the answer lies in history," but "there are other factors which have in part made history. These are the facts of geography."[51] He had no qualms about arguing that the British climate, relative to climates elsewhere, represented "an indestructable long term asset," fostering a highly variable and adaptable society.[52] He was, however, conscious of entering a controversial field. In notes he drafted for his preface to Emmanuel Le Roy Ladurie's 1967 *Histoire du climat depuis l'an mil*, he argued, "Some like to stress the cumulative effect of these extraneous factors, some prefer to minimise their importance and are wont to growl about would be determinist philosophies."[53] He highlighted the need for "caution in representing climate as only one of many factors affecting human life," which may have helped him avoid the criticisms that were leveled at determinists like Huntington.[54]

The remainder of this article considers Manley's approach to studying the climate through places, people, and their experiences of the British weather. Although his professional and academic standing was well established among both geographers and meteorologists, I focus on his frequent and regular articles in the *Manchester Guardian* (after 1959, the *Guardian*).[55] As is true of work by other twentieth-century climate scientists including C. E. P. Brooks, Alexander McAdie, and William Jackson Humphreys,[56] Manley's newspaper articles are woven with "apposite quotations from literature or from the ancient classics of meteorology" and with irony, humor, and historical knowledge,[57] and they might help inform a number of key issues highlighted in recent climate literature. Through his many detailed accounts of the

[48] Manley, "Revival of Climatic Determinism" (cit. n. 3), 98–105.

[49] See Manley's draft review of Tony John Chandler and Stanley Gregory, eds., *The Climate of the British Isles* (New York, 1979), Manley Papers, Department of Manuscripts and University Archives, Cambridge University (hereafter cited as "Manley Papers"), box 17, file 8.

[50] Fleming, *Historical Perspectives* (cit. n. 2), 103.

[51] Manley Papers, box 23, file 9.

[52] Manley, "Effective Rate of Altitudinal Change" (cit. n. 46), 412.

[53] Manley Papers, box 24, file 11.

[54] L. C. W. B., "Review of *Climate and the British Scene*," *Geogr. J.* 119 (1953): 99–100, on 100.

[55] A list of Manley's articles published in the *Manchester Guardian* between 1954 and 1961 appears in the back of Tooley and Sheail, *Climatic Scene* (cit. n. 22), but there does not appear to be a comprehensive collection of all his newspaper articles.

[56] Brooks, *Climate in Everyday Life* (London, 1950); McAdie, *Man and Weather* (Cambridge, Mass., 1926); Humphreys, *Ways of the Weather* (Tempe, Ariz., 1942).

[57] Marren, *New Naturalists* (cit. n. 27), 167.

regional idiosyncrasies of Britain's weather, his association of particular weather with specific people, spaces, and places, and his interest in local weather lore and proverbs, Manley illustrated the importance of considering climate's culturally and spatially variable dimensions and interpretations, as well as the relationships between local weather, cultural practice, and belief systems. The discussion below focuses on Manley's emphasis on the "idea" of climate as a phenomenon experienced through weather events.

BRITISH WEATHER, "WEATHER TALK," AND IDENTITY

Fascination with the weather has arguably reached its zenith in the United States, where, in addition to weather magazines, there is a twenty-four-hour Weather Channel with constantly updated forecasts and feature shows dealing with extreme weather.[58] It is fair to say, however, that the British also have "a peculiar outlook on the weather" and are "notorious" for engaging with this apparently safe and impersonal conversational topic.[59] This fact did not escape Manley, who noted that "if a census were taken of common topics of conversation amongst British people, it is very probable that the weather would take first place."[60] British "weather talk" has been understood as a form of "phatic" communication, an act of social bonding rather than simple transmission of information.[61] This unusual level of interest, which Trevor Harley considers to be verging on an "obsession,"[62] may be a function of the idiosyncrasy of the British climate, which, in Manley's opinion, was "paradoxical" and tended "to defy definition."[63] "There are indeed times," he noted in an article from the late 1950s, "when the sequence of meteorological events appears to develop in such an erratic manner that many may be tempted to retire from the field and opine that . . . 'things just happen.'"[64] This paradox, he argued, had been inscribed into British attitudes toward the weather and the "capacity for making ironic remarks" about it, especially when there are "pleasing departures from the normal," which "help to keep us lively."[65] While he was intrigued by the range of weather that could be experienced across the United Kingdom,[66] it is clear that Manley also recognized that there were "certain limits" beyond which British weather was unlikely to stray.[67] His argument was based on his meticulous reconstruction of long-term histories of temperatures, rainfall, and snowfall, in some cases extending back several centuries and often drawing on historical weather diaries. In an article from the mid-1960s, for example, he suggested that "200 years of records tell us that rain . . . must fall within well es-

---

[58] Steve Rayner, "Domesticating Nature: Commentary on the Anthropological Study of Weather and Climate Discourse," in Strauss and Orlove, *Weather, Climate, Culture* (cit. n. 17), 277–90, on 281.

[59] Jan Golinski, "Time, Talk and the Weather in Eighteenth Century Britain," in Strauss and Orlove, *Weather, Climate, Culture* (cit. n. 17), 17–38, on 17; Trevor A. Harley, "Nice Weather for the Time of Year: The British Obsession with the Weather," in Strauss and Orlove, 103–18.

[60] Manley, *Climate and the British Scene* (cit. n. 30), 13.

[61] Golinski, "Time, Talk and the Weather" (cit. n. 59); Rayner, "Domesticating Nature" (cit. n. 58), 277–90.

[62] Harley, "Nice Weather for the Time of Year" (cit. n. 59), 103.

[63] Manley, *Climate and the British Scene* (cit. n. 30), 295

[64] Manley, "An Umbrella in Reserve," *Manchester Guardian*, April 25, 1959.

[65] Manley, "Swinging Weather," *Guardian*, September 3, 1966.

[66] Manley, "The Range of Variation of the British Climate," *Geogr. J.* 117 (1951): 43–65.

[67] Manley, "Summer of 1954" (cit. n. 45).

tablished limits,"[68] and elsewhere he identified conditions that were apparently typical of the British weather at particular times of the year, reporting in one article from September 1962 on "typical bank holiday weather" in August—that is to say, rainy weather that made a "thorough nuisance of itself" for British holiday makers.[69]

Many of the attitudes toward weather that are common in Britain today have a long history. Sarah Warnecke has demonstrated that as early as the sixteenth century links were being drawn between the English air and a fascination with novelty, or "newfangledness."[70] Jan Golinski suggests that, as was happening elsewhere,[71] the weather became a topic of more continuous public interest and conversation during the Enlightenment, when the British "forged a sense of their national weather, its peculiarities and regularities" and attitudes to weather became even more strongly linked to a sense of national identity.[72] A familiarity with a long record of climate variability is also thought to be a key component in the framing of identities,[73] while, as Manley pointed out, distinctive regional variations in the weather can give rise to spatially distinctive identities and stereotypes.

Having been brought up in the north of England, Manley strove to highlight the climatic differences between the metropolitan hub of the southeast and the more peripheral regions further north. He focused particular attention on the northern Pennines, his home for many years, for which he harbored particular affection. Among the most notable regional distinctions across the British Isles that he discussed, perhaps reflecting a degree of tension over the divide between the metropole and more northerly provinces in scientific discussion about the weather, is a north-south divide in typical weather and associated weather talk. In an article from March 1968 he observed: "Sitting on the train on the way from London to Lancashire and onward, one is indeed aware that there is a North west and a South east; and so often the weather of the one just feels different from that of the other."[74] He also noted, however, that it was accepted wisdom that summer weather in the south should be better than that experienced in the north of the country according to long-standing regional weather stereotypes, which, he argued, "helped to build up that widespread notion strongly encouraged by the advertisers of Southern resorts that 'it is no good going North.'"[75] Manley drew on specific examples to illustrate this point. In talking of what he referred to as a "satisfactory" weather day in May 1964, which was "the warmest May day for five years, we were told," with a temperature of 78°, he noted that "for the whole of the South east, it was the perfect day indeed, when the reports came in that Manchester and other remote parts . . . had received what Londoners consider they

[68] Manley, "After the Monsoon, the Drought?" *Guardian*, July 21, 1965.

[69] Manley, "Work to Rule August Weather," *Guardian*, September 1, 1962.

[70] Warnecke, "A Taste for Newfangledness: The Destructive Potential of Novelty in Early Modern England," *Sixteenth Cent. J.* 26 (1995): 881–96.

[71] Bewell, "Jefferson's Thermometer" (cit. n. 39), 113.

[72] Golinski, "Time, Talk and the Weather" (cit. n. 59), 18; Golinski, *British Weather* (cit. n. 17), quotation on 57.

[73] Strauss and Orlove, "Up in the Air" (cit. n. 17), 10; Wendy L. Gardner, Cynthia L. Pickett, and Marilynn B. Brewer, "Social Exclusion and Selective Memory: How the Need to Belong Influences Memory for Social Events," *Personality and Social Psychology Bulletin* 26 (2000): 486–96, cited in Mike Hulme, Suraje Dessai, Irene Lorenzoni, and Donald R. Nelson, "Unstable Climates: Exploring the Statistical and Social Constructions of Normal Climate," *Geoforum* 40 (2009): 197–206, on 201.

[74] Manley, "Mainline from Corn to Grass," *Guardian*, March 7, 1968.

[75] Manley, "The Wettest Place," *Manchester Guardian*, December 29, 1945.

ought to have: that is rain."[76] Yet such perceived "norms," which in this case may reflect a metropolitan provincialism, were and are occasionally challenged. May Bank Holiday two years earlier in 1962, for example, had presented a diversion from the "acceptable" weather for this time of year. "Last year's sunny bank holiday at Blackpool," on the north west coast of England in Lancashire, "was almost unpardonable in the opinion of the rain swept South, where the legend of better weather is so strenuously upheld."[77]

Manley made efforts to illustrate the variety of weather that could be experienced in Britain, together with the diversity of local cultural characteristics across the country, drawing parallels between these factors. Weather was in this way used as a means of mapping and understanding regionalisms, while local cultural distinctions in turn influenced how weather was experienced. One of his anecdotes, discussed in an article from December 1971, illustrates this point: "It was an American visitor sitting one evening in a Langdale climbing hotel, listening to the varied accents and cheerful banter of Lancashire, Yorkshire and Tyneside, and London and Somerset who broke out 'Why you're all so different I can't understand in so small a country.'" To some extent, social differences are rendered more acute on a small island, but this diversity, Manley argued, arises from "3000 years of immigration," "the slow evolution of the mixture upon geological structures of remarkable variety within short distances," and, most significant, "a farming practice linked with those remarkably rapid changes of effective climate and of vegetation with altitude."[78] The link between climate, topography, and land use alluded to here was another of Manley's particular research interests and an issue about which he wrote extensively, but clearly even his work in this arena drew links between climate and regional character.[79]

Given that he spent his formative years in Lancashire, northwest England, and returned there regularly throughout his career, it is perhaps no surprise that many of Manley's reports focus on this particular region's apparently distinctive weather, landscape, and people. He argued in 1963 that "Lancashire weather has an individual eccentric quality. . . . It does not always fit with what goes on elsewhere."[80] This eccentricity had, he asserted, in part nurtured a competitive spirit in the monitoring of the impacts of local weather: "Lancashire men are disposed to take some pride in those winds that leave their mark on the trees throughout the county. Grandfathers recall those Edwardian days when . . . the old Southport Observatory and the Bidston Observatory . . . used to vie with each other when it was a question of who measured the highest gusts!"[81] Moreover, manifestations of distinctive local weather characteristics in northern England were ascribed a cultural significance. Writing again of Blackpool, he noted that "to be able to scrape salt off the window was almost as good a status symbol as having a billiard room; for it meant that you might have a

[76] Manley, "Fair Westerlies," *Guardian*, May 19, 1964.

[77] Manley, "The Cold Standard," *Guardian*, February 28, 1963.

[78] Manley, "Vive la Difference," *Guardian*, December 28, 1971.

[79] Manley, "Topographical Features and the Climate of Britain: A Review of Some Outstanding Effects," *Geogr. J.* 103 (1944): 241–58; Manley, "Further Climatological Averages for the Northern Pennines with a Note on Topographical Effects," *Quarterly Journal of the Royal Meteorological Society* 65 (1943): 251–61.

[80] Manley, "Cold Standard" (cit. n. 77).

[81] Both observatories are in the northwest of England, Bidston on the Wirral Peninsula and Southport on the coast north of Liverpool; Manley, "The Winter of Our Discontent?" *Guardian*, November 20, 1965.

house with a view of Blackpool Tower."[82] Further idiosyncrasy could be found at the subregional level. Thus, he argued in 1945, "among the fragments of geography recollected by older generations is the solemn statement of the Victorian textbook: 'Seathwaite [in Cumbria] is the wettest inhabited place in England.'"[83] Judging by a recent letter published in *Weather* magazine, a publication of the Royal Meteorological Society that was established while Manley was president,[84] this fascination with identifying places of extremes prevails among weather enthusiasts in the United Kingdom today.[85]

Adapting Latour's terminology, Golinski argues that "we have never been completely enlightened" in relation to the weather.[86] Notwithstanding the professionalization of meteorological science in the nineteenth century,[87] an interest in weather proverbs, superstition, and weather lore has proved persistent. Many folk and rural genres, for example, "express knowledge regarding the full range of weather and climatic conditions" based on familiarity with long-term climate variability.[88] Certainly, farming communities have long been associated with generating weather lore in order to predict and plan activities around future weather.[89] Reflecting his combined interests in meteorological science, cultural history, and vernacular law, Manley recognized the importance of proverbs, narratives, and "weather wisdom." He had access to an "extensive collection of proverbial sayings" thanks to the bequest to the Royal Meteorological Society by one of its former presidents, Richard Inward, of his book on weather lore. Such publications, along with others compiled by early twentieth-century meteorologists, such as Humphreys, who became head of research at the United States Weather Bureau,[90] enabled meteorologists to refer "to the wisdom of our forefathers."[91] Many of the proverbs and seasonal maxims date back many centuries and, as Manley suggested, "reflect the salutary caution with which Englishmen should ever be furnished: 'As the day lengthens, the cold strengthens'; . . . 'there are a hundred days of east wind in the first half of the year'; 'March dust on an apple leaf brings all kinds of fruit to grief.'"[92] It is clear that Manley regarded such popular knowledge as a manifestation of a long-standing engagement with the weather and as an example of weather talk. He drew on numerous regional examples of weather-related sayings and proverbs that have helped to define local, predominantly rural, attitudes toward seasonal and unseasonable weather. He noted, for example, that "lambing storms" were "proverbial in the north" and associated

[82] Ibid. Blackpool Tower in Lancashire is a tourist attraction that was opened to the public on May 14, 1894. It was inspired by the Eiffel Tower in Paris and rises to 158 meters.

[83] Manley, "Wettest Place" (cit. n. 75).

[84] Tooley and Sheail, "Life and Work" (cit. n. 22), 4.

[85] F. G. Thomas, letter, "The Wettest British Cities," *Weather* 59 (2004): 112.

[86] Golinski, "Time, Talk and the Weather" (cit. n. 59), 32; Latour's precise phrase was "we have never been modern." Discussed in James Fleming, review of Golinski's *British Weather and the Climate of Enlightenment*, *Amer. Hist. Rev.* 114 (2009): 830–1.

[87] Jankovic, *Reading the Skies* (cit. n. 18); Simon Naylor, "Nationalising Provincial Weather: Meteorology in Nineteenth-Century Cornwall," *Brit. J. Hist. Sci.* 39 (2006): 1–27; Strauss and Orlove, *Weather, Climate, Culture* (cit. n. 17); Golinski, "Time, Talk and the Weather" (cit. n. 59).

[88] Strauss and Orlove, "Up in the Air" (cit. n. 17), 7; Katherine Anderson, "Looking at the Sky: The Visual Context of Victorian Meteorology," *Brit. J. Hist. Sci.* 36 (2003): 301–32.

[89] Golinski, *British Weather* (cit. n. 17), 91.

[90] Humphreys, *Weather Proverbs and Paradoxes* (Baltimore, 1923).

[91] Manley, "Onward from the Fog: Cold Winter in Prospect?" *Manchester Guardian*, December 13, 1958.

[92] Ibid.; Golinski, "Time, Talk and the Weather" (cit. n. 59), 27.

with particular spring months: "The Scots have their 'lambing storms' in mid April, while Westmorland [Cumbria] knows 'the gesling blast' that harries the young birds near the end of the month."[93] In northern Lancashire, meanwhile, Manley reported on a tendency for a cold spell in mid-May, referring to it as "the cowquake," which, he argued, is "a term worth putting on record before our Londonised government stamps it out of memory, although it may linger in the grandeur of the Oxford Dictionary."[94] Such statements again seem to reflect a tension that Manley felt between the metropolitan focus of scientific discussion about the climate and what he referred to as "'the outlying foci' of early scientific progress" in the "provinces."[95]

Manley thus regularly drew on evidence from particular places to illustrate how British weather experiences framed and informed national, regional, and local identities and also how such identities in turn helped shape these experiences.[96] He was keen to demonstrate the long-term overall predictability of the British climate, but also the spatial variability of the British weather and the history of public engagement with both. Given recent calls to reestablish "what climate means for people and places and the relationships between people and places over time,"[97] it could be argued that Manley's approach to studying particular people and places and their experiences of the climate through weather, though now several decades old and arguably somewhat controversial amid fears of a revival of climatic determinism, has resonance and broader application today. This is particularly true in light of the fact that "everyday experiences and locality," or the situated nature of climate, are increasingly being recognized as fundamental to understanding how the public perceives, responds, and adapts to climate change.[98]

WEATHER, MEMORY, AND THE CULTURAL FABRIC OF THE BRITISH "SCENE"

The weather has been woven into our experiences of modern life in many ways. It "punctuates our daily routines" and has been inscribed into the social memory and cultural fabric of communities in the form of oral history, myth, tradition, folklore, technological adaptation, or narrative.[99] These different ways of recounting the past represent key means by which information on long-term climate change, and on short-term weather events, is gathered and transmitted across generations. According to what Harley refers to as the "recency effect,"[100] dramatic events tend to seize

---

[93] Manley, "Meteorologist Makes a March Journey: No Escaping the Weather," *Manchester Guardian,* April 3, 1958. The English term "gesling blast" is thought to come from "gosling blast" and means a sudden squall of rain or sleet.

[94] Manley Papers, box 24, file 17.

[95] Tooley and Sheail, "Life and Work" (cit. n. 22), 2.

[96] Hulme et al., "Unstable Climates" (cit. n. 73), 197–206; Strauss and Orlove, *Weather, Climate, Culture* (cit. n. 17).

[97] Hulme, "Geographical Work at the Boundaries" (cit. n. 16), 7. See also Hulme, *Why We Disagree about Climate Change* (Cambridge, 2009).

[98] Irene Lorenzoni and Nick F. Pigeon, "Public Views on Climate Change: European and USA Perspectives," *Climatic Change* 77 (2006): 73–95, on 80; see also J. P. Palutikof, M. D. Agnew, and M. R. Hoard, "Public Perceptions of Unusually Warm Weather in the UK: Impacts, Responses and Adaptations," *Climate Research* 26 (2004): 43–59.

[99] Golinski, "Time, Talk and the Weather" (cit. n. 59), 17; Robert De Courcy Ward, *The Essential Characteristics of United States Climates* (New York, 1920); Humphreys, *Weather Proverbs and Paradoxes* (cit. n. 90); Hassan, "Environmental Perception and Human Responses" (cit. n. 14).

[100] Harley, "Nice Weather for the Time of Year" (cit. n. 59).

popular attention more than "normal" conditions and expected seasonal variability, and, therefore, memories tend to be distorted with respect to extreme weather events. Defoe recognized this problem in his study of the storm event that affected much of Britain between November 26 and 27, 1703.[101] Extreme events can claim priority in people's memories as idealized stereotypes of seasonal conditions, while at the same time providing "anchors for personal memory."[102] Indeed, there is a tendency to remember the distant past with more accuracy if it is related to unusual or outstanding events, especially if those events are rare.[103] Weather can in this way play a metacognitive role in organizing memories.[104]

In accordance with his statements on the apparent "limits" of the British weather, Manley argued that Britain tended to be subject only to "gentle extremes,"[105] notwithstanding the devastating floods in Lynmouth, Devon, in August 1952, the North Sea floods during the winter of 1953, and the extreme winter that affected Britain ten years later.[106] Yet he did consider the significance of such events in the context of long-term weather observations and hinted at the potential for their historical distortion. Writing about the unusually cool summer of 1954, for instance, he noted that "so many meteorological events have found their way into the news or have affected our immediate comfort, that there are those who will emphatically declare that they have never known anything like the weather of 1954. But most memories are short and impressions can be deceptive."[107] Manley noted that the summer of 1954 was in fact, at the time, only the twelfth in order of coolness since 1700.[108]

There tends to be a collective rather than an individual nostalgia in terms of how people recall the weather of the past.[109] White Christmases represent a case in point. As Mike Hulme and his coauthors have suggested, "The expectation of a white Christmas in the UK is culturally perpetuated by the imagery and expectation reiterated every festive season, yet the likelihood of snow at Yuletide in most parts of England is now less than one in 10 years."[110] People's nostalgia for such weather events is often based on "statistically incorrect data," and also on popular "prominent but unrepresentative examples." Wistful recollections of a white Christmas may thus have a cultural rather than a memorial origin.[111] As Manley illustrated, however, cultural memory and statistical evidence can work in tandem. While "the notion of a snowy Christmas . . . owes far more to the early Victorian Romantics than to any real

[101] See Hamblyn, "Introduction" (cit. n. 18), xxx–xxxi.

[102] Hulme, *Why We Disagree* (cit. n. 97), 12.

[103] Hulme et al., "Unstable Climates" (cit. n. 73), 200; Harley, "Nice Weather for the Time of Year" (cit. n. 59).

[104] A. Karmillof-Smith, "From Meta-processes to Conscious Access: Evidence from Children's Metalinguistic and Repair Data," *Cognition* 23 (1986): 95–147, cited in Harley, "Nice Weather for the Time of Year" (cit. n. 59), 114.

[105] Manley, *Climate and the British Scene* (cit. n. 30), 273.

[106] There has been speculation that the unusually heavy rainfall that caused the flooding in Devon was associated with Royal Air Force cloud-seeding experiments; James Rodger Fleming, *Fixing the Sky: The Checkered History of Weather and Climate Control* (New York, 2010). *Climate and the British Scene* was reprinted five times and appeared in paperback, but was never revised in light of such events; Tooley and Sheail, "Life and Work" (cit. n. 22), 13.

[107] Manley, "The Weather of 1954: A Meteorological Retrospect," *Manchester Guardian*, December 31, 1954.

[108] Manley, "Summer of 1954" (cit. n. 45).

[109] Harley, "Nice Weather for the Time of Year" (cit. n. 59), 109.

[110] Hulme et al., "Unstable Climates" (cit. n. 73), 20.

[111] Harley, "Nice Weather for the Time of Year" (cit. n. 59), 113.

characteristics of the English climate,"[112] he did highlight support for such nostalgia, pointing toward statistical evidence that our Victorian ancestors experienced more extreme cold years, particularly during the 1880s and 1890s. Reflecting upon the unusually cold winter that affected much of Great Britain in 1962/3,[113] for example, he noted that "it will count as the coolest year since 1917 and 1919. But it will not fall to the level of 1892, 1888, 1887, 1885 or 1879."[114] In many of Manley's articles "Victorian weather" is used as a kind of standard. That era was, after all, a time when "our Victorian grandfathers wore boots . . . [and they have] told us of ice thatching on Windermere in 1895, of the coach on the Oxfordshire Thames in 1891, of the historic blizzards in 1886 in the North-east and 1881 in the South west, of the prolonged ordeal by cold from November 1878 to May 1879."[115] Such extremes have become inscribed into social memory and are used as benchmarks against which later cold events are compared.

There are other manifestations of the inscription of weather "memories" into the cultural fabric of British society. Steve Rayner, for example, argues that British weather has been "domesticated" through modernization, investments in material culture, and developments in infrastructure,[116] while William B. Meyer has presented similar arguments for American society.[117] Thus, housing, drainage, roads, and clothing have all evolved to mediate people's direct experiences of the weather. Manley was particularly fascinated with the way in which weather "vagaries enter into everyday life of the people" of Britain in these ways and regularly commented on changing attitudes toward and customary practices associated with the weather over time as a result of such developments.[118] It is possible to identify in his popular articles what Rayner refers to as a "climatological competence" or "ethnometeorological capacity" derived from experience of past weather events.[119] Manley referred, for example, to "native skees" (the local spelling of "skis") in the northeast of England that were "home made" by lead miners in the region to cope with heavy snowfalls; "according to an old inhabitant who was one of the last to use them, their use died out about 1900, partly with the decline of lead mining, partly on account of the predominance of milder winters during the first four decades of this century."[120]

Manley acknowledged that extremes did occasionally challenge the capacity for adaptation. "From time to time," he noted in 1969, "this idyllic island of ours becomes quite formidable and every year brings its toll of damage and difficulty." While he felt that this was something that could generally be accommodated, he also displaced blame for any shortcomings in this respect, arguing that the ability to cope was being compromised by "imported" trends, that "we should know enough to take precau-

[112] Manley, "Odds against a White Christmas," *Guardian*, December 22, 1962.

[113] The same cold event would unsettle Guy Stewart Callendar's theories about climatic warming at this time; see James Rodger Fleming, *The Callendar Effect: The Life and Times of Guy Stewart Callendar, the Scientist Who Established the Carbon Dioxide Theory of Climate Change* (Chicago, 2007), 31.

[114] Manley, "Good for Us," *Guardian*, January 2, 1964.

[115] Manley, "Cold Standard" (cit. n. 77).

[116] Rayner, "Domesticating Nature" (cit. n. 58), 280.

[117] Meyer, *Americans and Their Weather* (Oxford, 2000).

[118] Manley, *Climate and the British Scene* (cit. n. 30), xiii.

[119] Rayner, "Domesticating Nature" (cit. n. 58), 277.

[120] Manley, "Winds Blowing Hot and Cold: The Erratic British Winter," *Manchester Guardian*, February 15, 1958; Manley, "November Snowfalls," *Guardian*, November 24, 1962.

tions and learn to live comfortably in Britain in our own way,"[121] and that "many . . . troubles arise from our slavish following of designs and fashions derived for very different climates." The "narrow drainpipe trouser" came in for particular condemnation, becoming a "horror after wind-driven sleet," as did "the architectural folly" of dispensing with eaves and windowsills on buildings in Britain's rainy climate.[122] While "fashions change . . . the weather goes on," however, and continues to foster flexibility and adaptability.[123] In fact, in an article in 1962, Manley indicated that this capacity to deal with the unexpected was something that had become firmly embedded within British society, whose "prudent opportunism, . . . readiness to change the plan, to deviate from the policy, to refrain from putting all the eggs in one basket," was an "immemorial asset" and one that he attributed directly to the country's variable and unpredictable climate.[124] It was "a national characteristic to be adaptable," though this is obviously true of many nations.[125]

This "opportunism" extended to capturing the fleeting changes in weather in visual and literary media.[126] Weather has long attracted the attention of artists, poets, and writers intrigued by the aesthetic qualities of often rapidly changing atmospheric conditions, of clouds, and of sunsets, and by their implications for the appearance of the landscape.[127] John Thornes and Gemma Metherell, for example, argue that many artists have been attracted by the weather's very transitory nature, "which constantly challenges the ability to catch the 'atmosphere' on canvas."[128] Manley demonstrated how the features of the British landscape not only affected, but also contributed to, the diversity of local weather and took on a different appearance depending on the changeable weather.[129] He wanted to prove that even the most routine of seasonal weather conditions provided inspiration. The inclusion of a suite of plates, some in color, in *Climate and the British Scene*, depicting scenes of a hoarfrost on a birch wood in Derbyshire, "going to work in a Midland city in January," and an Edinburgh street in spring, one contemporary reviewer argued, provided much for the artist.[130] Manley's writings are also infused with literary "cullings," a reflection of the influence of the British weather on other parts of the creative imagination.[131] In a number of his articles in the *Manchester Guardian*, Manley drew the reader's attention to "the effect on the English mind that has begotten poets and scientists alike,"[132] and he expressed the feeling that "the inspiration born of the mood of the weather has set alight the imagination and creative intelligence" of poets and artists.[133] He considered the

[121] Manley, "Freak Weather a Test for Alien Fashions," *Guardian*, January 1, 1969.

[122] Manley, "Winds over Europe," *Guardian*, September 4, 1965.

[123] Manley, "After the Monsoon, the Drought?" (cit. n. 68).

[124] Manley, *Climate and the British Scene* (cit. n. 30), 277.

[125] Tooley and Sheail, "Life and Work" (cit. n. 22), 12.

[126] Arden Reed, *Romantic Weather: The Climates of Coleridge and Baudelaire* (London, 1983).

[127] Strauss and Orlove, "Up in the Air" (cit. n. 17), 9; Hulme, *Why We Disagree* (cit. n. 97), 2; Marilyn Gaull, *English Romanticism: The Human Context* (New York, 1988).

[128] Thornes and Metherell, "Monet's London Series and the Cultural Climate of London at the Turn of the Twentieth Century," in Strauss and Orlove, *Weather, Climate, Culture* (cit. n. 17), 141–60, on 142.

[129] Perry, "Classics in Physical Geography" (cit. n. 31).

[130] R. M. P., "Review of *Climate and the British Scene*," *Quarterly Journal of the Royal Meteorological Society* 79 (1953): 177–8.

[131] Hare, "Review of *Climate and the British Scene*" (cit. n. 27), 324.

[132] Manley, "Umbrella in Reserve" (cit. n. 64).

[133] Manley, *Climate and the British Scene* (cit. n. 30), 276.

powerful mental and physical impressions associated with the elemental imprint on various iconic British locations, such as

> the setting February sun on Crossfell; the superb brilliance of a May evening in Somerset after heavy showers; the fresh west wind on a June morning in Piccadilly; and even the hard January north wind above Haslingden with the green sky behind a snow covered Bowland or the recent delights of lunching on the Roman wall in the April sunshine while magnificent polar air showers swept over Newcastle and Carlisle.[134]

Manley also explored "British weather from within, and from the point of view of those who have lived in it," including those who, like himself, recorded its regularities and inconsistencies on a daily basis.[135] Golinski focuses specifically on the self-dedication of such weather observers and diarists in the late seventeenth and eighteenth centuries, for whom the "regular and exacting ritual" of keeping a diary helped frame their own personal identities.[136] As early as 1723, there were attempts to coordinate and standardize these often highly idiosyncratic meteorological observations. Andrea Rusnock demonstrates how, under James Jurin, the Royal Society acted as a "legitimating body" for the reports compiled by geographically dispersed individuals and also points to institutions elsewhere that were established for this purpose, including the Academia del Cimento in Italy and the Paris Académie des Sciences.[137] Other societies, though often ephemeral, such as the Meteorological Society of London, founded in 1823, or the Smithsonian Meteorological Project directed by Joseph Henry (1849), had a similar rationale.[138] Manley was acutely aware of the potential of the rich legacy of these individuals for the reconstruction of long-term temperature series and rainfall and snowfall records, and he received a research award from Shell to study these sources.[139]

He seems to have been equally fascinated by the individuals themselves, their practices, and their instruments, arguing that "we cannot fail to be impressed by the assiduity and regularity of some of those amateur enthusiasts whose records remain,"[140] which "testify to the existence of energy to spare, of enterprise, of persistence of nature worship, of that wish to acquire an orderly and precise set of measurements."[141] Manley discussed numerous examples of historical "amateur" weather diarists, as well as key figures in the history of meteorological observation, throughout his many articles.[142] He referred specifically to the British Rainfall Organisation (BRO), a society founded as a voluntary body in 1858 by George Symons following debates over a series of dry years in the mid-1850s and inspired by the apparent inadequacy of

[134] Manley, "Umbrella in Reserve" (cit. n. 64).
[135] Manley, *Climate and the British Scene* (cit. n. 30), 24.
[136] Jan Golinski, "Putting the Weather in Order: Narrative and Discipline in Eighteenth Century Weather Diaries," paper delivered at the William Andrews Clark Memorial Library, Univ. of California, Los Angeles, May 16, 1998, available online at http://www.unh.edu/history/golinski/paper3.html (accessed 23 December 2010).
[137] Rusnock, "Correspondence Networks and the Royal Society, 1700–1750," *Brit. J. Hist. Sci.* 32 (1999): 155–69.
[138] J. M. Walker, "The Meteorological Societies of London," *Weather* 48 (1993): 364–72.
[139] Manley Papers, box 24, various files.
[140] Manley, "The Countryman with a Logical Mind," *Guardian*, September 19, 1966.
[141] Manley, "For the Rain It Raineth Every Day," *Guardian*, November 4, 1967.
[142] Manley, "Weather and Diseases" (cit. n. 23).

available observations of rainfall up to that point.[143] He argued that "no better testimony to the great 'amateur tradition' of British scientific enquiry can be found than this noticeably successful enterprise for the collection and dissemination of rainfall statistics."[144] Amateur recorders were invited to send in rainfall readings, which Symons compiled into a bulletin, *British Rainfall*, that ran until 1968, though the results submitted were often far from trustworthy.[145] Part of the problem, as Symons noted, was the lack of consistency of weather-recording instruments and sites. He had seen "results published as air temperatures obtained from thermometers inside a hen house" and a rain gauge "under the eaves of a cottage and another under a tree."[146] While Manley acknowledged the "splendid conversations," "concealed rivalries," and "learned discussions" that lay behind those "carefully assembled and published measurements,"[147] he himself was all too aware of problems with their reliability, arguing in 1948 that considerable caution was needed when using early historical instrumental records.[148] Nevertheless, he regularly drew on the BRO data, as well as other amateur records and diaries, in his articles concerning the spatial and temporal variations in rainfall across the country in order to better understand contemporary weather anxieties, including those associated with the drought that affected much of the country in the mid-1950s and that resulted in the threat of "a national water shortage," of which there was significant media coverage at the time. It was to the amateur record that Manley turned to contextualize these concerns in "A *Manchester Guardian* Survey of a Public Service," published in November 1956, and he noted in a separate article that "thanks to the British Rainfall Organisation we have the finest set of rainfall statistics in the world: a good basis from which to start" in attempting to understand the implications of such droughts.[149]

In this "Survey of a Public Service," he commented that while "past records indicate that our climate is not likely to alter—save that we may run into a decade in which wetter or drier years predominate . . . we do know that it has occurred and that the consequences of such a succession of dry years today would be extremely serious."[150] The suggestion that Britain's climate was "not likely to alter" was no insignificant statement, given the controversial debates of the period over an apparent "climatic amelioration" or recession between the 1920s and 1950s.[151] Various climatic change theories had by this stage been forwarded.[152] Guy Stewart Callendar, a British steam-power engineer, for example, had revived Svante Arrhenius's theory linking the combustion

[143] D. E. Pedgley, "A Short History of the British Rainfall Organisation" (Occasional Papers on Meteorological History, no. 5, Specialist Group for the History of Meteorology and Physical Oceanography, Royal Meteorological Society, Reading, 2002), 1.

[144] Manley, *Climate and the British Scene* (cit. n. 30), 13.

[145] Fleming, *Historical Perspectives* (cit. n. 2), 35.

[146] George J. Symons, "History of English Meteorological Societies, 1823–1880," *Quarterly Journal of the Royal Meteorological Society* 7 (1881): 65–98, cited ibid.

[147] Manley, "For the Rain It Raineth Every Day" (cit. n. 141).

[148] Manley, Nelson Johnson, C. E. P. Brooke, Professor Hollingsworth, C. D. Ovey, Professor Hawkes, H. H. Lamb, G. A. A. Grant, D. J. Shove, and L. C. W. Bonacina, "Discussion," *Geogr. J.* 117 (1951): 65–8.

[149] Manley, "This Year's Dry Spring" (cit. n. 44).

[150] Manley, "Water: The National Shortage; A *Manchester Guardian* Survey of a Public Service," *Manchester Guardian*, November 12, 1956.

[151] Manley, "Range of Variation" (cit. n. 66), 44.

[152] Sverker Sörlin, "Narratives and Counter-narratives of Climate Change: North Atlantic Glaciology and Meteorology, c. 1930–1955," *J. Hist. Geogr.* 35 (2009): 237–55.

of fossil fuels with the enhanced greenhouse effect, arguing for anthropogenic global warming in an article published in 1938.[153] Callendar offered "a more detailed understanding of the spectrum of infrared absorption and emission" that became known as "the Callendar Effect."[154] Although his theory was supported by a number of high-profile scientists, because he was not a meteorologist, Callendar's arguments were not taken seriously by the broader meteorological community until much later, and other viewpoints proved to be more influential at the time, at least within British meteorology. For example, drawing on empirical data collection in the 1920s and 1930s, Swedish-born earth scientist Hans Ahlmann had found "overwhelming evidence" to demonstrate that Arctic glaciers were shrinking.[155] This, he argued, was not necessarily an irreversible trend, and could be attributed to polar, rather than global, warming.

Manley corresponded extensively with Callendar and referred to and greatly respected his work.[156] He was, however, also a particular supporter of Ahlmann and on a number of occasions cited the Swedish scientist's theories.[157] Manley's G. J. Symons Memorial Lecture, delivered on April 19, 1944, for example, focused on recent contributions to the study of climatic change and was heavily influenced by Ahlmann's work on glacial variations,[158] while his inaugural lecture paid tribute to the fieldwork ethic and exploratory tradition of Ahlmann, among others. Ahlmann's "comprehensive account" of climatic change published in the *Geographical Journal* in April 1949 also influenced Manley's own acknowledgment of a period of climatic amelioration between 1925 and 1950.[159] As an empiricist and one acutely aware of the "deeply conflicting nature of environmental evidence,"[160] however, Manley was skeptical of anthropogenic climate change theories, urging caution in assuming that any warming trend would continue.[161] Indeed, in his draft paper entitled "The Changeability of the British Climate," he expressed his conviction that "further study is necessary before we can express opinions regarding its possible continuance."[162]

(RE)PARTICULARIZING CLIMATE: FROM THE GLOBAL TO THE LOCAL

There have been different ideological and symbolic constructions of climate at different points in history, together with a dynamic set of associated anxieties.[163] The

---

[153] Callendar, "The Artificial Production of Carbon Dioxide and Its Influence on Climate," *Quarterly Journal of the Royal Meteorological Society* 64 (1938): 223–40. It is clear, however, that Callendar believed that this warming trend might bring opportunities and benefits; Hulme, *Why We Disagree* (cit. n. 97), 53.

[154] James Rodger Fleming, "Climate Change and Anthropogenic Greenhouse Warming: A Selection of Key Articles, 1824–1995, with Interpretive Essays," Primary Articles Learning Environment, *NSDL Classic Articles in Context*, no. 1 (April, 2008), http://wiki.nsdl.org/index.php/PALE:ClassicArticles/GlobalWarming (accessed 23 December 2010).

[155] For more on Ahlmann, see Sverker Sörlin in this volume.

[156] See Manley's letter of condolence upon Callendar's death, in Fleming, *Callendar Effect* (cit. n. 113), 89–90.

[157] Manley, "Range of Variation" (cit. n. 66); Manley et al., "Discussion" (cit. n. 148), 65–8.

[158] Tooley and Sheail, "Life and Work" (cit. n. 22), 4.

[159] Manley Papers, box 18, file 5; Manley, "Revival of Climatic Determinism" (cit. n. 3), 100.

[160] Frank Oldfield and Simon G. Robinson, "Geomagnetism and Palaeoclimate," in Tooley and Sheail, *Climatic Scene* (cit. n. 22), 186–205, on 186.

[161] Manley, "Range of Variation" (cit. n. 66), 64.

[162] Manley Papers, box 18, file 5. Judging by a letter included in box 18, file 3, this paper was written in 1950 and broadcast by the British Broadcasting Corporation on May 31, 1950, 9:30–9:50 p.m., and June 1, 7:15–7:45 p.m.

[163] Hulme, *Why We Disagree* (cit. n. 97).

dominant climate idea, or anxiety, of the twenty-first century focuses on the threats posed to humanity at large by global climate change. The rhetoric of climate change has become inseparable from the complicated and sometimes contradictory scientific evidence, and political debate has been articulated in language that, to some extent, is "impenetrable to the uninitiated."[164] The political and media focus on the possible implications of climate change, the predominantly scientific discourse in which this is couched, and the increasingly global scale of climate thinking have obscured the culturally specific and spatially distinctive meanings of climate.[165] Thus, as climate change has been elevated to the top of the global political agenda, climate and its cultural significance have become decoupled, and popular conceptualizations and discourses of climate, and its manifestations through local weather, have been replaced by a global, and mainly scientific, metanarrative.

To the general public, climate change remains a "rather abstract issue in a world full of pressing social and environmental concerns."[166] As a phrase, it "means different things to different people in different contexts, places and networks." In order to better understand these distinctive meanings, it has been argued that there is a need to reintroduce particularity to the debate.[167] Recent geographical scholarship, for example, has called for research on the "idea" of climate as a "hybrid phenomenon" that can and should be constructed not only through the use of meteorological statistics but also "inside the imagination," through "sensory experiences, mental assimilation, social learning and cultural interpretations."[168] Such work would investigate climate as a function of personal memory, experience, and intergenerational transfer of "climate knowledge" and by definition demands a more intimate spatial resolution.[169]

This article has attempted to illustrate the perspectives of one geographer and meteorologist, Gordon Manley, whose goal was to encourage such a culturally driven and local-scale consideration of climate as an expression of the "integrated experiences" of weather in Britain in the mid-twentieth century.[170] In the preface to *Climate and the British Scene*, Manley suggested that climate should be "apprehended as a whole and through several senses. Let the reader therefore try to recall not merely the meteorological situation, but all the feelings and associations of the landscapes at various seasons."[171] People's understandings and perceptions of the weather, and their personal appreciation and social memory of the weather, were his quarry. He was keen to produce the "home picture" of the weather, and in so doing he contributed to a lively geographical imagination of British identity at the time.

Some may consider that Manley's work bears resemblance to a determinism long since discredited. Moreover, he was clearly skeptical of any long-term trends in climate change, albeit at a time when such skepticism was perhaps less controversial. Yet what Manley tried to achieve was to infuse studies of meteorology with what

---

[164] Clair Gough and Simon Shackley, "The Respectable Politics of Climate Change: The Epistemic Communities and NGOs," *International Affairs* 77 (2001): 329–45, on 330.

[165] Andrew Ross, "Is Global Culture Warming Up?" *Social Text* 28 (1991): 3–30, cited in Mike Hulme, "The Conquering of Climate: Discourses of Fear and Their Dissolution," *Geogr. J.* 174 (2008): 5–16, on 13.

[166] Gough and Shackley, "Respectable Politics" (cit. n. 164), 331.

[167] Hulme, *Why We Disagree* (cit. n. 97), 325, 330.

[168] Hulme et al., "Unstable Climates" (cit. n. 73), 197.

[169] Hulme, *Why We Disagree* (cit. n. 97), 330.

[170] Manley, *Climate and the British Scene* (cit. n. 30), 1.

[171] Ibid., xvi.

has been lost as a function of increasingly global and scientific perspectives on climate—that is, "a sensory, aesthetic experience, philosophical endeavour, or more latterly a literary enterprise."[172] For Manley, the "fascinating complexities" of the British climate could only really be understood through the local physical subtleties of the weather and its spatially variable cultural manifestations.[173] His perspectives also have particular currency today, when the "relational context" of climate—"the places people live, their histories, daily lives, cultures or values"—is being identified as critical for understanding how different groups of people in different places comprehend and respond to climate change.[174]

As Hans Von Storch and Nico Stehr have suggested, there are still only a handful of researchers "engaged in studying the social and cultural processes of speaking about climate change, of the formation and using of lay knowledge, of the formation and social functioning of mental images, icons and popular explanations of climate and its interaction with people."[175] This may, perhaps, be a function of the pervasive reluctance to engage in work at the interface of climate and society lest that work, and the scholars who produce it, be branded deterministic. Notwithstanding anxieties over the "changing yet lingering role of deterministic ideas,"[176] however, there is still much to be done at the "fluid boundaries between climate, space and culture."[177] Exploring Manley's approach and applying it elsewhere may serve us well in our efforts to reframe, "particularize," or "reculture" climate at this scale.[178]

[172] Hulme et al., "Unstable Climates" (cit. n. 73), 197.

[173] Manley, *Climate and the British Scene* (cit. n. 30), xv–xvi.

[174] Rachel Slocum, "Consumer Citizens and the Cities for Climate Protection Campaign," *Environment and Planning A* 36 (2004): 763–82, discussed in Hulme, "Conquering of Climate" (cit. n. 165), 5–16.

[175] Von Storch and Stehr, "Anthropogenic Climate Change: A Reason for Concern since the 18th Century and Earlier," *Geografiska Annaler, Series A: Physical Geography* 88 (2006): 107–13, on 112.

[176] McGregor, "Huntington and Lovelock" (cit. n. 3), 237.

[177] Ian Bailey, "Geographical Work at the Boundaries of Climate Policy: A Commentary and Complement to Mike Hulme," *Trans. Inst. Brit. Geogr.* 33 (2008): 420–3, on 420.

[178] Hulme, *Why We Disagree* (cit. n. 97), 330.

INTERNATIONAL TO GLOBAL

Concentrating on CO_2:
The Scandinavian and Arctic Measurements

by Maria Bohn[*]

ABSTRACT

This article concerns atmospheric carbon dioxide (CO_2) measurements made in Scandinavia and in the Arctic region before measurements started at Mauna Loa, Hawaii, in 1958. The CO_2 hypothesis of climate change was one reason to measure atmospheric CO_2 in the mid-1950s. The earlier history of CO_2 measurements—for instance, the work of the chemist Kurt Buch—was also influential in this period. It is unclear when the CO_2 hypothesis of climate change began to provide sufficient motivation for measurements, and the measurements may relate in a nonlinear way to the growth in popularity of the hypothesis. Discussions between meteorologist Carl-Gustaf Rossby at Stockholm Högskola and scientists in America reveal how different kinds of CO_2 studies varied with regard to precision.

INTRODUCTION

Some historical accounts of Charles David Keeling's measurement series of atmospheric carbon dioxide (CO_2) at Mauna Loa, Hawaii, describe his work as "the first" of its kind.[1] We may look upon this single individual, the single place of measurement, and the single curve of the measurement series as a cultural expression that reduces a complex history. In this article I will depict the broader landscape that existed around Keeling.

Measurements of atmospheric CO_2 concentration were made starting in the late eighteenth century.[2] Guy Stewart Callendar drew on older records of measurements when he presented, in 1938, the hypothesis that fossil fuel combustion had caused

---

[*] Division of History of Science and Technology, Royal Institute of Technology, SE-100 44 Stockholm, Sweden; bohn@kth.se.

I thank the editors and my supervisors for comments on this article. I also appreciate the comments from the "Climate Anxieties" workshop at Colby College in April 2009.

[1] Spencer R. Weart, *The Discovery of Global Warming* (Cambridge, Mass., 2008), also includes Keeling's Antarctic measurements in his narrative; see also David M. Hart and David G. Victor, "Scientific Elites and the Making of US Policy for Climate Change Research, 1957–74," *Soc. Stud. Sci.* 23 (1993): 643–80. The measurement series from Mauna Loa are also commemorated by former vice president of the United States Al Gore, who, in his famous book *An Inconvenient Truth*, portrays Mauna Loa as the main research station in a pioneering study of atmospheric CO_2; Al Gore, *An Inconvenient Truth: The Planetary Emergence of Global Warming and What We Can Do About It* (New York, 2006), 31, 38.

[2] F. B. Mudge, "The Development of the 'Greenhouse' Theory of Global Climate Change from Victorian Times," *Weather* 52 (1997): 13–7. In 1952 an annotated bibliography of atmospheric CO_2 studies from 1756 appeared: Nina Stepanova, "Part II: A Selective Annotated Bibliography on Carbon Dioxide in the Atmosphere," *Meteorological Abstracts and Bibliography* 3 (1952): 137–70.

---

mean temperatures to increase.[3] Although some interwar climate researchers recognized the need to address the abrupt, secular climatic change, the idea of humans as causal agents in the process, as expressed by Callendar in 1938, was not widely accepted.[4] However, his work influenced some scientists.[5]

A longer, more detailed, and international history of CO_2 measurements could be crucial to better understanding the early studies of human-caused climatic change. This article addresses the following questions: How did it become important to study CO_2 in relation to burning fossil fuels and climate change? Who was involved in this shift in emphasis and in the resulting studies? What was the disciplinary background of the scientists involved in these investigations? What was the difference between the study of variation in the level of CO_2 and the study of the background value? Finally, can we discern a chronology of events and voices, as well as periods of silence?

To answer these questions, I focus on CO_2 measurements in Scandinavia.[6] In the 1950s a few scientists in Scandinavia began to study the circulation of elements in the atmosphere.[7] A program for the measurement of CO_2 was also set up.[8] We know that research in this part of the world was prominent in the fields of meteorology and glaciology, which were relevant to climate studies.[9] Strong scientific ties across the Atlantic were conducive to the Scandinavian work in climate change, as was geographical proximity to the Arctic and a tradition of Arctic field research. Indeed, Svante Arrhenius's and Arvid Högbom's work on CO_2 and climate change in the late nineteenth century was motivated by the drive to understand the ice age, which had turned out to be a primary factor in the economic development of Sweden.[10]

The pattern of measuring CO_2 in relation to fossil fuel combustion and climate that emerges in the mid-1950s in Scandinavia is a meandering trail of different measure-

[3] Callendar, "The Artificial Production of Carbon Dioxide and Its Influence on Temperature," *Quarterly Journal of the Royal Meteorological Society* 64 (1938): 223–40, repr. in *The Papers of Guy Stewart Callendar*, eds. James Rodger Fleming and Jason Thomas Fleming (Boston, 2007), DVD archive.
[4] Sverker Sörlin, "The Global Warming That Did Not Happen: Historicizing Glaciology and Climate Change," in *Nature's End: Environment and History*, eds. Sörlin and Paul Warde (London, 2009), 93–114.
[5] His work on infrared absorption convinced David Brunt and Sydney Chapman about the importance of further study of the subject in 1941, and later he also influenced Harry Wexler and Gilbert Plass; James Rodger Fleming, *The Callendar Effect: The Life and Work of Guy Stewart Callendar (1898–1964)* (Boston, 2007).
[6] This is partly an unexplored history. Spencer R. Weart has brought attention to the Scandinavian net of measurement stations, but not through Scandinavian archival material; Weart, *Global Warming* (cit. n. 1); Weart, "Global Warming, Cold War, and the Evolution of Research Plans," *Hist. Stud. Phys. Biol. Sci.* 27, pt. 2 (1997): 319–56. The Scandinavian work is also part of Keeling's recollections; Keeling, "Rewards and Penalties of Monitoring the Earth," *Annual Review of Energy and Environment* 23 (1998): 25–82; Keeling, "The Influence of Mauna Loa Observatory on the Development of Atmospheric CO_2 Research," in *Mauna Loa Observatory, a 20th Anniversary Report*, ed. John Miller (Silver Spring, Md., 1978), 36–54, available on the "Histories on SIO" Web page, University of California San Diego libraries, http://scilib.ucsd.edu/sio/hist/Keeling_Influence_of_Mauna_Loa.pdf (accessed 5 January 2011).
[7] Bert Bolin, "Carl-Gustaf Rossby: The Stockholm Period, 1947–1957," *Tellus* A–B 51 (1999): 4–12, on 10.
[8] Stig Fonselius and Folke Koroleff, with a preface by Kurt Buch, "Microdetermination of CO_2 in the Air, with Current Data for Scandinavia," *Tellus* 7 (1955): 258–65, on 259.
[9] Kristine C. Harper, *Weather by the Numbers: The Genesis of Modern Meteorology* (Cambridge, 2008). See also Sverker Sörlin in this volume.
[10] Christer Nordlund, "'On Going Up in the World': Nation, Region and the Land Elevation Debate in Sweden," *Ann. Sci.* 58 (2001): 17–50; Sverker Sörlin, "Rituals and Resources of Natural History: The North and the Arctic in Swedish Scientific Nationalism," in *Narrating the Arctic: A Cultural History of Nordic Scientific Practices*, eds. Michael T. Bravo and Sörlin (Canton, Mass., 2002): 73–122.

ment studies motivated by questions of climate, ocean chemistry, and geochemistry. There is an interest in the change over time of the atmospheric concentration of CO$_2$ before the measurements on Hawaii. Moreover, Hawaii was not a given as a place to study secular trends in CO$_2$, and series of observations of background CO$_2$ from other locations also now exist—for example, those taken in Barrow, Alaska, and in American Samoa starting in the 1970s.

EXAMPLES OF MEASUREMENT BEFORE THE 1950S AND THEIR MOTIVATION

Callendar presented in a 1938 article the hypothesis that fossil fuel combustion had contributed to an increase in the atmospheric content of CO$_2$, which, through its effect on "sky radiation," had caused mean temperatures to increase.[11] A leading British steam and combustion engineer, Callendar worked on the hypothesis from his home, not from within a scientific institution, and presented his work to the Royal Meteorological Society.[12]

In a number of his articles Callendar presented the historical development of the level of CO$_2$ in the atmosphere.[13] In the 1958 article "On the Amount of Carbon Dioxide in the Atmosphere," he discussed reliability and revealed something about the purpose of the measurements he used; he excluded for instance those "intended for special purposes, such as biological, soil air, atmospheric polution [sic], etc." since these were not "representative of the free air."[14] In his 1938 article Callendar used measurements taken outside the Jodrell Laboratory at Kew Gardens by the biochemist Horace T. Brown and his collaborator Ferguson Escombe around the turn of the twentieth century.[15] Brown and Escombe measured atmospheric CO$_2$ because they wanted to find the rate of photosynthesis in a leaf surrounded by "a concentration not far removed from the normal amount of 0.03 per cent."[16] The absolute variations in atmospheric CO$_2$ that they found were small, but still significant given what they knew about the relation between the rate of the leaf's assimilation of CO$_2$ and the partial pressure of CO$_2$ in the air.[17] In 1940, Callendar drew on several measurement series from the late nineteenth century, the earliest from 1872–80, and on a smaller number from the twentieth century. He was again interested in an increase in the

[11] Callendar, "Artificial Production" (cit. n. 3).

[12] Fleming, *Callendar Effect* (cit. n. 5).

[13] See Callendar, "Artificial Production" (cit. n. 3), in which he uses a measurement series from the turn of the nineteenth century and derives future concentrations theoretically; Callendar, "Variations of the Amount of Carbon Dioxide in Different Air Currents," *Quarterly Journal of the Royal Meteorological Society* 66 (1940): 395–400, repr. in Fleming and Fleming, *Papers* (cit. n. 3); Callendar, "Can Carbon Dioxide Influence Climate?" *Weather* 4 (1949): 310–4, repr. ibid.; Callendar, "On the Amount of Carbon Dioxide in the Atmosphere," *Tellus* 10 (1958): 243–8. Callendar's use of existing atmospheric CO$_2$ measurements received some criticism from a statistical point of view in Giles Slocum's review of the field, "Has the Amount of Carbon Dioxide in the Atmosphere Changed Significantly since the Beginning of the Twentieth Century?" *Monthly Weather Review* 83 (1955): 225–31. Harry Wexler, who is discussed below, directed Slocum's review.

[14] Callendar, "On the Amount of Carbon Dioxide in the Atmosphere" (cit. n. 13), 244.

[15] Callendar, "Artificial Production" (cit. n. 3).

[16] Horace T. Brown and Ferguson Escombe, "Researches on Some of the Physiological Processes of Green Leaves, with Special Reference to the Interchange of Energy between the Leaf and Its Surroundings," *Proceedings of the Royal Society of London* B 76 (1905): 29–111, on 29.

[17] Horace T. Brown and Ferguson Escombe, "On the Variations in the Amount of Carbon Dioxide in the Air of Kew during the Years 1898–1901," *Proceedings of the Royal Society of London* B 76 (1905): 118–21, on 119.

proportion of CO_2 in the air over time, but also in how it varied with air currents of different geographic origin. He compared old measurements to historical weather maps and laid the data out next to each other in his notebook.[18]

In Scandinavia some measurements of carbon dioxide in the late nineteenth century were made by Axel Hamberg and Augusta Palmqvist. Hamberg studied hydrology in the seas surrounding Greenland as part of the expedition there in 1883 led by Adolf Erik Nordenskiöld, and he later studied glaciology and meteorology in Sarek in northern Sweden. He became professor of geography at Uppsala University in 1907.[19] During the Greenland expedition, Hamberg collected samples of ocean water in order to study its CO_2 concentration and variations in that concentration. He was responding to other studies of CO_2 in ocean water that addressed questions of chemistry.[20] Palmqvist analyzed air samples in flasks from a number of localities in Norway, Sweden, Greenland, and surrounding seas. Otto Pettersson, professor of chemistry at Stockholm Högskola, had initiated the measurements and supervised Palmqvist's analyses.[21] She was concerned with questions that, as she wrote, had been studied since the beginning of the nineteenth century: "What is the atmosphere's normal content of carbonic acid, and what variations is this content subordinate to?"[22] Palmqvist especially mentioned Swiss botanist Théodore de Saussure's study of the subject, which he undertook from 1809 to 1830. Palmqvist studied possible variation with, for instance, vegetation, winds, time of day, and ocean temperature. Her conclusion also included an average of her results.[23] Saussure, in a study undertaken in the years 1827–9, was interested in how the level of CO_2 in the atmosphere varied with factors like the wind, rain, terrain, and time of day. He saw the virtue of studying variations of CO_2 in, for example, its relation to plant nutrition, its provision of new observations to meteorology, and its revelation of the mixing of the layers of the atmosphere.[24]

Svante Arrhenius, in his article "On the Influence of Carbonic Acid in the Air upon the Temperature of the Ground,"[25] used his theory to explain the ice ages.[26] Arrhenius, born in 1859, made important contributions to the field of physical chemistry and later took an interest in cosmic physics. He began to work at Stockholm Högskola in 1891 and then created the Stockholm Physics Society, where, among many other things, CO_2 measurement data were presented, for example by Otto Petters-

[18] Callendar, "Variations" (cit. n. 13); 1939–40 notebook, box 2, in Fleming and Fleming, *Papers* (cit. n. 3).

[19] *Svenskt Biografiskt Lexikon*, ed. Erik Grill (Stockholm, 1971), s.v. "Hamberg, Axel."

[20] Axel Hamberg, "Hydrografisk-kemiska iakttagelser under den svenska expeditionen till Grönland," appendix to *Kungliga Svenska vetenskapsakademiens handlingar* 10 (1885): 1–57, on 34–5. He also studied other constituents of ocean water.

[21] Augusta Palmqvist, "Undersökningar öfver atmosferens kolsyrehalt," appendix to *Kungliga Svenska vetenskapsakademiens handlingar* 18, sec. 2 (1892): 3–39.

[22] Ibid., 3; all translations in this article are my own unless otherwise indicated.

[23] Ibid.

[24] Théodore de Saussure, "Mémoire sur les variations de l'acide carbonique atmosphérique," *Annales de chimie et de physique* 44 (1830): 5–55. Saussure also did experiments on carbonic acid in plant tissue; *Larousse Dictionary of Scientists*, ed. Hazel Muir (New York, 1994), s.v. "Saussure, Nicolas Théodore de."

[25] Arrhenius, "On the Influence of Carbonic Acid in the Air upon the Temperature of the Ground," *Philosophical Magazine*, ser. 5, vol. 41, no. 251 (1896): 237–76, repr. in *The Legacy of Svante Arrhenius: Understanding the Greenhouse Effect*, eds. Henning Rhode and Robert Charlson (Stockholm, 1998), 173–212.

[26] Elisabeth Crawford, "Arrhenius' 1896 Model of the Greenhouse Effect in Context," in Rhode and Charlson, *Legacy* (cit. n. 25), 21–32.

son.[27] Palmqvist attended a few meetings of the society.[28] "On the Influence of Carbonic Acid" included several pages of an article on the carbon cycle by the Stockholm geologist Arvid Högbom that had originally appeared in *Svensk kemisk tidskrift* (*Swedish Chemical Review*) in 1894.[29] In his article Högbom was interested in the secular change in atmospheric CO_2 content. He did not refer explicitly to any scientists, but did mention work on the amount of CO_2 absorbed in ocean water, possibly similar to Hamberg's research.[30]

Callendar wrote several articles that set forth the theory that an increase in the content of CO_2 in the air caused temperature to increase.[31] The link was also noted by the Norwegian Knud Georg Meldahl in 1948. Meldahl's suggestion that fossil fuel combustion, as opposed to greater solar radiation, was causing a warming was discussed in a newspaper clipping attached in a letter from Harry Wexler, who was then chief of the Scientific Services Division of the U.S. Weather Bureau and later became deeply involved in measurements of CO_2 during the International Geophysical Year, to Carl-Gustaf Rossby, director of the Institute of Meteorology at Stockholm Högskola. Wexler was then involved in studies of the effect of volcanic eruptions on solar radiation and on "extra-terrestrial control of long period weather anomalies."[32] He was interested in, among other things, air mass development, radiative transfer, and climate.[33]

SCANDINAVIAN STUDIES ON CO_2 IN THE MID-1950s

Rossby was a dynamic meteorologist who after a long career in America returned to his home country, Sweden, in the late 1940s. It is to the Institute of Meteorology, which he directed, that I now wish to turn, to examine in detail the unfolding of research on CO_2 and the many people and places involved.

Callendar used measurements by the Finnish chemist Kurt Buch because they were among the most recent values of atmospheric CO_2 content.[34] Buch also took an interest in Callendar's work. He had asked a question in 1917 about the relation over time between the levels of CO_2 in the atmosphere and in the sea. He could not answer

[27] Ibid.

[28] Fysiska Sällskapet i Stockholm [Stockholm Physics Society], Protokollsbok 1 [Book of minutes 1] (21 October 1891–10 December 1898), Swedish Royal Academy of Sciences Archives, Stockholm.

[29] Crawford, "Arrhenius' 1896 Model" (cit. n. 26); Arrhenius, "Carbonic Acid" (cit. n. 25); Högbom, "Om sannolikheten för sekulära förändringar i atmosfärens kolsyrehalt," *Svensk kemisk tidskrift* 6 (1894): 169–76.

[30] Högbom, "Om sannolikheten" (cit. n. 29).

[31] Callendar, "Artificial Production" (cit. n. 3); Callendar, "The Composition of the Atmosphere through the Ages," *Meteorological Magazine* 74 (1939): 33–9, repr. in Fleming and Fleming, *Papers* (cit. n. 3); a little less confidently in Callendar, "Can Carbon Dioxide Influence Climate?" (cit. n. 13).

[32] Wexler to Rossby, 16 December 1948, in Rossbys korrespondens (hereafter cited as Rossby papers), internal archive of the Department of Meteorology, Stockholm University. The clipping Wexler attached was "House Heating Helps Change World Climate, Dane Reports," *Washington Star*, November 26, 1948. Meldahl had written a letter about his views and calculations to a Swedish daily, *Svenska Dagbladet*: "Ökad förbränning av kol höjer luftens temperatur: Jorden synes gå mot en ny stenkolsålder" [Increased combustion of coal raises the temperature of the air: Earth appears headed toward a new carboniferous age], August 21, 1948.

[33] James Rodger Fleming, "Polar and Global Meteorology in the Career of Harry Wexler, 1933–1962," in *Globalizing Polar Science: Reconsidering the International Polar and Geophysical Years,* eds. Roger D. Launius, Fleming, and David H. DeVorkin (New York, 2010), 225–41.

[34] Callendar, "Variations" (cit. n. 13). Buch, born in 1881, was professor of chemistry at Åbo Academy 1932–42 and at Helsinki University 1942–9.

the question, but Callendar's review of the content of CO_2 in the atmosphere had "shed new light on the question." Taking this evidence together with his own Atlantic observations from 1935 of a difference between the CO_2 tension of ocean water and the atmosphere, Buch suggested that increasing industrial combustion had upset a possible previous equilibrium and that CO_2 was now moving from the atmosphere to the sea.[35]

At a conference on atmospheric chemistry at the Institute of Meteorology in May 1954, opened by Rossby, one of the chemicals discussed was CO_2, including its possible accumulation in the atmosphere as a result of fossil fuel combustion and the effect of this on climatic conditions. Buch spoke on the subject.[36] Rossby and Buch had been in contact in the mid-1930s, and Rossby had then suggested that Buch take certain air samples of CO_2 over the sea "to test a theory that the CO_2-content changes proportionately to the logarithm of the height."[37] Buch had written to the institute in 1953 about Callendar's work on the effect of fossil fuel combustion on the level of CO_2 in the atmosphere and about Callendar's suggestion that this influenced temperature. The CO_2 hypothesis itself was hence one reason why the question of the content of CO_2 in the air was important to Buch, and to Rossby, who, as we will see, greeted his proposal very positively.[38] In the letter Buch also noted his earlier finding:

> Furthermore I have found that the CO_2 partial tension of ocean water on average appears to be the same as that of the atmosphere at the turn of the century . . . or in other words that the sea was at that time in CO_2 equilibirum with the atmosphere, but that the resulting movement of carbon dioxide now is in the direction of atmosphere → sea.[39]

In order to examine the question of an increase in the level of CO_2 in the atmosphere, Buch suggested they make some new measurements: "The question is undeniably interesting, but the material evidently too scant for a reliable judgment. However, now that twenty years have passed since the last CO_2 determinations, it appears to me that the question of renewed investigations is truly called for. That brings us to the practical circumstances concerning their realization; primarily, that of the localization of the sampling."[40] His proposal was met with interest at the institute in Stockholm. Following a visit by Buch, Erik Eriksson at the institute was making plans for places at which to take samples.[41] Eriksson visited Abisko Research Station in the far north of Sweden. The superintendent promised to participate. The janitor would do

---

[35] Kurt Buch, "Kolsyrejämvikten i Baltiska Havet," *Fennia* 68 (1945): 1–208, on 151–3.

[36] Erik Eriksson, "Report on an Informal Conference in Atmospheric Chemistry Held at the Meteorological Institute, University of Stockholm, May 24–26, 1954," *Tellus* 6 (1954): 302–7, on 303–4.

[37] Kurt Buch, English summary in "Beobachtungen über das Kohlensäuregleichgewicht und über den Kohlensäureaustausch zwischen Atmosphäre und Meer im Nord-Atlantischen Ozean," *Acta Academiae Aboensis Mathematica et Physica* 11 (1939): 28–31, on 30.

[38] Buch is also cited as being concerned with the hypothesis by the Finnish National Committee for the International Geophysical Year; E. Sucksdorff, "Report of the Finnish National Committee for the Année Géophysique Internationale," in Kommissionen för det Geofysiska Året 1957–1958: Handlingar 1952–1954 [Commission of the Geophysical Year 1957–1958: Documents, 1952–1954], internal archive of the Department of Meteorology, Stockholm University.

[39] Buch to "Ärade Kollega" (probably Rossby), 22 November 1953, Papers of Meteorologiska Institutionen, Stockholm University Archives (hereafter cited as MI), vol. E1A:1.

[40] Ibid.

[41] Eriksson to Buch, 21 January 1954, MI, vol. E1A:4.

the sampling; he took some observations at the station. "There is yet another possibility to sample in the Arctic," in Greenland, Eriksson wrote, if the Scandinavian
Airlines System opened the planned regular flight service with which one could carry
samples back from there.[42] Buch liked these ideas. Spitsbergen might be difficult to
get to in the winter, yet "samples at the highest latitudes might be the most important" in winter, since the "heat-delivering task" of CO$_2$ was more important then.[43]
The Oceanographic Institute in Finland planned sampling of CO$_2$ at Jungfruskär, located at the "shift" between the Åland and Åboland archipelagoes in the Baltic Sea.
Buch noted that "the air there is of course mixed, but ought to be free from industrial
smoke." Jungfruskär was already an observation station for hydrological and meteorological studies, with observations made at least as often as the first, tenth, and
twentieth of each month.[44]

Later in the spring Buch was invited to come to the conference on atmospheric
chemistry at the Institute of Meteorology and speak on CO$_2$.[45] In the general discussion of that conference it was concluded that a program for CO$_2$ analysis of the air
"must serve three separate purposes":

> One is to establish the geochemical distribution of CO$_2$ over a certain area, one is to
> study the modifications of the air with respect to CO$_2$ as it is carried from the oceans in
> over land areas and the last is to study secular trends. For the last purpose places with
> slight climatic variations and remote from any kind of atmospheric disturbances which
> may give rise to local changes are most suitable. Probably the most ideal place would be
> somewhere in the middle of the Sahara desert, and it was recommended to plan for such
> a station in the future. Another possible place would be Hawaii if samples were taken
> above the trade-wind inversion.[46]

Buch had initiated one project on geochemical distribution, covering Scandinavia,
when visiting the institute in January 1954, and Eriksson had now drafted its program. At the conference it was agreed that air samples would be taken at a few places
in Scandinavia on the first, tenth, and twentieth of each month. It was also suggested
that trajectories be established to determine the origin of the air mass sampled. Further, it was agreed that additional samples would be taken upon telegraphic order if
interesting weather situations arose, to study modifications in air masses. The group
of sampling locations was called a "Scandinavian net."[47] Rossby described it as a
"synoptic net."[48] Possibly, Rossby also regarded it as useful for studying the cycle of
carbon.[49]

A second project on geochemical distribution discussed at the conference aimed to
estimate the flux of CO$_2$ from the equator to the poles. Eriksson and Wendell Mordy,
meteorologist at the Pineapple Research Institute, Hawaii, hoped to have air samples
taken around the North Pacific Current.[50] Buch had earlier made air analyses in

[42] Ibid.
[43] Buch to Eriksson(?), 29 January(?) 1954, Rossby papers.
[44] Ibid.
[45] Buch to Eriksson(?), 15 April 1954, Rossby papers.
[46] Eriksson, "Informal Conference" (cit. n. 36), 305.
[47] Ibid., 305–6.
[48] Rossby to Lloyd V. Berkner, 13 April 1955, MI, vol. E1A:8.
[49] See Eriksson's account of Rossby's introduction to the conference; "Informal Conference" (cit. n. 36), 303.
[50] Ibid., 306.

Spitsbergen that indicated that CO_2 was absorbed by polar sea water.[51] Estimating the flux was one way to investigate the role of the ocean as a sink for CO_2.[52] Air sampling of CO_2 in a net of fifteen Scandinavian stations, as illustrated in figure 1, had begun already by November of 1954.[53]

The Swedish sampling stations were placed at Abisko Research Station, at Ultuna Agricultural College and Plönninge Agricultural School, at experiment stations in Öjeby and Flahult, the latter an old agrichemical station, and in the village of Bred-kälen. In Finland the stations were placed at Kauhava Military Training School, at Rissala and Luonetjärvi aerodromes, and at the zoological station in Tvärminne. In Norway they were placed at the experiment station in Vågönäs near Bodö, at Klones Agricultural School in Vågåmo, and at Bergen University. In Denmark one was placed at Ödum Agricultural Experiment Station and the other at Askov Agricultural Experiment Station.[54] In Sweden the actual air sampling was done by persons who worked at the sampling stations. Remunerations to the CO_2 samplers in Sweden were paid by the Meteorological Institute at Stockholm Högskola over the period November 1, 1955–June 30, 1956.[55] The location of sampling stations in the Scandinavian net was described in the sampling technique: "The air samples are taken on open places without foliage, such as a lake- or sea-shore, or on a rock without forest. Towers, or roofs without chimneys, are also suitable places."[56] Indeed, the samples were taken in places such as the roof of Abisko Research Station, in a tower at Bergen University, on airfields and graveled grounds, on a railway embankment, on the shore of the sea, and on the shore of a lake.[57] The stations chosen also lay at a distance from industry, and samples were taken at the time of day when the ground-air exchange was best developed.[58]

The method of analyzing the air samples was a modification of Danish physiologists August Krogh and Poul Brandt Rehberg's microtitration technique from 1929. Since Buch had used the same modified technique in a study in Scandinavia and the North Atlantic in 1935, the results could be compared.[59] Measurements were given to 1 part per million (ppm), and the principle was "to pass a definite small amount of air through a solution of barium hydroxide and then to titrate the excess of the hydroxide with hydrochloric acid."[60]

[51] Ibid., 303.

[52] Eriksson described two ways of determining how much of the CO_2 released in combustion was taken up by the sea rather than accumulated in the atmosphere: "One is to estimate the yearly flux of carbon dioxide from the equator towards the poles, and from the magnitude of this stationary process to estimate the effective volume of the sea exposed to the atmosphere yearly. Another way of approach is to map the age of the ocean water at different depths and positions by the C^{14} method" (ibid., 304).

[53] Fonselius, Koroleff, and Buch, "Microdetermination" (cit. n. 8), 258.

[54] Ibid., 260.

[55] Bert Bolin to Fröken T. Reinius, 19 December 1956, MI, vol. E1A:16.

[56] Fonselius, Koroleff, and Buch, "Microdetermination" (cit. n. 8), 259.

[57] Ibid., 260.

[58] Walter Bischof, "Periodical Variations of the Atmospheric CO_2-Content in Scandinavia," *Tellus* 12 (1960): 216–26, on 224.

[59] Stig Fonselius, Folke Koroleff, and Karl-Erik Wärme, "Carbon Dioxide Variations in the Atmosphere," *Tellus* 8 (1956): 176–83, on 177–8. In Fonselius, Koroleff, and Buch, "Microdetermination" (cit. n. 8), the source cited for the technique was August Krogh and Poul Brandt Rehberg, "CO_2-Bestimmung in der atmosphärischen Luft durch Microtitration," *Biochemische Zeitschrift* 205 (1929): 265–72. Krogh had studied CO_2 at the very beginning of the century, motivated by the possible influence of fossil fuel combustion on climate; *Dansk Biografisk Leksikon*, ed. Svend Cedergreen Bech (Copenhagen, 1981), s.v. "Krogh, Schack August Steenberg."

[60] Fonselius, Koroleff, and Buch, "Microdetermination" (cit. n. 8), 260.

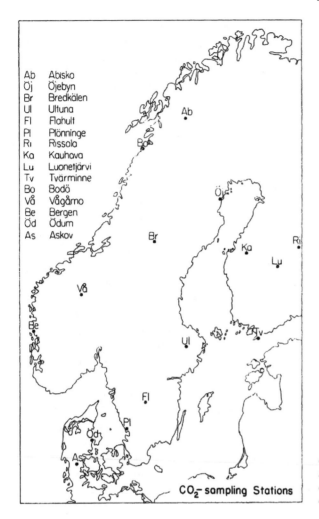

Ab Abisko
Öj Öjebyn
Br Bredkälen
Ul Ultuna
Fl Flahult
Pl Plönninge
Ri Rissola
Ka Kauhava
Lu Luonetjärvi
Tv Tvärminne
Bo Bodö
Vå Vågåmo
Be Bergen
Öd Ödum
As Askov

CO₂- sampling Stations

The original caption read, "Map showing the distribution of stations." Reprinted by permission from Fonselius, Koroleff, and Buch, "Micro-determination" (cit. n. 8), 260.

In an article on their sampling results, Stig Fonselius, Folke Koroleff (from the Institute of Marine Research, Helsinki), and Karl Erik Wärme wrote that they regretted the geographic limitation of their sampling net and that "the difference between Buch's and our values suggest [*sic*] an increase of the CO_2 content in Scandinavia, but it is impossible to say at present whether this increase is just a fluctuation in the regional CO_2-climate or if it represents a steady increase since 1935."[61] Their interest in CO_2 seemed very broad. They noted that knowledge about the concentration, distribution, and monthly and yearly variation of CO_2 and its cycle was poor. They expressed their hope that studies in other countries could be done during the International Geophysical Year (IGY): "The only way to investigate this problem is to carry out sampling and analyses over a large area simultaneously and consider all the meteorological factors."[62]

Rossby wrote that the data from the Scandinavian net had been found to vary

[61] Fonselius, Koroleff, and Wärme, "Carbon Dioxide" (cit. n. 59), 178.
[62] Ibid., 182.

considerably with air mass origin and that these measurements, taken in limited areas, were not reliable for estimating the "atmospheric carbon-dioxide reservoir and its secular changes."[63] He sympathized with the idea that measurements in synoptically inactive areas might be more useful for that inquiry.[64] A more extensive planned observation program for the IGY did not present a sure solution to Rossby: "It is very likely that great difficulties will be encountered in every attempt to compute the content of carbon dioxide in the atmosphere and its secular changes from such scattered observations."[65] He suggested that perhaps the method using radioactive carbon, first employed in 1953 by the chemist Hans Suess, held out great promise. It was also "very likely to eliminate local synoptic variations" in the CO_2 content of the air.[66]

The relationship between CO_2 and Earth's heat balance, and hence the CO_2 hypothesis, interested Rossby and Bert Bolin, also at the Institute of Meteorology. In an application for work in atmospheric chemistry to the Knut and Alice Wallenberg Foundation in 1956, Bolin, on behalf of the institute and Rossby, noted, "Intimately connected with the chemical composition of precipitation and the atmosphere is the atmosphere's content of carbon dioxide. The amount of carbon dioxide is of fundamental importance for the amount of heat radiation that leaves Earth and thus for Earth's entire heat balance."[67] With the IGY, they saw a possibility to do more on CO_2 measurement. The application (which was granted) was for chemistry work on atmosphere and precipitation during the IGY and included "CO_2 work."[68] They wished for a multifaceted program that included a northward expansion of the Scandinavian net, determination of the amount of CO_2 in the air from measurements taken aboard ships in the North Atlantic to study the synoptic distribution in the level of CO_2 and daily changes, and studies at North East Land, Spitsbergen, of the variation of CO_2 and its vertical transport to the sea surface.[69] When the IGY was starting, Eriksson expressed the further idea that CO_2 could be useful as a tag on air masses (possibly only those at high altitude).[70]

In the summer of 1956 the Soviet icebreaker *Ob* traveled by Spitsbergen, and some Norwegian and Swedish geophysicists were allowed to join the voyage and make observations. The Swedish National Commission for the Geophysical Year, of which Rossby was then chairman, instructed Eriksson at the Institute of Meteorology and glaciologist Valter Schytt at the Department of Geography at Stockholm Högskola, who was Ahlmann's student, to try to join the *Ob* and make preliminary observations for the IGY. According to Rossby, Eriksson's studies would concern "the

[63] Carl-Gustaf Rossby, "Current Problems in Meteorology," in *The Atmosphere and the Sea in Motion: Scientific Contributions to the Rossby Memorial Volume*, ed. Bert Bolin (New York, 1959), 9–50, on 15. Translated by staff at the International Meteorological Institute, Stockholm, from the Swedish original, "Aktuella meteorologiska problem," *Svensk naturvetenskap: Statens naturvetenskapliga forskningsråds årsbok* 10 (1956): 15–80. According to the preface to the memorial volume, Rossby finalized this Swedish version in 1957.

[64] Synoptic activity is, for instance, variable weather on a regional scale caused by variation (via atmospheric circulation or transport) in the origin of arriving air masses.

[65] Rossby, "Current Problems" (cit. n. 63), 15.

[66] Ibid., 16.

[67] Bolin to the Knut and Alice Wallenberg Foundation, 29 November 1956, MI, vol. E1A:17.

[68] Ibid.; the Knut and Alice Wallenberg Foundation to the Institute of Meteorology at Stockholm Högskola (Oscar af Ugglas to Bolin), 26 January 1957, MI, vol. E1A:19.

[69] Bolin to the Knut and Alice Wallenberg Foundation (cit. n. 67).

[70] Erik Eriksson, "Atmosfärkemi under geofysiska året," *Svenska Dagbladet*, July 10, 1957. A tag on an air mass can be used to determine where an air mass that has moved originally came from.

carbon-dioxide content of the air," along with other subjects.[71] During the actual IGY Fonselius was to study the CO$_2$ content of the air vertically by taking samples from aircraft.[72] At Kinnvika in Murchison Bay, where the Swedish-Finnish-Swiss IGY expedition had its base, Wärme worked to determine the CO$_2$ content of the air by using continuous registration with infrared light and by conducting chemical analysis on samples taken with pipettes.[73] CO$_2$ was also studied at the glacier station outpost Vestfonna.[74]

Atmospheric chemistry provides context for the CO$_2$ studies at the Institute of Meteorology, but the CO$_2$ studies were distinct in important ways. During the IGY atmospheric chemistry was part of the meteorological research program. Stations were classified according to whether they conducted analysis of precipitation, analysis of air samples, study of atmospheric CO$_2$ content, or measurement of condensation nuclei.[75] The institute began to formally collaborate in 1953–4 with Hans Egnér, a laboratory scientist at the Royal Agricultural College in Ultuna, on precipitation and atmospheric chemistry.[76] As they initiated the study of the circulation of atmospheric salts, the circulation of CO$_2$ and water vapor provided analogies to Egnér and Rossby.[77] There were also continuities in personnel and sampling locations with the CO$_2$ sampling net. It was through the collaboration with Egnér that Eriksson, an agronomist who worked at the time with Egnér, was drawn to the Institute of Meteorology.[78] Rossby planned to start "some sort of enquête in *Tellus*" on nitrogen in precipitation and air.[79] Eriksson wrote the piece and discussed the manuscript in Stockholm with Rossby, who "learned a fair amount."[80] In early 1954 Eriksson was the person responsible for the work in precipitation and atmospheric chemistry at the institute, and Fonselius was his co-worker.[81] The study of precipitation and atmospheric chemistry used a geographically more expansive and, in Sweden, finer, mesh of stations than that used for CO$_2$ sampling, but ten stations were in both nets and Kurt Buch was in both instances a Finnish counterpart.[82]

[71] Rossby to Nicolai Herlofson, 4 June 1956, MI, vol. EIA:17. According to Gösta Liljequist, Eriksson did travel with the *Ob*. Liljequist describes his purpose as having been "to study the regional distribution of the CO$_2$ content of the air, and its content of dust and salt particles"; *High Latitudes: A History of Swedish Polar Travels and Research* (Stockholm, 1993), 544.

[72] "Meteorologiska världsd Teknologflyg mäter kolsyra första gången efter Andrée," *Dagens Nyheter*, June 27, 1957. It is interesting that this article, in which Fonselius was interviewed, linked studies of CO$_2$ in the Scandinavian net with industry, but not with climatic change.

[73] Gösta Liljequist, with contributions from Eric Dyring, Sveneric Molander, and Carl-Axel Bäckstedt, *Arktisk utpost: Berättelsen om den svensk-finsk-schweiziska expeditionen till Nordostlandet 1957–1958* (Stockholm, 1960), 111.

[74] Gösta Liljequist to Bert Bolin, 14 June 1958, Polarårskommissionen, vol. 5, internal archive of the Department of Meteorology, Stockholm University.

[75] *Annals of the International Geophysical Year*, vol. 13 (Oxford, 1959), 87.

[76] Maria Bohn, "Polarårskommissionen och det Geofysiska året 1957–1958," *Ymer* 129 (2009): 177–99.

[77] Application from Egnér and Rossby to the Agricultural Research Council, 27 December 1952, Rossby papers.

[78] Rossby to Egnér, 22 August 1952, Papers of Hans Egnér, Institutionen för Radioekologi (Department of Radioecology) archives, Statens Landbruksuniversitet, Ultuna, vol. Ö1A:1 (hereafter cited as Egnér papers).

[79] Anders Ångström to Egnér, 27 June 1952, Egnér papers.

[80] Rossby to Egnér, 22 August 1952, Egnér papers.

[81] Rossby to the Knut and Alice Wallenberg Foundation, 30 September 1954, Archive of the Knut and Alice Wallenberg Foundation, Stockholm.

[82] Hans Egnér and Erik Eriksson, "Current Data on the Chemical Composition of Air and Precipitation," *Tellus* 7 (1955): 134–9. The stations in the CO$_2$ net that were not used again for this purpose were Bergen, Bodö, Abisko, Luonetjärvi, and Rissala.

TRANSATLANTIC CONTEXT AND PRECISION

The story of CO_2 studies in Scandinavia partly unfolds in a transatlantic context. Rossby's presence in American meteorology had left traces in the form of contacts he maintained with American scientists.[83] Rossby came to the United States in 1926, when in his late twenties, strengthened the meteorology departments at the Massachusetts Institute of Technology (MIT) and the University of Chicago, and, having received American citizenship in 1939, was part of the American war effort.[84] Harry Wexler earned his PhD from MIT in 1939 under Rossby's supervision.[85]

American CO_2 studies for the IGY were addressed at the U.S. IGY geochemical discussions at the Weather Bureau in January 1956. In the discussions Wexler called attention to Callendar's hypothesis from 1938 and to CO_2 as an example of a gas for which there was an "absence of reliable information on the changes occurring in the atmosphere over a period of years." Hans Suess, by that time at the Scripps Institution of Oceanography, expressed Scripps's interest in a coordination of oceanographic CO_2 studies and air sampling.[86]

Rossby corresponded regularly with the Swedish scientist Gustaf Arrhenius, grandson of Svante Arrhenius, who was also at Scripps.[87] Arrhenius had left Sweden in 1952 at the invitation of Roger Revelle, the director of Scripps, who had taken an interest in his early research on climate recording in sediment core material collected in the east equatorial Pacific.[88]

In 1956 Arrhenius and Rossby exchanged ideas about studies for the IGY. They sent each other drafts of their programs.[89] Arrhenius wrote to Rossby in the spring of 1956 about their aim to take a "careful inventory," plus or minus 1 percent, of present atmospheric CO_2 content in order to provide a baseline for study of future release of CO_2 from fossil fuels, as well as their aim to study the mechanisms of exchange with the sea.[90]

Rossby replied that a study of exchange mechanisms through gradient determinations would evidently need a method more precise than the one Krogh and Buch had developed. For the "geographical-climatological" study in the Scandinavian net, he thought that greater precision than 1 ppm might be "synoptically meaningless." Perhaps he meant that levels varied so much from place to place that a more precise method had no purpose. He also wrote, "I would . . . appreciate a statement from

---

[83] Harper, in *Weather by the Numbers* (cit. n. 9), demonstrates the close cooperation between Stockholm meteorologists led by Rossby and American scientists in the field of numerical weather prediction.

[84] *Dictionary of Scientific Biography*, s.v. "Rossby, Carl-Gustaf Arvid."

[85] Fleming, "Career of Harry Wexler" (cit. n. 33).

[86] U.S. National Committee for the International Geophysical Year 1957–1958, IGY geochemical discussion, meeting minutes, January 31, 1956, on 8, in Rossby to Gösta W. Funke, 23 March 1956, Svenska Nationalkommittén för Geodesi och Geofysik (Swedish National Committee on Geodesy and Geophysics), Swedish Royal Academy of Sciences Archives, vol. E1:1A.

[87] Letters in MI.

[88] "Oral History of Gustaf Olof Svante Arrhenius," April 11, 2006, "Oral Histories" Web page, University of California San Diego libraries, http://scilib.ucsd.edu/sio/oral/Arrhenius.pdf (accessed 7 January 2011), on 6–7.

[89] Rossby to Arrhenius, 29 March 1956, MI, vol. E1A:17. Rossby thanked Arrhenius for a memo dated February 27 in which "La Jolla's large carbonic acid program was sketched" and sent him a draft program for work in atmospheric chemistry at Spitsbergen. I have not found the actual programs in the archive.

[90] Arrhenius to Rossby, 24 February–7 March 1956, MI, vol. E1A:13.

you, about how we should best adapt our work here in Sweden on carbonic acid to fit into your general program, and you ought not hesitate to criticize what you may find unsatisfactory or incomplete in our current observation program."[91] Arrhenius and Keeling suggested coordinating CO$_2$ measurements with Rossby's group during the IGY.[92] Rossby would then need the same instrument that they would use. Perhaps a copy could go to Spitsbergen?[93] Later in the year, Keeling wrote to Rossby along the same lines:

> It seems desirable to maintain continuous measurements at two or three stations in any case. . . . A station in the Arctic would also be particularly desirable, and we wonder whether a continuous recording analyzer could be accommodated at Spitzbergen. This would also serve to bridge the gap between the U.S. and Scandinavian programs and furnish a direct comparison of data.[94]

Keeling initially began to study background concentrations of carbon dioxide in the air as information he needed to determine the amount of carbonate in water. His original orientation toward the field of atmospheric CO$_2$ studies was hence related to and set in the context of an already existing field of study. He then saw a diurnal pattern in the atmospheric observations and continued with them. In 1956 he was confident that he had discovered a "near constancy" in the level of CO$_2$ in the atmosphere, as opposed to variability with air mass origin or photosynthetic activity.[95] That discovery made it important to measure with increased precision. The method used in the Scandinavian net was not very expensive, and it was simple.[96] But with Keeling's discovery it became insufficiently accurate.

There was a discussion in America in 1956 about the required precision of CO$_2$ measurements during the IGY.[97] Keeling explained to Rossby the need for increased precision in comparison to the Scandinavian net: "Our belief in the need for more accurate data is based upon recent measurements of maritime Pacific air. These indicate a concentration range in carbon dioxide roughly one power of ten less than currently observed in Scandinavia, nevertheless with a systematic variability appearing in measurements accurate to about +/−0.5 ppm."[98] Rossby was in favor of measuring with increased accuracy. He was interested in having measurements made at a few stations with the more accurate infrared analyzer, in order to study how fast CO$_2$ increased in the atmosphere as a result of industrial combustion.[99]

[91] Rossby to Arrhenius (cit. n. 89).

[92] Arrhenius to Rossby, 12 April 1956, MI, vol. E1A:13; Keeling to Rossby, 21 September 1956, ibid.

[93] Arrhenius to Rossby (cit. n. 92).

[94] Keeling to Rossby (cit. n. 92).

[95] Keeling, "Rewards and Penalties" (cit. n. 6). See also Weart, *Global Warming* (cit. n. 1).

[96] Rossby to Arrhenius (cit. n. 89). Joseph Priestley valued simple experimental methods because they could be replicated; Jan Golinski, "The Nicety of Experiment," in *The Values of Precision,* ed. M. Norton Wise (Princeton, N.J., 1995), 72–91. The measurements in the Scandinavian net were similarly useful in their simplicity.

[97] Norris W. Rakestraw to Maurice Ewing, 23 August 1956, in Rossby to Arrhenius, 13 September 1956, MI, vol. E1A:17.

[98] Keeling to Rossby (cit. n. 92).

[99] Rossby to Edward Georg Bowen, 3 October 1956, Rossby papers; compare this to the position Rossby expressed in 1957. In the letter a station in Western Australia is suggested by "us"—probably the Scandinavian group, Keeling, and Wexler (who are copied): "All of us who are concerned with the CO$_2$ program . . ."

Keeling started to work at Scripps in August 1956 on a program of CO_2 studies. It included continuous measurements with an infrared gas analyzer at the South Pole starting in 1957 and at Mauna Loa starting in 1958. The meteorological observatory at Mauna Loa had recently been built by the Weather Bureau, and Wexler supported the Scripps program; in fact, Wexler had previously offered Keeling a position to take continuous measurements at Mauna Loa and Little America, but Keeling preferred the working environment at Scripps.[100] Wexler gave initial funding for an infrared gas analyzer at Mauna Loa in October 1956.[101] Listed in the *Annals of the International Geophysical Year* are sixty-two stations where measurements of the amount of carbon dioxide in the air were taken, located in seven of eight regions defined by geomagnetic latitude. One of the fifteen stations in the Scandinavian net, Abisko, is on this list. Other stations included Little America, Arctic Ice Floe Station A, and Mauna Loa.[102]

CONCLUSIONS

In his letter to Rossby in 1953 Buch had set forth Callendar's calculation of the relation between the level of CO_2 and temperature, and Callendar's hypothesis was clearly a reason for the institution of the Scandinavian net;[103] at the 1954 conference Earl Barett of the Institute of Meteorology at Stockholm Högskola had spoken on work on the heat balance by Lewis Kaplan, Callendar, and Gilbert Plass, suggesting a connection between increased CO_2 levels in the atmosphere and "the observed change in climatic conditions during the last half century which has manifested itself for instance in the recession of glaciers";[104] and again at the conference in 1955, Buch "mentioned the influence of an increase of carbon dioxide in the atmosphere on the heat economy of the earth. The greatest effect would occur in polar regions, where the amount of water vapor is low."[105] Yet the CO_2 hypothesis of climate change was not named as a motivation in the article that described the Scandinavian net and its first results. In a preface Buch introduced the problem as the possible increase of CO_2 in the atmosphere. The "principal objection" to an increase had to do with the exchange process between the atmosphere and the sea, which, he noted, "may be regarded as a separate problem with its own intrinsic interest."[106]

The hypothesis expressed by Callendar arrived into a community of scientists that had methods and reasons to study CO_2. The hypothesis was one reason why CO_2 measurements were taken in Scandinavia in the 1950s, but it was not necessarily a sufficient reason. Keeling's Mauna Loa measurement series of annual increases in CO_2 levels itself made the CO_2 hypothesis more plausible.

---

[100] Keeling, "Rewards and Penalties" (cit. n. 6).

[101] James Rodger Fleming, *Historical Perspectives on Climate Change* (New York, 1998), 126. During the IGY Wexler "supported a program of worldwide CO_2 measurements"; Fleming, *Callendar Effect* (cit. n. 5).

[102] *Annals of the International Geophysical Year* (cit. n. 75), table 2-D, "Atmospheric Chemistry: Number of Stations," on 88; the list of stations is on 144–8.

[103] Buch to "Ärade Kollega" (cit. n. 39).

[104] Eriksson, "Informal Confererence" (cit. n. 36), 304.

[105] Erik Eriksson, "Report on the Second Informal Conference on Atmospheric Chemistry, Held at the Meteorological Institute, University of Stockholm, May 31–June 4, 1955," *Tellus* 7 (1955): 388–94, on 390.

[106] Fonselius, Koroleff, and Buch, "Microdetermination" (cit. n. 8), 258–9.

The deeper history of CO$_2$ measurements provides a contrast to the current cultural singularity of Keeling's measurement series. Perhaps this singularity is related to the fact that the curve of the measurement series from Mauna Loa became an icon of the greenhouse effect or, more broadly, an environmental icon.[107] A recent piece of science journalism called it "a symbol of our times."[108] The fact that the curve shows not only a rise but also, in the annual indents, the seasonal cycle of vegetation, has caused some to look upon it aesthetically.[109] The place, Mauna Loa, where the measurements were taken can be a vivid presence in a narrative about the measurements.[110] But not only was the choice of Hawaii nonevident, there are also other places and measurements that relate historically—as did Hawaii at the time—in a pre-iconic way to the CO$_2$ hypothesis of climate change.

[107] Weart, *Global Warming* (cit. n. 1), 35; Fleming, *Historical Perspectives* (cit. n. 101), 126.

[108] Helen Briggs, "50 Years On: The Keeling Curve Legacy," BBC News Online, December 2, 2007, http://news.bbc.co.uk/2/hi/science/nature/7120770.stm (accessed 3 January 2011).

[109] Tim Flannery, *The Weather Makers: The History and Future Impact of Climate Change* (Melbourne, 2005), 25.

[110] Ibid.

Melting Empires?

Climate Change and Politics in Antarctica since the International Geophysical Year

*by Adrian Howkins**

ABSTRACT

This article examines the relationship between climate change and politics in Antarctica since the International Geophysical Year of 1957–8, paying particular attention to the work of the British Antarctic Survey. Research conducted in Antarctica has played an important role in the understanding of climate change on a global scale. In turn, fears about the consequences of global climate change have radically changed perceptions of Antarctica and profoundly shaped scientific research agendas: a continent that until fifty years ago was perceived largely as an inhospitable wilderness has come to be seen as a dangerously vulnerable environment. This radical shift in perception contrasts with a fundamental continuity in the political power structures of the continent. This article argues that the severity of the threat of climate change has reinforced the privileged political position of the "insider" nations within the Antarctic Treaty System.

INTRODUCTION

In 1965, four years after the ratification of the Antarctic Treaty, the British Antarctic Survey (BAS) faced closure.[1] Before the signature of this innovative international treaty, the conduct of science had been a central tool in the promotion of British sovereignty in Antarctica. But article 4 of the Antarctic Treaty had suspended (or "frozen," in the official pun of the Washington conference at which the treaty was negotiated) all sovereignty claims to the continent, significantly reducing international tensions in Antarctica, including a costly dispute between Britain, Argentina, and Chile over the ownership of the Antarctic Peninsula. In theory at least, the terms of the Antarctic Treaty meant that no nation could make new sovereignty claims or do anything to strengthen or weaken existing claims or rights. For a brief moment, British Antarctic science seemed to have lost its political raison d'être. One strategy employed at the time by British Antarctic scientists to defend their institution was

* Department of History, Colorado State University, B368 Clark Building, 1776 Campus Delivery, Fort Collins, CO 80523-1776; howkins@mail.colostate.edu. The author would like to thank Ellen Bazeley-White and Joanna Rae at the British Antarctic Survey Archives for their helpful and friendly assistance during the research for this article. He would also like to thank James Fleming, Vladimir Jankovic, and the anonymous reviewers of the article for their useful comments on earlier drafts.

[1] For a full discussion of the closure debate, see "Policy: Future of Survey," AD3/1/AS/164/1(1), British Antarctic Survey Archives, Cambridge (hereafter cited as BASA).

to focus on the threat that melting ice in Antarctica would lead to rising sea levels around the world. "If an unstable ice surge, such as occurs on smaller glaciers, were to develop," Gordon Robin, a prominent Antarctic meteorologist, argued in his case against closing the BAS, "its effect on sea level and climate would be world-wide and drastic."[2] Antarctic science, Robin implied, mattered to Britain because the ice of Antarctica was linked to the sea level and climate of the rest of the world.

The British government spared BAS from closure for a number of reasons. Perhaps most pressing was the realization that Antarctic science remained strongly connected to Antarctic politics, and that to abandon scientific research in Antarctica would be to surrender hard-won political influence and prestige in the continent.[3] The use by British scientists, in their arguments for continued Antarctic research, of the threat of melting ice and rising sea levels offers an early example of the connection between politics and climate change in Antarctica. In the 1960s, the threat of global climate change remained a distant concern, and British scientists did not directly connect the possibility of ice sheets breaking away in Antarctica to rising global temperatures. But over the next few decades, as the perceived threat from global warming increased, the connection between climate change and politics in Antarctica would grow ever stronger.

Since the late 1950s, the instrumental period of Antarctic science, research conducted in Antarctica has made a significant contribution to the development of climate change science at a global scale. Measurements taken by Charles Keeling at the South Pole and published in 1960 suggested increasing levels of atmospheric carbon dioxide (CO_2).[4] Antarctic ice-core research revealed a possible correlation between high levels of atmospheric CO_2 and high temperatures in Earth's past. Global climate models have been calibrated using data from Antarctica. Although it is impossible to quantify the precise contribution of Antarctic science to climate research, measurements and experiments conducted in Antarctica have substantially increased scientific understanding of anthropogenic climate change.

Increasing fears of climate change on a global scale have led to major shifts in perceptions of the Antarctic environment. In a neat circularity, climate change research conducted in Antarctica has contributed to these changes in perception, and they in turn have shaped the scientific research agenda there. On the eve of the International Geophysical Year (IGY) of 1957–8, Antarctica continued to be seen largely as an impenetrable wilderness: nobody had set foot on the South Pole since the famous race between Amundsen and Scott of 1911–2. Today, as a result of fears of climate change, the continent has come to be seen as a dangerously vulnerable environment. A recently published 500-page report by the Scientific Committee on Antarctic Research (SCAR) entitled *Antarctic Climate Change and the Environment* comprehensively sets out contemporary scientific understanding of the past, present, and future of Antarctic climate change.[5] According to the report, ice-core reconstructions

[2] D. C. Martin [executive secretary of the Royal Society] to Sir Vivian Fuchs, 11 November 1965, in "Policy: Future of Survey" (cit. n. 1). The letter includes an enclosure by Robin, "Outline of the Importance of Antarctic Research."

[3] See, e.g., Executive Secretary of the Royal Society to Mr. R. M. K. Slater, 17 September 1965, in "Policy: Future of Survey" (cit n. 1).

[4] Charles D. Keeling, "The Concentration and Isotopic Abundances of Carbon Dioxide in the Atmosphere," *Tellus* 12 (1960): 200–3.

[5] John Turner et al., *Antarctic Climate Change and the Environment: A Contribution to the International Polar Year, 2007–2008* (Cambridge, 2009).

suggest that Antarctic temperatures have increased on average by about 0.2°C since the late nineteenth century.[6] These changes, however, have not been uniform across the continent, with the Peninsula region showing the highest temperature increases: the Faraday/Vernadsky Station on the Antarctic Peninsula, for example, has undergone a decadal increase of +0.53°C over the period 1951–2006, and 87 percent of the Peninsula's marine glaciers have shown overall retreat since 1953.[7] The report stresses that the ozone hole over Antarctica has shielded large parts of East Antarctica, by far the largest part of the continent, from much of the effect of global warming. But it notes that if ozone concentrations above Antarctica recover as they are expected to do over the next century, and greenhouse gas concentrations increase at the present rate, then "temperatures across the continent will increase by several degrees and there will be about one third less sea ice."[8] The consequences of global warming in Antarctica could have devastating global implications. Research directly related to climate change has come to dominate other fields of Antarctic science, such as glaciology and biology, as scientists seek to understand these implications. Substantial scientific resources, for example, have been put into understanding the West Antarctic Ice Sheet for fear that its collapse could have disastrous effects around the world.[9]

These dramatic shifts in perceptions of Antarctica's environment over the last fifty years contrast starkly with a fundamental continuity in the political structures of Antarctica from the IGY up to the present. By showing how Antarctic science has been coproduced with politics of the Southern Continent from the mid-twentieth century to the present, this article argues that the threat of climate change in Antarctica has largely acted to reinforce the political status quo in the continent.[10] This argument runs in some ways counter to what might be labeled the "official narrative" of the Antarctic Treaty System (ATS), which suggests that scientific idealism has largely trumped political rivalry.[11] But it is in broad agreement with recent research analyzing the connections between Antarctic science and geopolitics during and after the IGY in such fields as echo sounding and surveying.[12] As imperial powers such as Britain once did in order to justify their sovereignty claims to Antarctica, the members of the ATS today draw upon a benevolent paternalism to argue that they are conducting "science for the good of humanity" in seeking to understand climate change and potentially mitigate the worst of its consequences. Because of the threat of climate change, any challenge to the ATS, its members implicitly argue, represents a threat to humanity by disrupting and downgrading Antarctic science.[13] While the

[6] Ibid., xvii.

[7] Ibid., vi–xviii.

[8] Ibid., xiii.

[9] David G. Vaughan, "West Antarctic Ice Sheet Collapse—the Fall and Rise of a Paradigm," *Climate Change* 91 (2008): 65–79.

[10] Sheila Jasanoff, ed., *States of Knowledge: The Co-production of Science and Social Order* (London, 2004).

[11] Paul A. Berkman, *Science into Policy: Global Lessons from Antarctica* (San Diego, Calif., 2001).

[12] See, e.g., Simone Turchetti, Simon Naylor, Katrina Dean, and Martin Siegert, "On Thick Ice: Scientific Internationalism and Antarctic Affairs, 1957–1980," *History and Technology* 24 (2008): 351–76; Dean, Naylor, Turchetti, and Siegert, "Data in Antarctic Science and Politics," *Soc. Stud. Sci.* 34 (2008): 571–604; Naylor, Dean, and Siegert, "The IGY and the Ice Sheet: Surveying Antarctica," *J. Hist. Geogr.* 34 (2008): 574–95; Turchetti, Dean, Naylor, and Siegert, "Accidents and Opportunities: A History of the Radio Echo-Sounding of Antarctica, 1958–79," *Brit. J. Hist. Sci.* 41 (2008): 417–44.

[13] For early examples of the scientific defense of the ATS, see United Nations Organization, *Question of Antarctica: Study Requested under General Assembly Resolution 38/77; Report of the Secretary General* (New York, 1984).

word "imperialism" may not be appropriate in describing contemporary Antarctic politics, the ATS remains deliberately exclusive. Importantly, the ATS makes an explicit connection between science and politics in its requirement that a country be conducting scientific research in Antarctica before it is granted a place at the political table.[14] This requirement is expensive and works to exclude poorer nations from the science and politics of the Antarctic continent. South Africa, for example, is the only African country to be a member.

While the article takes a broad approach to the relationship between politics and climate change in Antarctica since the IGY, it pays particular attention to the work of the BAS. As it is the British government's official institution for Antarctic research, the history of the BAS highlights the development of climate change research, perceptual change, and political continuity over a substantial period of time. Before it took its present name on January 1, 1962, the BAS was known as Operation Tabarin and the Falkland Islands Dependencies Survey.[15] The survey boasts the longest history of continuous research on the Antarctic continent, dating back to 1944.[16] Over this time, the BAS has played a leading role in the conduct of climate change research in Antarctica, and its scientific agenda has changed to reflect broader preoccupations with climate change. Because of its intrinsic connections with British Antarctic policy, the history of the BAS also helps to demonstrate the fundamental continuity of political structures in Antarctica from before the IGY up to the present.

FROZEN EMPIRES

By the early 1950s, Antarctica was the stage for a number of geopolitical rivalries. In the Antarctic Peninsula region the claims of Great Britain, Argentina, and Chile overlapped, and an active three-way sovereignty dispute had been raging since the late 1930s.[17] In other parts of the continent four other countries—New Zealand, France, Australia, and Norway—asserted territorial claims, leaving only West Antarctica unclaimed. The United States and the Soviet Union refused to recognize any claims to Antarctic sovereignty.[18] Instead both superpowers reserved their own right to claim any part of the continent, and by the mid-1950s it looked increasingly likely that the tensions of the cold war would spread to Antarctica. Almost all of these claims and assertions of rights could be described as "imperial" in one sense or another, and they tell us much about the expansionist mentality of the first half of the twentieth century. The continent of Antarctica offered nations with this imperial mentality an opportunity to "conquer nature" in one of its most extreme forms.[19]

As various countries sought to promote their competing claims and rights to Antarctica, they used their self-proclaimed ability to understand and control the Antarctic

[14] See article 9 of the treaty, the full text of which is available on the U.S. Department of State Web site, http://www.state.gov/www/global/oes/oceans/antarctic_treaty_1959.html (accessed 4 January 2011).

[15] For an early history of the BAS, see Vivian Fuchs, *Of Ice and Men: The Story of the British Antarctic Survey, 1943–73* (Oswestry, 1982).

[16] Klaus Dodds, *Pink Ice: Britain and the South Atlantic Empire* (London, 2002).

[17] Adrian Howkins, "Icy Relations: The Emergence of South American Antarctica during the Second World War," *Polar Record* 42 (2006): 153–65.

[18] Frank Klotz, *America on the Ice: Antarctic Policy Issues* (Washington, D.C., 1990).

[19] For a fuller discussion, see Adrian Howkins, "Appropriating Space: Antarctic Imperialism and the Mentality of Settler Colonialism," in *Making Settler Colonial Space: Perspectives on Race, Place and Identity,* eds. Tracey Banivanua-Mar and Penelope Edmonds (Basingstoke, 2010), 29–52.

environment to support their respective cases. These assertions can be thought of as a form of "environmental authority" that both facilitated and legitimized political influence in the continent.[20] In making their case against Chile and Argentina over sovereignty in the Antarctic Peninsula, the British, for example, drew upon their research into whaling and their attempts to regulate the industry.[21] In a broadcast address to the people of the Falkland Islands made in 1948, the governor, Sir Miles Clifford, argued that because of its ability to regulate the whaling industry, Britain's motive for claiming Antarctica was purely unselfish: "to conserve the harvest of these seas for the benefit of mankind as a whole."[22] The origination of the BAS during the Second World War can also be seen as an assertion of environmental authority, alongside an attempt to occupy the region.[23] In opposition to such claims, the Argentines and Chileans built their own form of environmental authority into their case for ownership, arguing, for example, that the geological extension of the Andes into the Antarctic Peninsula gave them rights to sovereignty.[24] Interestingly, one Chilean author, perhaps with his tongue in his cheek, even brought a vision of climate change into his positive analysis of the future value of the Antarctic continent: "With the centuries that polar spring could return. And maybe the great-great-grandchildren of our great-great-grandchildren will be able to drink tea comfortably under giant palms, in order discreetly to escape the polar heat."[25]

The contest for environmental authority in Antarctica culminated in the IGY of 1957–8.[26] The IGY had something of a Jekyll-and-Hyde personality.[27] On the one hand, this major international research effort, which had a particular focus on Antarctica, sought to coordinate geophysical research around the world in order to improve scientific understanding of Earth's natural systems. It involved unprecedented levels of international cooperation, even between such bitter rivals as the United States and the Soviet Union. On the other hand, the IGY witnessed continued geopolitical competition as rival powers sought to outdo each other in the quality and logistical complexity of their scientific contributions. In Antarctica, the United States and the Soviet Union competed to establish bases in the most remote regions, and the British Commonwealth completed the first land-based crossing of the whole continent.[28]

[20] The use of "environmental authority" was common throughout histories of European imperialism and the cold war. For a recent overview of this growing field of scholarship from the British perspective, see William Beinart and Lotte Hughes, *Environment and Empire* (Oxford, 2007). For a discussion of the relationship between science and politics during the cold war, see Ronald E. Doel and Kristine C. Harper, "Prometheus Unleashed: Science as a Diplomatic Weapon in the Lyndon B. Johnson Administration," *Osiris* 21 (2006): 66–85.

[21] Ann Savours and Margaret Slythe, *The Voyages of the Discovery: The Illustrated History of Scott's Ship* (London, 2001).

[22] Clifford, "Broadcast Address by His Excellency the Governor," February 22, 1948, MISS. Brit Emp.s 517 4/1, "Falkland Islands 1946–1957: Clifford," Rhodes House Library, Oxford.

[23] Dodds, *Pink Ice* (cit. n. 16).

[24] Enrique Cordovez Madariaga, *La Antártida sudamericana* (Santiago, 1945).

[25] Juan Bardina, quoted in Oscar Pinochet de la Barra, *La Antártica chilena* (Santiago, 1948), on 166.

[26] Simon Naylor, Martin Siegert, Katrina Dean, and Simone Turchetti, "Science, Geopolitics and the Governance of Antarctica," *Nature Geoscience* 1 (2008): 143–5.

[27] Adrian Howkins, "Science, Environment, and Sovereignty: The International Geophysical Year in the Antarctic Peninsula," in *Globalizing Polar Science: Reconsidering the Social and Intellectual Implications of the International Polar and Geophysical Years*, eds. Roger D. Launius, James Rodger Fleming, and David H. DeVorkin (New York, 2010), 245–64.

[28] For a discussion of the U.S.-Soviet rivalry, see Oscar Pinochet de la Barra, *Medio siglo de recuerdos antárticos: Memorias* (Santiago, 1994). For an analysis of the trans-Antarctic expedition,

When the Soviet Union discovered that the United States had decided to build a base at the South Pole, for example, Soviet planners decided to switch their attentions to the so-called Pole of Relative Inaccessibility, the place on the continent furthest from any coastline. In undertaking such a difficult task the Soviet Union hoped to demonstrate its logistical superiority to the United States and all other Antarctic powers. Despite their occasionally farcical nature, such feats might be thought of as assertions of environmental authority similar to those made earlier by the British Empire.

Despite their complex motivations, the scientific achievements of the IGY were impressive, both in Antarctica and around the world.[29] At the end of the eighteen-month enterprise, a Russian scientist boldly declared: "Mankind has learned more about Antarctica in the last three or four years than in all the one hundred thirty years since the day of discovery."[30] The scientific results of the IGY radically changed environmental perceptions of the Southern Continent. These changed perceptions had important political consequences, particularly the realization that the short-to-medium-term economic potential of Antarctica was limited.[31] Despite the unprecedented levels of scientific activity in Antarctica, no economically worthwhile mineral deposits were found. On the contrary, the IGY confirmed the inhospitable nature of the Antarctic environment: a place with extraordinarily cold temperatures, strong winds, and deep ice. Fantastical dreams of Antarctica as a frozen El Dorado, brimming with mineral resources, were shattered, at least temporarily. This negative discovery served to weaken the attachment of countries such as Britain and the United States to exclusive political sovereignty, opening the way for a limited internationalization of the continent.

In 1956, India had suggested that Antarctica should become an international territory under the authority of the United Nations.[32] All seven claimant countries rejected this idea as a violation of their sovereign rights. Instead, in the aftermath of the IGY, the twelve nations that had participated in research in Antarctica met in Washington, D.C., to negotiate the Antarctic Treaty between themselves.[33] Signed on December 1, 1959, the Antarctic Treaty suspended all sovereignty claims to the continent and sought to promote ongoing scientific cooperation in Antarctica as begun during the IGY, thereby creating what came to be known as "a continent for science."[34] The preamble to the treaty expresses the conviction that a continuation of the IGY's policy of scientific freedom "accords with the interests of science and the progress of all mankind."[35] The framers of the treaty deliberately sought to present their work as a result of the overflowing scientific idealism fostered by the IGY. International disagreements could, they implied, be solved through the pursuit of a common scientific purpose, and this claim formed the basis of what might be termed the "official narrative" of the IGY and the Antarctic Treaty.

In reality, the origins of the Antarctic Treaty were less idealistic. The ATS did not develop as an alternative to great-power politics, but was constructed by the great

---

see Klaus Dodds, "The Great Trek: New Zealand and the British/Commonwealth 1955–58 Trans-Antarctic Expedition," *Journal of Imperial and Commonwealth History* 33 (2005): 93–114.

[29] Launius, Fleming, and DeVorkin, *Globalizing Polar Science* (cit. n. 27).

[30] Quoted in Philip W. Quigg, *A Pole Apart: The Emerging Issue of Antarctica* (New York, 1983), 40.

[31] Howkins, "Science, Environment, and Sovereignty" (cit. n. 27).

[32] Adrian Howkins, "Defending Polar Empire: Opposition to India's Proposal to Raise the 'Antarctic Question' at the United Nations in 1956," *Polar Record* 44 (2008): 35–44.

[33] Peter Beck, *The International Politics of Antarctica* (London, 1986).

[34] Richard S. Lewis, *A Continent for Science: The Antarctic Adventure* (New York, 1965).

[35] See the text of the treaty on the State Department Web site (cit. n. 14).

powers of the late 1950s to preserve their political interests in the continent.[36] Imperial claims to Antarctica remain "frozen" by article 4 of the Antarctic Treaty, and the exclusive political arrangements of the ATS might themselves be thought of as a form of "frozen empire" in a postcolonial world. Weary of the conflict with Argentina and Chile, and increasingly aware of the limited economic potential of Antarctica, British policy makers pushed for a solution to the "Antarctic problem" that would preserve their interest without the need for them to exercise formal control.[37] The United States and the Soviet Union could largely agree that they did not want the tensions of the cold war spreading to Antarctica, although both superpowers wanted to retain their political interests in the region. In order to achieve their objectives, the principal architects of the treaty—most notably Britain and the United States—used the rhetoric of scientific cooperation to bring their disparate interests together. Science also offered the signatory powers a very useful means of retaining their political interests to the exclusion of other nations. Article 9 explicitly stated that a country could only become a full consultative member of the treaty if it could demonstrate "its interest in Antarctica by conducting substantial research activity there, such as the establishment of a scientific station or the dispatch of a scientific expedition."[38] In private, British officials stated explicitly that the purpose of this provision was to exclude the "troublemakers," meaning the newly independent third world states and Soviet satellites.[39]

With the addition of a number of supplementary treaties and measures, including the 1991 Madrid Environmental Protocol, which introduced a number of conservation measures such as the prohibition of mineral prospecting for at least fifty years, the ATS remains the governing structure of the Antarctic continent. Arguably, in no other part of the world does scientific research enjoy such status and political influence as it does in Antarctica. With the support of SCAR—perhaps not coincidentally housed at the Scott Polar Research Institute in Cambridge, England—the ATS continues to draw its political legitimacy from the claim that it encourages "science for the good of humanity." Over time, the ATS has increasingly come to focus its scientific research efforts on the question of climate change.[40] As the threat of climate change has grown, fears of the potential consequences have further reinforced the ATS by strengthening its claims to be acting in the best interests of all humanity.

A GLOBAL LABORATORY

By promoting political stability and international cooperation, the ATS has done much to advance scientific research in Antarctica. With the exception of Hans Ahl-

[36] Adrian Howkins, "Reluctant Collaborators: Argentina and Chile in Antarctica during the IGY," *J. Hist. Geogr.* 34 (2008): 596–617.

[37] The expression "Antarctic problem" comes from Eric William Hunter Christie, *The Antarctic Problem: An Historical and Political Study* (London, 1951). The argument for the preservation of British interests is similar to that set out in W. R. Louis and R. Robinson, "The Imperialism of Decolonization," *Journal of Imperial and Commonwealth History* 22 (1994): 462–511.

[38] Treaty text (cit. n. 14).

[39] British Embassy, Washington, to Foreign Office, 14 January 1958, FO 371/131905, National Archives, Kew, London.

[40] SCAR paper of May 1998 on global change and the Antarctic (AT29051998C), in W. M. Bush, *Antarctica and International Law: A Collection of Inter-state and National Documents* (Dobbs Ferry, N.Y., 2003), 21–6.

mann and a handful of others, scientists had not considered Antarctica's role as a laboratory for global environmental science prior to the IGY of 1957–8.[41] It was the global nature of IGY science that was truly innovative, and the late 1950s marked a distinctive turning point in the history of Antarctic research, linking it to research in the rest of the world.[42] Writing in 1962, Phillip Law, Australia's leading Antarctic specialist, described the continent as an "untapped reservoir of knowledge, and a unique uncontaminated laboratory."[43]

Antarctica's role as a laboratory has been particularly important in the field of climate research. Meteorological research during the IGY filled in many of the theoretical gaps in the understanding of Antarctica's weather and climate and its influence on global atmospheric systems.[44] This research was internationally coordinated and overseen by such prestigious scientists as Harry Wexler, director of Meteorological Research at the U.S. Weather Bureau and chief scientist of the U.S. National Committee for the IGY Antarctic Program. Data from across the continent fed into "Weather Central" at the U.S. Little America base, where it was processed by a team of international meteorologists.[45] Atmospheric science played an important role in encouraging international cooperation during the IGY and in creating an increasingly global perspective within the earth sciences.

In retrospect, among the most important projects conducted in Antarctica as part of the IGY were Keeling's CO_2 measurements at the South Pole Station.[46] Over the two-year cycle of measurements at the pole, CO_2 readings demonstrated a perceptible upward trend that would be repeated at Keeling's other test site, in Mauna Loa, Hawaii (where measurements continued past the IGY). Shortly after the end of the IGY Keeling published his results from Antarctica in *Tellus*, bringing his findings to the wider scientific community.[47] Building on the work of Svante Arrhenius, Guy Stewart Callendar, Roger Revelle, and others, Keeling believed that rising atmospheric CO_2 levels would cause a rise in world temperatures through the so-called greenhouse effect.[48] By showing an increase in atmospheric greenhouse gases, Keeling's work added urgency to what had until then been a largely theoretical problem. If CO_2 did indeed contribute to global warming, then these measurements from Antarctica suggested that the world might be heating up. Over the next fifty years, scientists would continue to monitor atmospheric CO_2 levels in Antarctica, which continued to

[41] Peder Roberts, "När polarforskaren blev professionall, 1900–1960" [The professionalization of polar research, 1900–1960], *Ymer*, 2009, 129–50. See also Sverker Sörlin, "Narratives and Counter-narratives of Climate Change: North Atlantic Glaciology and Meteorology, c.1930–1955," *J. Hist. Geogr.* 35 (2009): 237–55; Sörlin in this volume.

[42] James Rodger Fleming, "Polar and Global Meteorology in the Career of Harry Wexler, 1933–1962," in Launius, Fleming, and DeVorkin, *Globalizing Polar Science* (cit. n. 27), 225–41.

[43] Quoted in Lorne K. Kriwoken, Julia Jabour-Green, Alan D. Hemmings, and Stuart Harris, eds., *Looking South: Australia's Antarctic Agenda* (Leichhardt, 2007), on 168.

[44] Adrian Howkins, "Political Meteorology: Weather, Climate and the Contest for Antarctic Sovereignty, 1939–1959," *History of Meteorology* 4 (2008): 27–40.

[45] Dian Olson Belanger, *Deep Freeze: The United States, the International Geophysical Year, and the Origins of Antarctica's Age of Science* (Boulder, Colo., 2006).

[46] See Maria Bohn in this volume; Spencer R. Weart, *The Discovery of Global Warming* (Cambridge, Mass., 2008).

[47] Keeling, "Concentration and Isotopic Abundances" (cit. n. 4).

[48] James Rodger Fleming, *Historical Perspectives on Climate Change* (New York, 1998); Fleming, *The Callendar Effect: The Life and Times of Guy Stewart Callendar (1898–1964), the Scientist Who Established the Carbon Dioxide Theory of Climate Change* (Boston, 2007).

rise.[49] While these measurements alone could not demonstrate a causal connection between CO_2 and warming temperature trends, they did suggest that human activity was changing the atmosphere through the burning of fossil fuels.

Since the early 1960s, ice-core science in Antarctica has helped to demonstrate the connection between atmospheric CO_2 and temperature. Antarctica's ice provides scientists with a frozen archive of Earth's climatic past. "Analysis of ice cores," a 1987 British scientific report noted, "yields evidence on climate, on environmental factors that can induce climate change, on the size of the ice sheet and the area of sea ice surrounding it, and in the upper part, evidence for the impact of Man."[50] Thus, Antarctic ice offered a key site for understanding global climate change. The theory and practice of Antarctic ice coring was international and collaborative from its very beginning. The Antarctic Treaty and SCAR provided a ready-made framework for international collaboration in this field and in climate research more generally.[51] The logistics of drilling for prolonged periods of time in some of the world's remotest places made the collection of ice cores dauntingly expensive, and cooperation through the ATS sometimes allowed these costs to be shared.[52]

Techniques for deep ice coring were initially pioneered at the opposite end of the planet: at Camp Century in Greenland, where drilling began in 1961.[53] One of the major scientific-technical problems to be solved at this early stage was how to use the ice to measure changes in Earth's past climate. The initial solution to this problem was pioneered by the Danish glaciologist Willi Dansgaard, who looked at the ratio of the isotopes ^{18}O to ^{16}O in the ice: the more ^{18}O, he proposed, the warmer the temperature when the ice was formed.[54] Scientists drilled the first significant Antarctic ice core in 1968 at Byrd Station in West Antarctica. Less than a year later, the publication of their results demonstrated a strong continuity with results from Greenland, suggesting that climate operated as a global system and regional conditions alone were unlikely to cause dramatic warming or cooling on a regional scale.[55]

By the late 1980s, new techniques had been developed for extracting the air from the tiny bubbles contained in the ancient ice.[56] Scientists carefully cleaned ice samples, crushed them in a vacuum, and then quickly measured the gas. Crucially, this allowed for measurements of the CO_2 content of the ancient air. This coincided with new developments in the technology and practice of extracting ice cores. In the late 1980s, a combined Russian-French team working at the Soviet Vostok base in East Antarctica produced an ice core that stretched back 400,000 years. The Vostok ice core covered four glacial cycles, and tests of the air it contained revealed that CO_2

[49] Turner et al., *Antarctic Climate Change and the Environment* (cit. n. 5), 213.

[50] "Strategy for Research in the Ice and Climate Division," undated, 60/98/1454, BASA. For a wider discussion of the history of ice-core research, see Richard B. Alley, *The Two-Mile Time Machine: Ice-Cores, Abrupt Climate Change, and Our Future* (Princeton, N.J., 2000).

[51] See, e.g., "SCAR Preliminary Report of Its Planning on Global Change Research in Antarctica at Bremerhaven, 18–20 September 1991" (AT20091991), in Bush, *Antarctica and International Law* (cit. n. 40), 22–3.

[52] Paul A. Mayewski and Frank White, *The Ice Chronicles: The Quest to Understand Global Climate Change* (Lebanon, N.H., 2002).

[53] Weart, *Discovery of Global Warming* (cit. n. 46), on 70.

[54] Ibid., 71.

[55] Ibid.

[56] Ibid.

levels had fallen during these cycles and risen during interglacial periods. In other words, CO_2 in the atmosphere rose during warm periods in Earth's history. These results demonstrated a correlation between CO_2 and climate change. Although the causal relationship remained unclear, these results and others like them added strong evidence to the suggestion that greenhouse gas emissions caused by human activities might warm the atmosphere.[57]

Perhaps the most important recent development in Antarctica's contribution to climate change science has been connected to the field of mathematical climate modeling.[58] Mathematical climate modeling became an important part of Antarctic research in the late 1970s.[59] With the development of its Ice and Climate Division in 1987, the BAS emphasized the importance of mathematical modeling to its work, although at the time many modeling centers acknowledged the poor performance of their models for Antarctica.[60] Since models have to be "calibrated" against past climates, modelers have worked closely with ice-core researchers. Mathematical climate modeling has become absolutely central to scientific understanding of climate change both in Antarctica and globally,[61] and this importance looks set to continue into the twenty-first century.[62]

Despite Antarctica's importance to climate change research on a global scale, understanding how global trends have affected the continent of Antarctica itself has not proved easy. From a historical perspective, the most significant problem in understanding temperature trends in Antarctica has been the relative scarcity of data. The BAS *Annual Report* from 1983–4 cited a lack of data and noted, "It is not possible to say at this stage whether recent apparent warming trends are more than part of the natural variability with a long period or Man's influence on the environment."[63] In order to demonstrate significant climatic trends, it is necessary to have at least several decades of meteorological data.[64] Until the IGY, the scant meteorological data that existed for Antarctica came from occasional expeditions and a very small number of scientific stations.[65] The locations of these stations were chosen primarily for geopolitical purposes, and meteorological considerations barely factored into the decisions. As a consequence, meteorological observing conditions were often flawed, and local conditions overrode large-scale ones. During the IGY systematic weather observations began to be taken.[66] But only a handful of stations have continuously collected meteorological data in Antarctica since then. Therefore, although SCAR, the Antarctic Treaty, and the World Meteorological Organization (WMO) have been successful in facilitating the exchange of the meteorological data that do exist, until

[57] More recent ice cores have gone back 800,000 years; see Turner et al., *Antarctic Climate Change and the Environment* (cit. n. 5), 126.

[58] J. C. King and J. Turner, *Antarctic Meteorology and Climatology* (Cambridge, 1997).

[59] BAS, *Annual Report, 1978–79* (Cambridge, 1979), 14.

[60] "Strategy for Research" (cit. n. 50).

[61] Clark A. Miller and Paul N. Edwards, *Changing the Atmosphere: Expert Knowledge and Environmental Governance* (Cambridge, Mass., 2001).

[62] Thomas J. Bracegirdle, William M. Connolley, and John Turner, "Antarctic Climate Change over the Twenty First Century," *J. Geoph. Res.—Atmospheres* 113 (2008), doi:10.1029/2007JD008933, http://www.agu.org/journals/jd (accessed 5 January 2011).

[63] BAS, *Annual Report, 1983–84* (Cambridge, 1984), 39.

[64] See Sörlin in this volume.

[65] Howkins, "Political Meteorology" (cit. n. 44).

[66] Harry Wexler, "Seasonal and Other Temperature Changes in the Antarctic Atmosphere," *Quarterly Journal of the Royal Meteorological Society* 86 (1959): 196–208.

recently there have been barely enough data to test models or theories, especially given the vast size of Antarctica.[67]

The accumulation of new meteorological data over recent years has made analysis more feasible. These data have come from new stations, unmanned observing systems, and satellites to produce records from a much wider area.[68] The recently published SCAR report *Antarctic Climate Change and the Environment* draws upon these developments to present a comprehensive overview of the current state of scientific understanding of the Antarctic climate.[69] Central to the report is the idea that over the last thirty years the hole in the ozone layer over Antarctica has "shield[ed] the continent from much of the effect of global warming."[70] But it also makes clear that this situation will not last, and that as the ozone layer repairs over future decades, the effects of climate change over Antarctica will become more pronounced. The report summarizes the recorded temperature trends since the IGY as follows:

> Surface temperature trends show significant warming across the Antarctic Peninsula and to a lesser extent West Antarctica since the early 1950s, with little change across the rest of the continent. The latest warming trends occur on the western and northern parts of the Antarctic Peninsula. There the Faraday/Varnadsky Station has experienced the largest statistically significant (<5% level) trend of +0.53°C per decade for the period 1951–2006. The 100-year record from Orcadas on Laurie Island, South Orkney Islands, shows a warming of +0.20°C per decade. The western Peninsula warming has been largest during the winter, winter temperatures at Faraday/Vernadsky increasing by +1.03°C per decade from 1950–2006. There is a high correlation during the winter between sea ice extent and surface temperatures, suggesting more sea ice during the 1950s–1960s and reduction since then. This warming may reflect natural variability.[71]

The report makes clear that within the unique conditions of the Antarctic continent, such temperature trends are consistent with global warming trends.

Antarctica's contribution to the science of climate change has taken place simultaneously with research in other parts of the world. It is impossible to quantify the influence of Antarctic research on the "discovery of global warming," or to say what would have happened if climatic research had not taken place in Antarctica.[72] But it is fair to conclude that research in Antarctica has played—and continues to play—a significant role in scientific understanding of global climate change. In a sense climate change research has become self-perpetuating: research in Antarctica has contributed to a growing understanding of climate change around the world, which in turn has led to more climate research being done in Antarctica.[73]

[67] For an example of information exchange within the context of the ATS, see BAS, *Annual Report, 1975–76* (Cambridge, 1976).

[68] King and Turner, *Antarctic Meteorology and Climatology* (cit. n. 58).

[69] Turner et al., *Antarctic Climate Change and the Environment* (cit. n. 5).

[70] Ibid., xiii.

[71] Ibid., xvi.

[72] Weart, *Discovery of Global Warming* (cit. n. 46); Fleming, *Historical Perspectives on Climate Change* (cit. n. 48).

[73] E.g., the establishment of the WMO's World Climate Program in 1980 indirectly stimulated the BAS's growing emphasis on climate research; see A. S. Loughton, Note on Climate Research, December 15, 1980, 158/70/19, Committees, N.E.R.C. Climate Research Policy, BASA.

CLIMATE ANXIETY IN ANTARCTICA

Early in 2002, the 3,250-square-kilometer Larsen B Ice Shelf, on the east coast of the Antarctic Peninsula, collapsed and surged into the sea, where it began to melt. The event made headlines around the world as an example of the consequences of global warming. Under the title "Large Ice Shelf in Antarctica Disintegrates at Great Speed," the *New York Times* reported that "many experts" attributed the collapse to global warming, as they ran out of other explanations.[74] Environmental commentators argued that this collapse was a foretaste of things to come as melting ice in Antarctica could have devastating consequences for the whole planet. The fears generated by the Larsen B collapse highlighted the central position that Antarctica had come to assume in discussions about the consequences of global warming. It also demonstrated how perceptions of the Antarctic environment had shifted dramatically over the previous fifty years. On the eve of the IGY, many people still considered Antarctica to be a virtually impenetrable wilderness and a symbol of nature's enduring strength; by the first decade of the twenty-first century, the continent had come to be seen as a dangerously vulnerable environment.

The focus on Antarctica as a victim of anthropogenic climate change is a relatively recent phenomenon. In 1974, with the creation of an Atmospheric Sciences Division in the BAS, the survey's annual report made no direct mention of climate change as an important area of study.[75] Instead there was an admission that climate research had not played a major role in the BAS's recent scientific programs. Thirteen years later, with the creation of an Ice and Climate Division to replace Atmospheric Science, there was a pronounced focus on the threat of global warming.[76] A noticeable shift in emphasis took place at the beginning of the 1990s. In the 1989–90 BAS handbook, the opening lines describing the Ice and Climate Division stated,

> All branches of Antarctic research must take account of the ice environment. The vast ice sheet, through its heat exchange with the atmosphere and its mass exchange with the ocean, influences conditions over the entire Earth. It is the principal factor controlling sea level around the coasts of Britain. Antarctica provides the only contemporary example of an ice sheet in an extreme glacial period.[77]

One year later, the same opening lines had become significantly more threatening:

> One of the potential disasters facing mankind over the next few centuries is an increase in global sea level included [*sic*] by the "greenhouse effect." Some general circulation models (GCMs) of the atmosphere predict that warming will be enhanced at higher latitudes. This suggests that polar ice sheets may be especially vulnerable to a climate change that could cause significant melting.[78]

Such a change in language suggests that a "tipping point" in scientific perceptions of the Antarctic continent was reached around 1990, and this is consistent with a more

---

[74] Andrew Revkin, "Large Ice Shelf in Antarctica Disintegrates at Great Speed," *New York Times*, March 20, 2002.

[75] BAS, *Annual Report, 1972–75* (Cambridge, 1975).

[76] "Strategy for Research" (cit. n. 50).

[77] BAS, *Handbook, 1989–90* (Cambridge, 1989), 9.

[78] BAS, *Handbook, 1990–91* (Cambridge, 1990), 9.

general increase in concerns about global climate change within the scientific community at this time.[79]

As the BAS handbooks made clear, melting ice in Antarctica mattered to the global community not because people cared particularly about what happened in some remote corner of the Weddell Sea, but because what happens in Antarctica might have consequences for the rest of the world. The most obvious threat is that a warming climate might melt the Antarctic ice and raise sea levels throughout the world. In another early statement of the relevance of Antarctic research, a BAS report from 1977 noted, "If the people of Britain have reason to be interested in storm surges and flooding of the Thames, they have reason to be interested in Antarctica. For the Antarctic ice sheet is the principal factor controlling sea-level on the coasts of Britain and throughout the world."[80] Experiments conducted during the IGY had demonstrated that the ice across much of Antarctica was several kilometers thick, causing estimates of the world's water budget to be revised upward significantly.[81] One scientist calculated that if all the Antarctic ice were to melt, sea levels around the world would rise by 59 meters.[82]

Almost all ice in Antarctica is lost though calving and disintegration at the continent's edges (as in the case of the Larsen B Ice Shelf), rather than melting. The reality, therefore, is more complicated than a simple "warming equals melting" scenario. Scientists have long realized that the key concept is glacial mass balance: how much ice goes into a system and how much goes out. By 1981 the BAS Atmospheric Research Division had teams working under glaciologist Charles Swithenbank on glacial mass balance and ice flow.[83] Some scientists have suggested that a slight warming over Antarctica might cause an increase in precipitation, leading to more ice formation, a net increase in mass balance, and potentially falling sea levels.[84] But if the rate of ice loss at the continent's edge increases, perhaps because of warming sea temperatures, without a significant increase in the precipitation over Antarctica, then it is indeed possible that there will be a net loss in the glacial mass balance of Antarctica, and as a consequence sea levels may rise.

The most dramatic scenarios for ice loss in Antarctica involve entire ice sheets breaking off. In the early 1970s, scientists began to focus on the West Antarctic Ice Sheet, and this has proven to be one of the most enduring subjects of study in Antarctic research.[85] Unlike the ice in East Antarctica, which rests firmly on the continental landmass, the West Antarctic Ice Sheet is only partially anchored to land. One of the leading scientists to focus on the potential consequences should the West Antarctic Ice Sheet break off was John Mercer, a glaciologist at Ohio State University.[86] He presented a doomsday scenario in which the entire West Antarctic Ice Sheet broke away after melting had destabilized its edges. Mercer predicted that the quantity of

[79] Weart, *Discovery of Global Warming* (cit. n. 46).
[80] BAS, "Strategic Research Requirements in the Antarctic," policy document, January 13, 1977, AD3/2/121/70/08 Administration NERC Strategic Research, BASA.
[81] Lewis, *Continent for Science* (cit. n. 34).
[82] Quoted in "Strategic Research Requirements in the Antarctic" (cit. n. 80).
[83] BAS to Natural Environment Research Council (NERC), 22 January 1981, AD3/2/121/70/08 Administration NERC Strategic Research, BASA.
[84] Briefing notes for a BBC World Service interview, June 2, 1989, 60/98/1735, Documents: Research BAS, Ice and Climate, BASA.
[85] Weart, *Discovery of Global Warming* (cit. n. 46).
[86] Ibid.

water contained in the West Antarctic Ice Sheet would be enough to raise global sea levels by around 5 meters, causing the flooding of coastal cities around the world.[87] Perhaps even more seriously, the presence of the West Antarctic Ice Sheet floating in the southern ocean could potentially lower global temperatures by increasing the albedo (reflectivity of light) over a significant proportion of Earth's surface and thereby reflecting more of the sun's radiation back into space. The possibility was considered that a breakaway West Antarctic Ice Sheet could even block the Drake Passage between South America and the Antarctica Peninsula, with major consequences for ocean circulation.[88] Policy makers, especially in the United States, focused on the national security implications of a potentially unstable West Antarctic Ice Sheet, and the severity of the predicted consequences encouraged them to spend large sums on research.[89]

In recent years, as various scientific disciplines have competed for funding and prestige, climate change research has assumed an increasing importance within Antarctic science. In Britain, the competition between scientific disciplines intensified when a more free-market approach to research funding was introduced in the 1990s. No longer were different disciplinary fields within the BAS guaranteed their own funds; instead, they had to compete openly with each other.[90] In this competition for funds, global relevance increasingly became one of the major criteria for funding awards, and climate change science has been ideally placed to claim relevance. The focus on climate change has spread into other disciplines. For example, marine biologists working in Antarctica now seek to ask questions about how climate change will influence the continent's biological life and to consider how changes in Antarctic biology are potential indicators of climate change.[91] These research agendas have become self-perpetuating and reinforce perceptions of Antarctica as a vulnerable environment.

MELTING EMPIRES?

In 1985, scientists working at Britain's Halley Base on the Weddell Sea observed a 40 percent increase in the size of the springtime hole in the ozone layer over Antarctica (compared to 1957).[92] In a letter to *Nature* published on May 16, 1985, BAS scientists J. C. Farman, B. G. Gardiner, and J. Shanklin argued that the hole's growth was caused by the large amounts of chlorofluoromethanes (mainly the halocarbons CFC-11 and CFC-12) released into the troposphere.[93] Since 1974, atmospheric chemists had been suggesting that the use of CFCs could damage the ozone layer, and scientific work in Antarctica later revealed that this was actually happening. Following corroboration by NASA satellites, the world community rapidly responded

[87] Vaughan, "West Antarctic Ice Sheet Collapse" (cit. n. 9).

[88] Kenzo Takano and Keiji Higuchi, "A Numerical Experiment on the General Circulation in the Global Ocean," in *Challenges from the Future: Proceedings of the International Future Research Conference*, ed. Nihon Mirae Gakkai (Tokyo, 1970), 361–3.

[89] Vaughan, "West Antarctic Ice Sheet Collapse" (cit. n. 9).

[90] See recent BAS five-year plans, AD2/2/26, BASA.

[91] Meredith Hooper, *The Ferocious Summer: Palmer's Penguins and the Warming of Antarctica* (London, 2007).

[92] BAS, *Annual Report, 1985–86* (Cambridge, 1986), 35.

[93] Farman, Gardiner, and Shanklin, "Large Losses of Total Ozone in Antarctica Reveal Seasonal C1Ox/NOx Interaction," *Nature* 315 (1985): 207–10.

and signed the Montreal Protocol in 1987, which dramatically reduced the production of CFCs. Although this response cannot be interpreted as an entirely altruistic policy, since the reduction of CFCs was a major economic opportunity, the history of the human response to the growing ozone hole has generally been presented as an environmental success story.[94] Recent research suggests that the ozone hole over Antarctica is no longer growing significantly, although ozone variation and projected recovery are both within the limits of error and variability and it will take many years to repair the damage already done. Scientists in turn predict that ozone recovery will have important consequences for the Antarctic climate.[95]

Among those most satisfied with the rapid action taken to save the ozone layer was the British prime minister at that time, Margaret Thatcher.[96] Following the Falklands War of 1982 between Britain and Argentina, Thatcher's Conservative government had increased the funding for the BAS in an effort to increase Britain's presence in the region.[97] She saw the "discovery" of the ozone hole as a vindication of her government's position:

> In the aftermath of the Falklands conflict we were able to strengthen Britain's presence in the South Atlantic by increasing our scientific effort. This paid off remarkably quickly in a totally unexpected way with the discovery by the British Antarctic Survey of the ozone "hole" over Antarctica in the austral spring. This brought home to the whole world the potentially dangerous changes in the environment which mankind's activities are bringing about and led to the first measures to control pollution on a global scale.[98]

The ability of British scientists to understand the Antarctic environment, Thatcher implied, justified Britain's leading position within the ATS and its continued occupation of the Falkland Islands. This reaction to the discovery of the ozone hole offers a salient reminder of the multiple political uses to which Antarctic science can be put.

At a national level, climate change science has been put to political use in order to win support for Antarctic research. Antarctic research institutions such as the BAS need to justify the relevance of their work to win national funding, not only in competition with non-Antarctic science such as medical research, but also against the threat of receiving no money at all. The threat of climate change provides Antarctic scientists and administrators with a useful tool to argue for the importance of their work to the "national interest" in making their case for funding. A 1977 BAS research document stressed the economic importance of climate change research in Antarctica to the rest of the world, hinting at possibilities for geoengineering: "Rapid technological progress has also made it possible to alter the atmospheric environment by design or accident with significant and possibly catastrophic economic and social implications."[99] More recently, Chris Rapley, head of the BAS from 1998 to 2007, has

[94] Richard Elliot Benedick, *Ozone Diplomacy* (Cambridge, Mass., 2007).

[95] Turner et al., *Antarctic Climate Change and the Environment* (cit. n. 5).

[96] Maureen Christie, *Ozone Layer: A Philosophy of Science Perspective* (Cambridge, 2001).

[97] BAS, *Annual Report, 1982–83* (Cambridge, 1983), 1.

[98] Margaret Thatcher, foreword, in G. E. Fogg, *A History of Antarctic Science* (Cambridge, 1992), xv.

[99] "Strategic Research Requirements in the Antarctic" (cit. n. 80); see also Matthias Dörries in this volume.

brought prominence to Antarctic science through his work as a high-profile climate change campaigner.[100]

A similar scenario has developed in Australia. Since the 1991 Madrid Protocol put an end to the potential for resource exploitation in Antarctica, at least in the medium term, the Australian Antarctica Division has increasingly turned to climate change research in Antarctica as a way of justifying its existence to the Australian people and the Australian government.[101] In addition to the generalized threat of melting ice leading to rising sea levels, the Australian Antarctic Division has been able to draw on the long-acknowledged connection between the weather and climate of Antarctica and those of Australia to argue that climate change in Antarctica could have a profound influence on Australia. In this way, Australian Antarctic scientists are able to draw upon climate anxieties in Australia itself to boost their case for Antarctic research.[102]

As noted above, the geopolitical rivalries of the 1940s and 1950s did not simply disappear with the signature of the Antarctic Treaty: all seven territorial claims to the continent remain "frozen" in a state of suspended animation, with the United States and Russia continuing to reserve rights to the whole continent. At an international level, scientific research in Antarctica offers rival powers an outlet for friendly competition—politics by other means.[103] The sense that good science promotes political interests is explicit in the BAS mission statement from 1997–8: "The mission of the BAS is to undertake a program of first class science through which an active and influential role can be sustained in the Antarctic region, giving the UK an authoritative voice in Antarctic affairs."[104] Despite the rhetoric of international scientific cooperation, in private British scientists and politicians continued to refer to their supposed scientific partners in Antarctica as "competitors."[105] Given the centrality that climate change science has assumed in the world of science and politics, it is particularly important to be seen to be leading this field. For Argentina and Britain, the two countries with the longest series of meteorological records from Antarctica, it does their political claims no harm at all when data from their stations are prominently displayed and analyzed in scientific papers on Antarctic climate change, such as SCAR's *Antarctic Climate Change and the Environment.*

Since the signature of the 1959 Antarctic Treaty, preexisting national interests in Antarctica have largely been preserved within the ATS. The great strength of the ATS is that is has almost always been in the interests of the member states to preserve and strengthen the system in order to preserve and strengthen their individual interests in Antarctica. It is at this global level, where there are "insider" and "outsider"

---

[100] See "Interview with Former BAS Director, Professor Chris Rapley—Nunatak," on the BAS-NERC Web site, http://www.antarctica.ac.uk/indepth/nunatak/interview/director.php (accessed 29 September 2009).

[101] Aynsley Kellow, "A Caution on the Benefits of Research: Australia, Antarctica and Climate Change," in Kriwoken et al., *Looking South* (cit. n. 43), 165–75.

[102] See Ruth Morgan in this volume.

[103] Aant Elzinga, "Antarctica: The Construction of a Continent by and for Science," in *Denationalizing Science: The Contexts of International Scientific Practice*, eds. Elisabeth Crawford, Terry Shinn, and Sverker Sörlin (Dordrecht, 1993), 73–106.

[104] BAS, *Handbook, 1997–98* (Cambridge, 1997), AD 2/2/24, BASA.

[105] Briefing note for the director of the BAS, E. M. Morris, March 22, 1993, 60/98/1454 Ice and Climate Research Policy Strategy Document, BASA.

nations, that the political implications of climate change science in Antarctica are perhaps most disturbing. Although the number of consultative parties to the Antarctic Treaty has increased from the initial twelve to twenty-eight today, and nineteen other nations have acceded to the treaty as nonvoting parties, the ATS continues to be criticized by some as an "exclusive club."[106] Following India's accession to the treaty in 1983, Malaysia played a leading role in challenging the exclusive nature of the ATS on behalf of the outsider countries, particularly in relation to the perceived unfairness of the minerals negotiations taking place among the member states during the 1980s. The signature of the 1991 Madrid Environmental Protocol significantly reduced criticism of the treaty by prohibiting all activities related to mineral exploitation for at least fifty years. Nevertheless, the ATS continues to exclude a large number of poor nations through the requirement that a country be conducting substantial scientific research in order to fully participate in the political decision-making process.

Given the important role that Antarctica has come to assume in climate change debates, countries on the outside of the ATS are excluded from an important forum. It is often argued that the world's poorest countries will suffer the most from the consequences of a warming planet, and yet these countries often have no stake in Antarctic science and politics. Instead of direct participation in the ATS, nonmembers are left to rely on the goodwill and expertise of those member countries conducting research in Antarctica. This paternalistic arrangement has been reinforced by the predicted severity of the consequences of climate change. Member states can justify the exclusionary practices of the ATS on the basis that it offers the most efficient and perhaps least environmentally damaging means of conducting research for "the good of humanity." Meddling with the current state of affairs, the implicit argument goes, could jeopardize important research and put the whole of humanity at risk.

CONCLUSION

The ATS has proved remarkably successful in preserving peace, promoting science, and protecting the environment in the Southern Continent. The history of Antarctica suggests that the possibility of the continent once again becoming the object of international discord would be very real if the treaty should ever lapse or break down. If international conflict were to replace cooperation in Antarctica as a result of a challenge to the ATS, scientific research would almost certainly suffer. Given the contribution that Antarctic science has made to scientific understanding of climate change, and given the fundamental severity of the threat of climate change, any risk to Antarctic science can also be legitimately presented as a risk to the world much more broadly.

But at the same time as this article acknowledges the achievements of the ATS, it has also suggested that there is something a little disturbing in the fact that the threat of climate change is helping to preserve a political system that excludes large parts of the world. The history of the relationship between climate change and politics in Antarctica exemplifies contemporary assertions of environmental authority: the worse the threat of climate change becomes, the stronger the case of the ATS powers for looking after the interests of the world. The countries excluded from the science and politics of Antarctica by the requirement to be conducting expensive scientific

---

[106] See, e.g., the Malaysian reply in United Nations Organization, *Question of Antarctica* (cit. n. 13).

research are often the places where the consequences of climate change are predicted to be most severe. Such exclusion from the science and politics of Antarctica perpetuates a sense of international paternalism and prevents any solutions to the "problem" of climate change from ever being truly inclusive.[107]

This article has focused on the history of the relationship between climate change and politics in Antarctica precisely because of the uniqueness of this history. In few other parts of the world is the requirement to be conducting scientific research in order to participate in political discussions quite so explicit. But this does not mean that Antarctica should be seen as a "pole apart."[108] In fact, the history of climate change and politics in Antarctica can be seen in some ways as a microcosm of the relationship between climate change and politics at a global scale. In particular, the case of Antarctica highlights the difficulties involved in reconciling good science with political inclusiveness. While the requirement to be conducting scientific research is not a formal barrier to entry into international political negotiations on climate change, the conduct of significant climate change research clearly adds prestige and influence to a country's negotiating position. The benevolent paternalism exercised by those countries conducting global climate change research is very similar to that exercised by the members of the ATS. Expert knowledge about climate change—often discussed in terms of grants, scientific papers, and research institutions—can be thought of as a form of modern-day environmental authority. Within global debates about climate change, access to scientific knowledge creates hierarchical relationships. These relationships are self-reinforcing: economically powerful countries have money and expertise to invest in climate change research, while economically weaker countries do not. Without suggesting that it will be easy to conduct good science in a politically inclusive way, it is important to acknowledge that the threat of climate change might be reinforcing the political status quo not only in Antarctica but also at a global scale.

[107] On the "problem" of climate change, see Mike Hulme, *Why We Disagree about Climate Change: Understanding Controversy, Inaction and Opportunity* (Cambridge, 2009).
[108] Quigg, *A Pole Apart* (cit. n. 30).

The Politics of Atmospheric Sciences:
"Nuclear Winter" and Global Climate Change

by Matthias Dörries[*]

ABSTRACT

This article, by exploring the individual and collective trajectories that led to the "nuclear winter" debate, examines what originally drew scientists on both sides of the controversy to this research. Stepping back from the day-to-day action and looking at the larger cultural and political context of nuclear winter reveals sometimes surprising commonalities among actors who found themselves on opposing sides, as well as differences within the apparently coherent TTAPS group (the theory's originators: Richard P. Turco, Owen Brian Toon, Thomas P. Ackerman, James B. Pollack, and Carl Sagan). This story foreshadows that of recent research on anthropogenic climate change, which was substantially shaped during this—apparently tangential—cold war debate of the 1980s about research on the global effects of nuclear weapons.

INTRODUCTION

The phrase "nuclear winter" was coined in 1983 as a result of political pressure. During the Reagan years, the leadership of the National Aeronautics and Space Administration (NASA) was eager to avoid friction with Washington on sensitive issues such as nuclear conflict and was thus averse to seeing terms like "nuclear war" or "nuclear weapons" in the titles of its scientific publications.[1] Therefore the TTAPS group (an acronym for five researchers, Richard P. Turco, Owen Brian Toon, Thomas P. Ackerman, James B. Pollack, and Carl Sagan), some of whom were located at NASA Ames Research Center, searched for an alternative to "nuclear war" when they prepared their first article on this subject. Turco came up with "nuclear winter." In an article on the history of the nuclear winter theory in the political arena, Lawrence Badash has pointed out how much NASA's efforts "backfired, for the phenomenon became far more recognizable to a wide public" thanks to Carl Sagan's relentless efforts to disseminate it to a larger audience throughout the 1980s.[2]

The term "nuclear winter" linked nuclear and meteorological considerations, thus embedding climate concerns firmly within the sphere of military strategy and defense policy in ways that continue to reverberate in contemporary debates about anthropogenic climate change. This is not to suggest that these concerns were new in the

---

[*] IRIST, Université de Strasbourg, 7 rue de l'Université, 67000 Strasbourg, France; dorries@unistra.fr.

[1] Carl Sagan and Richard Turco, *A Path Where No Man Thought* (New York, 1990), 465.

[2] Badash, "Nuclear Winter: Scientists in the Political Arena," *Phys. Persp.* 3 (2001): 76–105, on 87.

1980s. There had been speculation that multiple nuclear explosions could lead to local, regional, or even worldwide climatic change from the first uses of the atomic bomb in 1945 and tests of the hydrogen bomb in the early 1950s. What was new, however, was that a single term that brought climate change and nuclear weapons together resonated within a growing and multifaceted scientific community by linking pressing cold war issues to worries over the planet's environmental future and stimulated a vigorous debate.

This happened because the TTAPS group had upset conventional military thinking by bringing the parameter of smoke into their calculations of the atmospheric consequences of nuclear war. Their results, first published in *Science* in 1983, indicated that particularly the black, sooty smoke from the burning of cities and industrial facilities in the aftermath of a major nuclear war would induce a significant diminution of global surface temperatures, leading to a global climate modification lasting several months, with disastrous consequences for people all over the planet.[3]

In the light of this theory, various research communities that had hitherto worked more or less independently, in organizations such as the Lawrence Livermore National Laboratory (LLNL), Los Alamos National Laboratory (LANL), the NASA research laboratories, the U.S. Weather Service and National Oceanic and Atmospheric Administration (NOAA), Department of Defense think tanks, and university astronomical and environmental science departments, all became interested in related and overlapping questions. These institutions had different missions, research traditions, and technical strengths, and they increasingly represented—especially after the later stages of the Vietnam War—diverging political orientations and worldviews. The nuclear winter theory was the culmination of a series of inquiries, originally speculative and partly secret, into the possibility of inadvertent anthropogenic climate modification via a powerful weapon. These studies had quietly been done for decades under the auspices of defense laboratories and think tanks, but now—conceived on the fringes of the civil NASA space research program and heavily promoted—they took place in public for the first time and became part of a politically charged environmental debate.

The intensity of the debate points to a profound transformation of the meaning of the term "environmental" in nuclear weapons studies during the 1960s and 1970s. The uses of the term in the defense studies of the 1950s and early 1960s pertained primarily to short-term side effects in geographical proximity to military or industrial activity; for example, the downwind propagation of radioactive particles toward civilian population centers.[4] If the term ever referred in this period to truly Earth-encompassing effects then it was mostly on a speculative, qualitative level, especially given the lack of global data. The outer-space tests of the late 1950s already brought with them a new geographical dimension, but it was only from the late 1960s that the use of "environmental" transgressed its original boundaries to include not only the whole Earth, but also an expanded time frame, now encompassing the systematic study of long-term consequences of human action over years or even decades. This epistemological double transformation was inseparable from contemporaneous

[3] R. P. Turco, O. B. Toon, T. P. Ackerman, J. B. Pollack, and Carl Sagan, "Nuclear Winter: Global Consequences of Multiple Nuclear Explosions," *Science* 222 (1983): 1283–92.

[4] For the link between military research and environmental sciences after 1945 and further references, see Ronald E. Doel, "Constituting the Postwar Earth Sciences: The Military's Influence on the Environmental Sciences in the USA after 1945," *Soc. Stud. Sci.* 33 (2003): 635–66.

political debates about the economic and political future of Earth and the threat of environmental disasters to long-term economic growth.[5]

Nuclear winter and anthropogenic climate change (through either nuclear weapons, cooling from dust, or heating from the production of carbon dioxide) thus emerged together as exciting and contested focal points for scientists from different backgrounds and institutions, military and civilian. Much energy was spent during the 1980s to defend both long-held beliefs and newly conceived ideas and theories, as environmental concerns came increasingly to be regarded as a threat to the economic and political status quo.

The history of nuclear winter has been largely neglected by historians. The only in-depth study is a recent book by Badash,[6] a thorough, carefully researched history of nuclear winter from its origins to the end of the cold war. Badash focuses on the conflict in the public and political arena, complementing this study by tracing the various historical lines of research that combined in the nuclear winter theory. His book closely follows the twists and turns of the debate during the 1980s, a difficult learning experience for everybody who participated, because it meant dealing in a public arena with the tricky issue of scientific uncertainty. The debate did not and could not reach definitive closure (happily, nobody has carried out the essential experiment to verify the theory). Badash shows how in the end, as the issue dragged out over years without resolution, this uncertainty played into the hands of those who had political power and advocated the scientific and political status quo.

This article examines the history of nuclear winter from a different angle. Rather than following the path and pace of the highly politicized scientific and public debate over the course of the 1980s, I study the period up to late 1983 and look at what made scientists, whether for or against the theory of nuclear winter, get interested in this research in the first place, and I explore the individual and collective trajectories that led to the nuclear winter debate. The story told here foreshadows the recent research on anthropogenic climate change that appears to have been substantially shaped during this—apparently tangential—cold war debate of the 1980s about research on the global effects of nuclear weapons. This point is easily obscured, since the most heated scientific and public debates on climate change came in 1990 with the first assessment report of the Intergovernmental Panel on Climate Change (IPCC), after the fall of the Berlin Wall, and apparently unconnected to nuclear arms issues. I by no means claim that the nuclear winter debate is the sole reason for the rise of interest in climate change.[7] However, the American debate about nuclear winter set the topic

[5] See, e.g., Barbara Ward and René Dubos, *Only One Earth: The Care and Maintenance of a Small Planet* (New York, 1972); Donella H. Meadows et al., *The Limits to Growth* (New York, 1972).

[6] Badash, *A Nuclear Winter's Tale: Science and Politics in the 1980s* (Cambridge, Mass., 2009). I am grateful to Badash for providing me with the proofs of his book as well as transcripts of interviews he conducted with various actors during the 1980s and 1990s. For the political context, see also Badash, "Nuclear Winter" (cit. n. 2); Paul Harold Rubinson, "Containing Science: The U.S. National Security State and Scientists' Challenge to Nuclear Weapons during the Cold War" (PhD diss., Univ. of Texas at Austin, 2008), esp. chap. 7, "An Elaborate Way of Committing National Suicide: Carl Sagan, Popularization, and Nuclear Winter." For nuclear winter within the history of climate change, see Spencer Weart, *The Discovery of Global Warming* (Cambridge, Mass., 2003), 144–6; see also the book's Web site at the Center for History of Physics of the American Institute of Physics, http://www.aip.org/history/climate (accessed 7 January 2011). For nuclear winter within the history of atmospheric research at NASA, see Erik M. Conway, *Atmospheric Science at NASA: A History* (Baltimore, 2008), 206–12.

[7] Other significant events include climate-related problems in agriculture during the 1970s, the oil crises, ozone depletion, and El Niño events.

and the tone for what was to follow. It hardened positions on scientific method and political practice that adumbrated the still-ongoing debate on climate change and global warming. Stepping back from the day-to-day action and looking at the larger cultural and political context reveals sometimes surprising commonalities among main actors who found themselves on opposite sides, as well as differences within the apparently coherent TTAPS group.

The patterns of the TTAPS members' scientific careers both reflect larger cultural and political transformations and indicate the range of options among which the actors had to choose at certain career stages. The actors not only contributed to framing the nuclear winter debate, they were themselves transformed by it. The biographical approach leaves room for the subtleties of intermediate, complex, and not necessarily internally coherent positions and avoids reducing the scientists to puppets deterministically defined by an all-powerful context. The remarkable cohesiveness of the TTAPS group was a result not only of intense external pressure, but also of the successful cooperation of quite different individual characters. My approach here is perhaps closest to Myanna Lahsen's in her recent article on a trio of prominent global climate change skeptics, part of a study of the "cultural and historical dimensions that structure US climate science politics." Her study of the three distinguished nuclear physicists, who all served as powerful policy advisers in Washington, shows that they acted in defense of their "preferred understandings of science, modernity, and of themselves as a physicist elite."[8]

My biographical analysis of the five TTAPS members focuses on three points. First is their choice of research in atmospheric science and the motivations that lay behind it. This field emerged as a key environmental science over the period I discuss, pushed to the forefront in part by the TTAPS members, who then found themselves, partly to their own surprise, in the midst of highly political and public debates, a learning experience for all members. Second, they chose a research topic that had been previously investigated only by military request and at military laboratories or think tanks. The question here is what kind of relationships the TTAPS members developed during their careers with the military and military institutions, especially given the larger issues of the Vietnam War that confronted all of them. The investigation of this question also provides additional hints on the "actual political views of scientists in disciplines other than physics during the Cold War," which Ron Doel has asked for.[9] Third, they settled on a particular tool kit, aerosol and climate modeling, as a way to predict the consequences of a nuclear war. Here I only examine the training that allowed the TTAPS members to handle these models with confidence.[10] I draw to a large extent upon interviews and e-mail exchanges carried out with the three surviving TTAPS members during 2009 and 2010.

WEATHER, CLIMATE, AND THE BOMBS

The study of the meteorological consequences of nuclear bomb explosions began with the observation of black rain after the bombing of Hiroshima. In the late 1940s,

---

[8] Myanna Lahsen, "Experiences of Modernity in the Greenhouse: A Cultural Analysis of a Physicist 'Trio' Supporting the Backlash against Global Warming," *Global Environmental Change* 18 (2008): 204–19, on 204.

[9] Doel, "Constituting the Postwar Earth Sciences" (cit. n. 4), 654.

[10] I plan to develop this point in a future article.

Harry Wexler (1911–62), the director of research at the U.S. Weather Bureau, became deeply involved in this study, for example in the classified Project Gabriel, submitted to the U.S. Atomic Energy Commission (AEC) in 1949.[11] Wexler published articles in the open scientific literature on the effects of volcanic dust on the atmosphere, the flip side of his mostly classified research on the longer-lasting effects of dust caused by nuclear bombs.[12] The Rand Corporation's Project Sunshine of 1953, sponsored by the AEC and the U.S. Air Force, looked at effects that geographically exceeded the immediate destruction area, within an expanded temporal span. The report was an outcome of a Rand meeting in the summer of 1953, in which more than seventy scientists participated, among them—to name only the most prominent ones or those who will reappear in this article—Wexler, Edward Teller (University of California), Lester Machta (U.S. Weather Bureau), who had been hired by Wexler to head the Special Projects Section, Roger Revelle (Scripps Institution of Oceanography), John von Neumann (Institute of Advanced Studies), William W. Kellogg (Rand), and F. J. Krieger (Rand). The report was mostly concerned with radioactive-biological hazards; the atmospheric section was limited to a discussion of the distribution of radioactive debris (especially its diffusion downwind). It mentioned, however, that long-range effects of nuclear bombs, such as the "loading of the atmosphere with particulate matter . . . may affect the weather of the earth" in a similar way to a volcanic eruption, and participants at the meeting discussed the effects of upper atmosphere ionization by radioactive debris.[13]

The Project Sunshine report was already, in its own words, "an historical document" when it appeared in 1953.[14] The American atmospheric hydrogen-bomb explosions in the Pacific, beginning in November 1952, led to further studies of their meteorological and climatic consequences, on regional and ultimately global levels. Japanese researchers, following public concerns in that country, worried in a 1954 article about extremely cold summers with excessive rains, leading to bad harvests, which had occurred several times over the twentieth century, and as recently as 1953 and 1954. The author, Hidetoshi Arakawa of the Japanese Meteorological Research Institute (who had invented the balloon bomb used by the Japanese military during the Second World War), discussed the role of dust in the upper atmosphere (stratosphere) caused by either volcanoes or the atmospheric nuclear explosions, particularly those between March and May 1954.[15] Another Japanese study, by the Observation Mission of the Central Meteorological Observatory, suggested that a series

[11] Parts of this classified study were published a year later in Los Alamos Scientific Laboratory, U.S. Department of Defense, and U.S. Atomic Energy Commission, *The Effects of Atomic Weapons* (Washington, D.C., 1950). See also U.S. Atomic Energy Commission and U.S. Air Force, *Worldwide Effects of Atomic Weapons: Project Sunshine; U.S. Atomic Energy Commission, Contract AT (11-1)-135; U.S. Air Force Project Rand, Contract AF 33(038)-6413, August 6, 1953* (Santa Monica, Calif., 1953), 1. On Wexler see *Complete Dictionary of Scientific Biography,* s.v. "Harry Wexler," by Sepideh Yalda, and James Rodger Fleming, *Fixing the Sky: The Checkered History of Weather and Climate Control* (New York, 2010), chap. 7.

[12] Harry Wexler, "On the Effects of Volcanic Dust on Isolation and Weather," *Bulletin of the American Meteorological Society* 32 (1951): 10–5; Wexler, "Spread of the Krakatoa Volcanic Dust Cloud as Related to the High-Level Circulation," *Bull. Amer. Meteor. Soc.* 32 (1951): 48–51.

[13] Atomic Energy Commission and Air Force, *Project Sunshine* (cit. n. 11), 4.

[14] Ibid., iii.

[15] Arakawa, "Possible Atmospheric Disturbances and Damages to the Rice-Crops in Northern Japan That May Be Caused by Experimentation with Nuclear Weapons," *Geophysical Magazine* (1954): 125–34.

of hydrogen-bomb experiments "might be able to throw as much fine dust as a volcanic explosion up to the stratosphere" and thus "affect the world-wide circulation through contaminating the higher atmosphere." These scientists relied exclusively on Japanese observations and data and expressed their desire to have further access to American or global research.[16]

In the United States, the new test sites in Nevada, which had been established in 1951, and a prolonged period of drought in the Midwest raised public fears of a modification of the weather due to atmospheric nuclear explosions. These issues were addressed for the first time in a public Congressional hearing in 1955 before the Joint Committee on Atomic Energy. Wexler presented parts of the results of a wide-range study on meteorological aspects of nuclear explosions, financed with some $100,000 by the Division of Biology and Medicine of the AEC, which indicated that any effect on the weather was "unlikely."[17] Questioned on long-term effects of nuclear weapons on the weather, John von Neumann suggested in a brief response of a speculative nature that given the cooling effect of the pall of dust, it would take a hundred of the largest atomic explosions (roughly one Krakatau eruption) per year over a period of ten or twenty years to bring back "the conditions of the last ice age."[18] However, the hearings excluded any discussion of the possible effects of hydrogen bombs, and in 1955 available data were not sufficient "to justify more than a tentative conclusion" that there had been "any obvious changes in the weather . . . outside of the test area."[19] The meteorological part of the final "study of the biological effects of atomic radiation" by the National Academy of Sciences (NAS) came to the same conclusion, providing, however, an interesting caveat:

> The failure to detect statistically significant changes in the weather during the first 10 years of the atomic age is no proof that physically significant changes have not been produced by the explosions, but it does show that a careful physical analysis of the effects of atomic and thermonuclear explosions on the atmosphere must be made.[20]

The authors discussed the possibility that "a series of explosions designed for the maximum efficiency in throwing debris into the upper atmosphere might significantly affect the radiation received at the ground," but admitted the difficulty of coming up with "a final authoritative evaluation."[21] The subsequent years, including the International Geophysical Year (IGY) in 1957, were characterized by Wexler's effort to better understand the general circulation of the lower and upper atmosphere.[22] In the late 1950s, he focused on expanding "the field program involving specially

[16] Akio Arakawa et al., "Climatic Abnormalities as Related to the Explosions of Volcano and Hydrogen-Bomb," *Geophysical Magazine* (1955): 231–55, on 252–3, 242.

[17] Joint Committee on Atomic Energy, *Health and Safety Problems and Weather Effects Associated with Atomic Explosions*, hearing, 84th Cong., 1st sess., 1955, 47. For a publication of these results, see Lester Machta and D. L. Harris, "Effects of Atomic Explosions on Weather," *Science* 121 (1955): 75–81.

[18] Ibid., 35–6.

[19] Machta and Harris, "Effects" (cit. n. 17), 80.

[20] Harry Wexler, D. Lee Harris, R. J. List, Lester Machta, and W. W. Kellogg, "Meteorological Aspects of Atomic Radiation," *Science* 124 (1956): 105–12, on 112.

[21] Ibid., 112.

[22] For Wexler's role as U.S. chief scientist for Antarctic programs in the IGY, see James Rodger Fleming, "Polar and Global Meteorology in the Career of Harry Wexler, 1933–1962," in *Globalizing Polar Science*, eds. Roger D. Launius, Fleming, and David DeVorkin (New York, 2010), 225–41.

equipped aircraft, balloons, and rockets to obtain stratospheric air samples and measurements," on accelerating data analysis with the help of "high-speed automatic electronic computers," and on developing mathematical models for stratospheric circulations.[23] Wexler also saw the need for having a highly qualified meteorologist heading these observational efforts.

Wexler died suddenly in August 1962 at the age of 51, and with him died many of his research initiatives. U.S. and Soviet atmospheric nuclear explosions also died down with the atmospheric test ban treaty of 1963. From then on, researchers had to content themselves with existing data, drawn mostly from the nuclear explosions of the 1950s and early 1960s. As the hydrogen-bomb explosions of the 1950s did not seem to have left any clearly decipherable mark on Earth's climate, there seemed no necessity for further debate or public reassurance, and the issue thus remained mostly out of public view.

THE RISE OF ENVIRONMENTAL SCIENCES DURING THE 1960s AND 1970s

However, this apparent silence did not mean that no one was interested. For example, military researchers at LLNL continued to look at the atmospheric impact of bombs. Edward Teller was the driving force here. Military strategists and researchers from the 1950s until the 1970s were keenly interested in weather and climate for numerous reasons: they wished to know both the impact of weather on nuclear explosions and the inverse, that of nuclear explosions on weather; they were eager to obtain long-term weather forecasts for better strategic planning; and they also considered the use of weather and climate modification (possibly with the help of nuclear bombs) as strategic tools.[24] Weather and climate change were first-rank defense issues, and the best minds of the time turned toward them, either in a speculative mode or by way of modeling complex weather and atmospheric systems with the help of computers.[25] Some of this research was classified at the time—for example, studies on the atmospheric dust due to nuclear explosions or on computational physics and modeling.[26] Teller also supervised a PhD student at the University of California at Davis, Michael MacCracken, who worked on the analysis of the ice age theory by computer simulation in the second half of the 1960s. MacCracken subsequently joined LLNL and did atmospheric science modeling.

A second line of research was pushed by Herman Kahn, who was interested in the survivability of nuclear war. Kahn, by then director of his own Hudson Institute (an "also-RAND"[27]), obtained a grant from the Federal Emergency Management Agency (FEMA) and hired a physics postdoc, Robert U. Ayres, to work on the problem. Ayres had obtained a PhD in mathematical physics from Kings College at the University of London, had operated some detection instruments during the last atmospheric nu-

[23] Joint Committee on Atomic Energy, *Fallout from Nuclear Weapons Tests: Hearings before the Special Subcommittee on Radiation*, vol. 3, 86th Cong., 1st sess., 1959, 2004–5.

[24] Fleming, *Fixing the Sky* (cit. n. 11); Paul Edwards, *The Closed World: Computers and the Politics of Discourse in Cold War America* (Cambridge, Mass., 1997).

[25] Kristine C. Harper, *Weather by the Numbers: The Genesis of Modern Meteorology* (Cambridge, Mass., 2008).

[26] R. G. Gutmacher, G. H. Higgins, and H. A. Tewes, *Total Mass and Concentration of Particles in Dust Clouds*, URCL-14397 rev. 1 (Livermore, Calif., 1965). An unclassified version was released in March 1983 as UCRL-14397 rev. 2.

[27] Martin Shubik, "Processing the Future," *Science* 166 (1969): 1257–8, on 1257.

clear test series at Johnson Island in the western Pacific, and had published a substantial three-volume report, *Environmental Effects of Nuclear Weapons*, of which certain parts also referred to atmospheric consequences.[28] However, the report did not present new research and seems not to have been read widely, either at FEMA (for which it was probably too technical) or by atmospheric scientists (who probably did not have easy access to it). Ayres himself was "unhappy" about doing this kind of work. He "did not want to be thought of as a nuclear war analyst" and did little to publicize his work to a larger audience.[29] A possible book contract with Princeton University Press did not materialize, and Ayres shifted his subsequent work toward themes more rewarding to himself, such as technological forecasting and environmental economics.[30] Similarly to Ayres's work, a Rand study prepared for the AEC discussed the effects of nuclear weapons in a qualitative way. It did not exclude the possibility of the modification of weather and climate by a nuclear war but pointed to the "complexity and the lack of thorough understanding of the interdependent meteorological processes" that made quantitative estimations impossible.[31]

A third line of research was stimulated by the rise of environmental science that was increasingly independent of military control and funding and led to atmospheric studies whose results challenged the assumptions prevalent within military circles. During the 1960s, environmental issues in the United States of local or regional origin were increasingly treated at the national level, as exemplified in the creation of government bodies such as the Committee on Environmental Quality of the Federal Council of Science and Technology, the Panel on the Environment of the President's Science Advisory Committee (PSAC), the White House Council on Environmental Quality, the NAS Environmental Studies Board, the Committee on Environmental Alterations of the American Association for the Advancement of Science, and other activities by nongovernmental organizations (e.g., the Conservation Foundation). The early 1970s also saw the creation of the Environmental Protection Agency (EPA, 1970) and the Office for Technology Assessment (OTA, 1972).

Zuoyue Wang has recently shown in detail the crucial role of PSAC in putting environmental issues on the political agenda, particularly in the aftermath of the publication of *Silent Spring* by Rachel Carson in 1962, which looked at the detrimental long-term effects of DDT. Wang attributes to PSAC a pervasive "technological skepticism" that extended from its evaluation of nuclear weapons to its view on pesticides.[32] A landmark in environmental history was the 1965 PSAC report *Restoring the Quality of Our Environment*, issued by the Environmental Pollution Panel with John W. Tukey of Princeton University as chairman. This wide-ranging report included a section on atmospheric carbon dioxide and a discussion of the consequences of global warming written by Roger Revelle (Harvard University), Wallace Broecker (Columbia University), Harmon Craig (University of California at San Diego),

---

[28] Robert U. Ayres, *Environmental Effects of Nuclear Weapons*, 3 vols. (New York, 1965); Ayres, "What about Nuclear Proliferation: Iran and North Korea" (unpublished manuscript, Paris, France, 2004).

[29] Robert U. Ayres, e-mail messages to author, 17 June and 27 November 2009.

[30] See, e.g., his widely read book, *Technological Forecasting and Long-Range Planning* (New York, 1969).

[31] E. S. Batten, *The Effects of Nuclear War on the Weather and Climate* (Santa Monica, Calif., 1966), v.

[32] Zuoyue Wang, *In Sputnik's Shadow: The President's Science Advisory Committee and Cold War America* (Piscataway, N.J., 2008), 200.

Charles D. Keeling (Scripps Institution of Oceanography), and Joseph Smagorinsky (U.S. Weather Bureau and Environmental Sciences Services Administration).

During the 1970s, study of environmental issues grew exponentially outside the military realm and reached a larger scientific and political community. A noted climatologist with government ties, Helmut E. Landsberg, professor at the University of Maryland, noticed a "new trend toward thinking in ecological terms" in a 1970 *Science* article entitled "Man-Made Climatic Changes."[33] The United Nations played a key role in that change. In 1968, UNESCO (the United Nations Educational, Scientific, and Cultural Organization) had organized a conference on man and the biosphere. The same year, the UN Economic and Social Council passed a resolution identifying an "urgent need for intensified action at the national and the international level, to limit and, where possible, to eliminate the impairment of the human environment."[34] Later in 1968, the UN General Assembly, with Sweden in the lead, decided to convene the Conference on the Human Environment in Stockholm for 1972. In the United States, there was also a rising interest in a global perspective. Already in 1968, S. Fred Singer, for example, then deputy assistant secretary of the Department of the Interior, had organized a conference on the global effects of environmental pollution. The chief motivation for pushing these global environmental studies was "sound economic and social development,"[35] to be guaranteed—it was hoped— by studying the possible environmental consequences of human action.

The unofficial summary report in preparation for the 1972 Stockholm UN meeting was *Only One Earth*, written by Barbara Ward and René Dubos, a book that drew together numerous contributions into "a unique experiment in international collaboration."[36] Ward and Dubos set the tone for a new global environmental consciousness. The atmosphere, shared by all nations, was being affected by "more and more of industrial man's activities."[37] Researchers should study "the points at which human actions, however miniscule their effects may seem when set against the total scale of the planet's energy system, may nonetheless trigger off one of those small but fateful changes which alter the balance of the seesaw."[38] And, "given the global interdependence of man's airs and climates," there was a need for "a new capacity for global decision-making and global care . . . a new commitment to global responsibilities."[39]

In preparation for the Conference on the Human Environment, two important reports were issued that had a strong impact on the field of atmospheric science: the 1970 SCEP (Study of Critical Environmental Problems) report, a result of a one-month interdisciplinary conference at the Massachusetts Institute of Technology (MIT), and the 1971 SMIC (Study of Man's Impact on Climate) report, from a conference sponsored by MIT and hosted by the Royal Swedish Academy of Sciences and the Royal Swedish Academy of Engineering Sciences.[40] Both reports were meant

[33] Landsberg, "Man-made Climatic Changes," *Science* 170 (1970): 1265–74, on 1273.

[34] "Constitution of the Conference," United Nations Environmental Programme Web site, http://www.unep.org/Documents.Multilingual/Default.asp?DocumentID=97&ArticleID=1496 (accessed 7 September 2009).

[35] Ibid.

[36] Ward and Dubos, *Only One Earth* (cit. n. 5), vii.

[37] Ibid., 191.

[38] Ibid., 192.

[39] Ibid., 195.

[40] SCEP, *Man's Impact on the Global Environment* (Cambridge, Mass., 1970); SMIC, *Inadvertent Climate Modification* (Cambridge, Mass., 1971).

as "useful reference works to researchers and students in the many disciplines involved in solving global environmental problems."[41]

The SMIC and SCEP reports, although out-of-date for leading atmospheric scientists when they appeared, served as important orientation points for young atmospheric scientists in the early 1970s. The reports were remarkable in that they addressed the issue of global climate modification exclusively from an environmental standpoint. They represented a clear break with the past: first, they hardly referred to previous military research, and, with the exception of Kellogg and Machta, none of the actors important in that sphere figured in the reports, and second, in contrast to the inaccessibility of the military research to the general public, their authors declared that it was now time to "raise the level of informed *public* and scientific discussion and action."[42] Furthermore, there was a clear push for environmental research on an international scale in order to respond adequately to the global challenges as laid out by the UN.

Another stimulus to long-term environmental studies came from a potent new tool, computer modeling, and wider access to increasingly powerful computers. Complex systems, such as ocean currents, ecosystems, and global atmospheric circulation, came under intense scrutiny. A climatology textbook of the early 1970s noted the transformation of climatology from an observational science into a mathematical and physical one and asserted that "climatology deals with numbers, and one of the first things a person with a scientific mind will want to do, on coming upon a set of climatic figures, is to see if he can use these figures to predict another set."[43]

THE 1975 NAS STUDY

At the beginning of the 1970s, ozone depletion was the most important and widely discussed issue in atmospheric science. In a 1970 article, Paul J. Crutzen, a postdoctoral fellow at Oxford University, had shown that nitrous oxide played an important role in destroying ozone, and in 1971 a professor of chemistry at the University of California at Berkeley, Harold Johnston, had pointed to the possible destruction of ozone in the stratosphere due to supersonic transport.[44] This debate on ozone (enhanced in 1974 by concern about the influence of chlorofluorocarbons on the ozone layer) had set a precedent, in that it energized the community of atmospheric scientists and strengthened the field in terms of funding. These issues did not escape the attention of Fred C. Iklé, the director of the U.S. Arms Control and Disarmament Agency (ACDA), who in April 1974 asked the NAS for a report on the long-term worldwide effects of nuclear weapons on ozone.

The 1975 NAS report was aimed at a larger audience; the possible environmental consequences of nuclear war were no longer enshrined in secrecy. However, the report did not consider the possible impact of fires and smoke in the aftermath of a

---

[41] William H. Matthews, William W. Kellogg, and G. D. Robinson, eds., *Man's Impact on the Climate* (Cambridge, Mass., 1971), xiii.

[42] SMIC, *Inadvertent Climate Modification* (cit. n. 40), xvi–xvii (italics mine).

[43] E. T. Stringer, *Foundations of Climatology* (San Francisco, 1972), xi; Harper, *Weather by the Numbers* (cit. n. 25); Edwards, *Closed World* (cit. n. 24).

[44] Crutzen, "The Influence of Nitrogen Oxides on the Atmospheric Ozone Content," *Quarterly Journal of the Royal Meteorological Society* 96 (1970): 320–5; Johnston, "Reduction of Stratospheric Ozone by Nitrogen Oxide Catalysts from Supersonic Transport Exhaust," *Science* 173 (1971): 517–22. See also Badash, *Nuclear Winter's Tale* (cit. n. 6), 11–4.

nuclear explosion. It confirmed the danger of ozone depletion (at least for nuclear explosions greater than 1 megaton) and drew comparisons to the effects of volcanic dust on climate. In sum, however, the report minimized the risk of major climatic consequences, though it explicitly did not exclude such a possibility, pointing to the uncertainty and insufficiency of present research.[45]

ACDA, under Iklé, also issued its own short summary report in 1975, based on the NAS report and other studies, under the title *Worldwide Effects of Nuclear War*. Several aspects of this report are remarkable and serve to distinguish it from earlier work. First, Iklé explicitly addressed a larger public than any previous report had done: "Realistic and responsible arms control policy calls for our knowing more about these wider effects and for making this knowledge available to the public."[46] Second, the report listed a series of previous scientific "surprises," such as the size of the Castle Bravo explosion of 1954, which led to the irradiation of the Japanese crew of the fishing vessel *Lucky Dragon* and the fallout of iodine-131, seriously affecting young Rongelapese children, and the 1962 explosion of a nuclear device called Starfish at an altitude of 250 miles (in the ionosphere), which produced an artificial belt of charged particles trapped in Earth's magnetic field for several years. Iklé concluded, "Much of our knowledge was thus gained by chance—a fact which should imbue us with humility as we contemplate the remaining uncertainties (as well as the certainties) of nuclear warfare."[47] He stressed the remaining uncertainties, which were of "such magnitude" that they worked as a "further deterrent to the use of nuclear weapons."[48] Third, the report represented a switch of perspective from the national to the international to consider the effects "from the standpoint of the countries not under direct attack" and also from the "less-developed countries."[49]

While some of the points mentioned above may have been merely rhetorical, the report nevertheless clearly shows that the global environmental and disarmament agenda set by the UN had to be addressed directly. A subsequent report in 1979, conducted by the OTA, adopted the NAS's concern about ozone depletion, while pointing out that potential damage would be less likely because of the expected future development of smaller warheads of less than 1 megaton.[50]

TTAPS: ELEMENTS OF A COLLECTIVE BIOGRAPHY

Behind the acronym TTAPS stood five planetary and atmospheric scientists with different education and worldviews. They also belonged to different generations: Carl Sagan was born in 1934, James B. Pollack in 1938, Richard P. Turco in 1943, Owen Brian Toon in 1947, and Thomas P. Ackerman in 1949. The group as a whole combined both seniority and experience. But the differences in age mattered. For example, Sagan's career took off around 1960, during tense years of the cold war, which influenced his early choice of work, drawing him into military research. On the other hand, the Vietnam War, with its severe repercussions for the American nation and

[45] National Research Council, *Long-Term Worldwide Effects of Multiple Nuclear-Weapons Detonations* (Washington, D.C., 1975).
[46] ACDA, *Worldwide Effects of Nuclear War: Some Perspectives* (Washington, D.C., 1975), 4.
[47] Ibid., 7.
[48] Ibid., 4.
[49] Ibid., 18.
[50] OTA, *The Effects of Nuclear War* (Washington, D.C., 1979), 114.

its institutions, significantly shaped the careers of at least two members of TTAPS (Toon and Ackerman), who completed their college educations under the shadow of the draft. Furthermore, during the late 1960s and early 1970s, the three younger scientists were drawn into the rapidly expanding field of atmospheric science, of whose existence they had had no knowledge at the beginnings of their careers. They entered—and ultimately also shaped—the field of environmental research without at first realizing it.

While they were involved to a lesser degree in the details of the nuclear winter research, Sagan and Pollack, the senior members of the group, played important roles in promoting, organizing, encouraging, reviewing, and even—in Sagan's case—hyping the issue. They lent their scientific capital to what would have otherwise been a brilliant but rather powerless group. Pollack focused within the group exclusively on the constructive critique of his younger colleagues, while Sagan became the public face of TTAPS, relentlessly defending the nuclear winter theory in public settings and debates, taking some of the intense political and public pressure off of the younger TTAPS members, whose careers were still at stake.

Carl Sagan

As a young man in the late 1950s Sagan had been a part of the research on the global effects of nuclear weapons. After receiving an MS in physics at the University of Chicago in 1956, he submitted in 1960 his dissertation in astronomy and astrophysics under the supervision of the planetary astronomer Gerard P. Kuiper (1905–73) of the Yerkes Observatory, part of the University of Chicago.[51] Kuiper's close relationship with the military led him to suggest to Sagan that he participate in a secret project with the code name A119, under the directorship of Leonard Reiffel of the Armour Research Foundation (ARF) at the Illinois Institute of Technology in Chicago. From 1949 to 1962 ARF conducted studies on the global environmental effects of nuclear explosions.[52] In 1958 the U.S. Air Force asked Reiffel for research into the visibility and effects of a hypothetical nuclear explosion on the moon. It was interested in a "surprise demonstration explosion, with all its obvious implications for public relations and the Cold War."[53] Via Kuiper, Reiffel hired Sagan to model "the expansion of an exploding gas/dust cloud rarifying into the space around the Moon."[54]

Sagan also worked as a consultant for the Rand Corporation while a postdoc at the University of California at Berkeley (1960–2).[55] He thought big and he fitted seamlessly—it seems—into the Rand model of an avant-garde think tank. For

[51] Carl Edward Sagan, "Physical Study of Planets" (PhD diss., Univ. of Chicago, 1960); Keay Davidson, *Carl Sagan: A Life* (New York, 1999); William Poundstone, *Carl Sagan* (New York, 1999). On Kuiper, see Ronald E. Doel, *Solar System Astronomy in America: Communities, Patronage, and Interdisciplinary Research, 1920–1960* (Cambridge, 1996).

[52] Leonard Reiffel, "Sagan Breached Security by Revealing US Work on a Lunar Bomb Project," *Nature* 405 (2000): 13.

[53] Ibid., 13; L. Reiffel, *A Study of Lunar Research Flight*, vol.1 (classified report, Air Force Special Weapons Center, Kirtland, N.M., June 19, 1959).

[54] Reiffel, "Sagan" (cit. n. 52), 13. Sagan accidentally listed two classified texts in his Miller Institute postdoctoral fellowship application: "Possible Contribution of Lunar Nuclear Weapons Detonations to the Solution of Some Problems in Planetary Astronomy" (1958) and "Radiological Contamination of the Moon by Nuclear Weapons Detonations" (1959).

[55] Carl E. Sagan, *Is the Martian Blue Haze Produced by Solar Protons?* (Santa Monica, Calif., 1961).

example, in a 1961 article, he considered "the prospect of microbiological planetary engineering" on the planet Venus.[56] Sagan's close relationship with military research soured, however, during the Vietnam War, while he was at the Smithsonian Astrophysical Observatory in Cambridge, Massachusetts, in the mid-1960s and a professor of planetary studies at Cornell University from 1968. His first step to distance himself was to resign from the Air Force Scientific Advisory Board's Geophysical Panel, a decision that many of his colleagues thought "was really dumb," given that the contacts provided by the position were invaluable for a career in science. In a 1991 interview, Sagan admitted that in his early career he had been "happy to get some military support," but said that after "a serious look at Vietnam," he "couldn't do it" any longer and claimed not to have accepted military funding ever since.[57] In the early 1970s, Sagan worked with Toon on planetary research. The *Mariner 9* mission to Mars in 1971–2 gave astronomers the first chance ever to observe an actual climatic change, caused by a planetary-scale dust storm. Planetary research fit nicely into the new global perspective on Earth laid out in the SMIC and SCEP reports, both in its emphasis on the smallness and fragility of Earth in space and in its operation on long-term timescales.

In 1982 Sagan was deeply involved with his popular television series *Cosmos*, a huge success that made him famous inside and outside the United States. He was driven in this series by a mission to counter pseudoscience and to promote the wonders of astronomy, particularly planetary science; it also gave him a platform to advocate political arguments such as disarmament.

The theory of nuclear winter suited Sagan's big thinking and his desire to shake up scientific tradition and open up new horizons. Sagan did no active research on nuclear winter; he "was not directly involved in building the models or running the simulations," Turco recalled, but "was responsible for suggesting scenarios, critically interpreting results."[58] Furthermore, "he meticulously reviewed all of the written material and greatly improved much of it," organized an external review of the initial TTAPS research report, and became "instrumental by making this review work."[59]

During the 1980s Sagan, because of his media and scientific omnipresence, became the preferred target of critics of the nuclear winter theory. One standard line of attack was to point to the multiple acknowledged uncertainties in the theory, then to single out Sagan and link him to the peace movement, and finally to dismiss the theory as part of a liberal political strategy aimed at subverting the status quo in the cold war. While Sagan had indeed taken an explicit anti-Vietnam and anti–nuclear weapons course, and preached on television in a quasi-religious way within a cosmological perspective, there was no straight path for him from political conviction to the scientific choice of an object of research. Always eager to pick up ideas, he probably saw immediately the politically subversive potential of the nuclear winter theory, but

[56] Carl Sagan, "The Planet Venus," *Science* 133 (1961): 849–58, on 857.

[57] Carl Sagan, oral history interview with Ronald E. Doel, August 27, 1991, quoted in Davidson, *Sagan* (cit. n. 51), 195. Concerning university protests against military research, see Stuart Leslie, *Cold War and American Science: The Military-Industrial Complex at MIT and Stanford* (New York, 1993), esp. chap. 9, "Days of Reckoning: March 4 and April 3."

[58] Richard P. Turco, e-mail message to author, 3 June 2009.

[59] Ibid.; interview with Owen Brian Toon, May 1, 2009.

he was also careful to have sufficient scientific support from colleagues and to have put the theory through quite an elaborate review before he decided to throw all his force behind it.

James B. Pollack

Pollack (1938–94) acted mostly in the background of TTAPS. He graduated from Princeton in 1960 and earned an MA in nuclear physics at the University of California at Berkeley in 1962, then wrote his PhD thesis at Harvard under Sagan in 1965 on the greenhouse effect on Venus.[60] After a few years with Sagan at the Smithsonian Astrophysical Observatory and then at Cornell, Pollack moved to NASA Ames in 1970 to continue his work on planetary science, creating models for the climate and meteorology of Venus, Mars, and the outer planets, modeling radiative transfer and measuring optical properties, and exploring the origins of planets. He also became interested during the 1970s in Earth's atmosphere, particularly the effects of volcanoes, and initiated large experimental efforts to use Ames airplanes to study volcanic clouds.

Pollack brought to the TTAPS group experience and thoroughness. He had previously worked with Turco on the applications of aerosol-radiation models for a variety of cases, including planetary studies, research on the extinction of the dinosaurs, and assessments of the emissions of supersonic transports and the space shuttle. Though he stayed "more at arm's length during most of the nuclear winter epoch,"[61] he protected the two youngest members of TTAPS (Toon and Ackerman), who also worked at Ames, by putting his name on the common paper and thus played an important role in keeping the group together, despite all the pressure.

Richard P. Turco

Turco, the first T in TTAPS, originally trained as an engineer and later worked in a military think tank. He did not hesitate to cross institutional and intellectual boundaries or to follow new directions whenever he saw the urgent need to do so. Strongly encouraged by Sagan, he took the scientific lead in the nuclear winter theory and subsequent debate.

Born in 1943, Turco studied electrical engineering at Rutgers University (BS, 1965) and the University of Illinois (MS, 1967). At Illinois he was drawn into the emerging science of solid-state physics as applied to semiconductors, a field that was developing between electrical engineering and physics. Turco took courses in both departments and sought advisers in both for his master's thesis on a solid-state laser. However, "the competition among the engineering and physics students for graduate research positions was fierce, jobs were scarce." To support himself he took a graduate job with the director of the Aeronomy Laboratory, Sidney A. Bowhill, professor of electrical engineering and the lab's director since 1962.

---

[60] James Barney Pollack, "Theoretical Studies of Venus: An Application of Planetary Astrophysics" (PhD diss., Harvard Univ., 1965). On Pollack, see Brian Toon, Jeff Cuzzi, and Carl Sagan, "In Memoriam: James B. Pollack (1938–1994)," *Icarus* 113 (1995): 227–31.

[61] Turco, 3 June e-mail (cit. n. 58).

> I drifted into "aeronomy" . . . mainly because I needed a graduate job. It was not an early conscious decision. Later, as graduation approached I recognized the tremendous opportunities in the emerging field of stratospheric chemistry. Accordingly, I moved ahead with this opening, dropped all interest in solid state physics, and never looked back on these decisions (looking back now [in 2009], however, I would say that this was one of the luckiest breaks in my career!).[62]

The assignment led Turco to work on the chemistry of the ionosphere and eventually to write a PhD thesis on the behavior of the sunrise D-region (the lower ionosphere) under the supervision of Chalmers F. Segrist, another professor at the Aeronomy Laboratory. During his time at the University of Illinois in Urbana-Champaign, Turco made use of the university's excellent computer facilities and was lucky to get in contact with C.W. Gear of the computer science faculty, who gave him his code for solving "stiff" nonlinear ordinary differential equations of the sort that Turco was dealing with in the ionosphere. This computational approach put Turco at the forefront of atmospheric research in the late 1960s.

While he was confronted with anti–Vietnam War demonstrations on campus, Turco remembers not being "really interested in the politics of the war."[63] He had taken the army physical, but obtained a PhD deferment. In 1971, he started work as a research scientist at R&D Associates (RDA), an offspring of the Rand Corporation in Marina del Rey, California.[64] Here he came in contact with leading nuclear-weapons specialists, such as the physicist Albert L. Latter, former head of the physics department of Rand and cofounder of RDA. He acquired intimate knowledge of nuclear weapons and was easily able to provide nonclassified data for the TTAPS scenarios.

Turco's early work on aerosol modeling was easily adapted to the stratospheric ozone problem, a line of research that he pursued in the 1970s and 1980s in close cooperation with the NASA Ames aeronomer Robert C. Whitten (who hired Turco to participate in research at Ames). In fact, the nuclear winter work was not the first research of Turco's to link nuclear weapons to atmospheric concerns. Already in 1975, Turco had published a paper with Whitten and W. J. Borucki on the impact of nuclear explosions on ozone.[65]

The work at Ames made Turco familiar with Toon, Ackerman, and Pollack, and they published articles together on a variety of research projects during the 1970s and early 1980s. Turco had switched disciplines and wandered between the two quite different worlds of the military think tank, RDA, and the civilian laboratory, Ames. The combination of bold, avant-garde scientific thinking at RDA with the well-established and well-funded practice of atmospheric and planetary observations at the NASA laboratory made it possible to assemble the multiple elements that entered into the complex nuclear winter theory. Turco brought his experience in atmospheric

---

[62] Ibid.

[63] Richard P. Turco, e-mail message to author, 17 July 2009.

[64] RDA provided high-tech electronics systems and research for the Defense Department. It was founded by former Rand researchers William R. Graham, Albert L. Latter, and Roland F. Herbst. All three would work later as policy advisers and staff for the Reagan administration.

[65] R. C. Whitten, W. J. Borucki, and R. P. Turco, "Possible Ozone Depletions following Nuclear Explosions," *Nature* 257 (1975): 38–9.

chemistry to this work and made use of an aerosol-radiation model that he had developed with Toon and others.

Owen Brian Toon

When Sagan, Pollack, and Turco were already at midcareer, Toon and Ackerman were at the beginning of theirs. Both came to do atmospheric environmental science without even having known the field existed at the beginning of their scientific studies. Toon (born in 1947) had flirted originally with political science and philosophy, before obtaining an MA in physics at the University of California at Berkeley in 1969. During the summers he worked in oceanography as part of the navy's Man-in-the-Sea Program, which from the mid-1960s had studied the effects on humans of long-term confinement in an undersea environment. After a year with the Environmental Research Division of the Naval Coastal Systems Center in Panama City, Florida, Toon opted for graduate school at Cornell University, not least to avoid the political tension and tear gas of the Vietnam demonstrations on the Berkeley campus. Cornell also attracted Toon because of its diverse physics department, as he was not yet quite sure what he wanted to do. However, when he applied, physics was in a recession. Toon recalls that solid-state research had suffered from a significant drop in interest; few students were accepted, job prospects were meager, and students were encouraged "to look for something that was not just straight physics."[66] He had thought originally to do something on gravitational waves, but took a planetary class with Sagan (of whose rising prominence he had little knowledge at the time), became intrigued by the exciting and rapidly moving research in the field, and volunteered to work for him for free.

However, the assigned experimental research, studying optical properties of dust and stratospheric aerosols (such as sulphuric acid), was too narrow and specialized for Toon. He preferred broad theoretical research, and following Sagan's advice, he looked at climate changes on Earth and Mars due either to volcanic eruptions or to dust storms, work that culminated in his 1975 PhD thesis.[67] In 1973, Toon's published review of the SCEP and SMIC reports pointed out the relevance of planetary studies for studying Earth's climate and developing models, laying out a future research program: "Changes in the Earth's global climate are far from being well understood. Man may be capable of causing irreversible changes without being aware of it until too late."[68] From 1975 to 1997, Toon did research at Ames, first as a research associate, then from 1978 on as a research scientist. His aerosol work touched upon a wide range of subjects: besides the climatic planetary studies, he worked on atmospheric pollution due to space shuttles, supersonic transports, and other high-flying aircraft, as well as the climatic consequences of volcanoes and asteroid impacts, in close cooperation with Turco, Pollack, and Ackerman. In contrast with Turco, Toon had no links with the military involving nuclear weapons or nuclear winter and "never did any classified research related to it."[69]

---

[66] Interview with Toon (cit. n. 59).

[67] Owen B. Toon, "Climatic Change on Mars and Earth" (PhD diss., Cornell Univ., 1975).

[68] Owen B. Toon, "Review: Man's Impact on the Climate; Inadvertent Climate Modification," *Icarus* (1973): 609–10, on 610.

[69] Interview with Toon (cit. n. 59); Toon, e-mail message to author, 16 December 2009.

Thomas P. Ackerman

Born in 1949, Ackerman obtained a BA in 1970 in physics at Calvin College, a Christian liberal arts school in Grand Rapids, Michigan, before moving to the University of Washington at Seattle for an MS in physics (1971). However, he ultimately had little interest in the experimental work on physical constants that some physicists were doing and looked instead for "exploratory sets of questions." When he took an independent study with Professor of Atmospheric Sciences and Geophysics Conway Leovy, he "fell in love with atmospheric science" and "entered the environmental sciences without knowing it."[70]

Leovy had obtained a PhD in meteorology from MIT in 1963, and he worked afterward at Rand on such topics as the modeling of the general circulation of the Martian atmosphere before moving to the University of Washington in 1968.[71] He and Ackerman developed a close and enduring relationship. Ackerman made the unfortunate discovery that his planned PhD research on the photochemistry of Mars's atmosphere had already been done and decided to study Earth rather than other planets, specializing in the issue of radiative transfer, to which he could happily apply his mathematical and computational skills. His thesis dealt with the effects of pollutants on meteorological variables in the Los Angeles Basin.[72] As a graduate student he read the SCEP and SMIC reports, which left a lasting impression on him, introducing him to the possibility of human-induced climate modification. Guided by his interest in climate change, he worked from 1976 to 1979 as a postdoc research scientist at the Australian Numerical Meteorology Research Center in Melbourne, where he spent most of his time learning how to develop global climate models and to write programs for cutting-edge computers. From 1979 Ackerman was at NASA Ames, first as a research associate, and from 1982 to 1987 as a research scientist, working closely with Pollack.

Ackerman recalls that in the midst of the nuclear winter work, after President Reagan's "Star Wars" speech in March 1983, he was offered a job tied to the Strategic Defense Initiative (SDI) at a considerably higher salary by representatives of Lockheed Missile and Space Division in Sunnyvale (next to Ames), who were interested in using his excellent knowledge of radiative transfer to detect the hot trails of missiles.[73] Ackerman decided against it. In this decision multiple factors were at play: the confrontations over the Vietnam War on the politically active campus of the University of Washington in the 1970s, memories of his fear of being drafted (made even more vivid by the war injuries of his older brother, who had been conscripted), and finally, the lies and inconsistencies of the official government statements of the time, which were in contradiction to what soldiers experienced on the ground. The war, he felt, was "morally wrong," and he wanted to have "nothing to do with the military and

[70] Interview with Thomas P. Ackerman, May 5, 2009.

[71] Conway B. Leovy and Yale Mintz, *A Numerical Circulation Experiment for the Atmosphere of Mars* (Santa Monica, Calif., 1966). Leovy was later one of the reviewers of the 1984 NAS study on nuclear winter.

[72] Thomas Peter Ackerman, "A Study of the Influence of Aerosols on Urban Boundary Layers with Particular Applications to the Los Angeles Basin" (PhD diss., Univ. of Washington, 1976).

[73] Thomas P. Ackerman, e-mail message to author, 30 April 2010; for the history of military uses of the transfer of radiation, see Spencer Weart, "Global Warming, Cold War, and the Evolution of Research Plans," *Hist. Stud. Phys. Biol. Sci.* 27 (1997): 319–56.

military research."[74] He considered the nuclear winter work something significant and unique that he could not just drop at any time, despite the political and economic pressure. It was "the moral thing to do," in line with the ethical values he had been taught through the Christian Reformed Church and his education at Calvin College. As a student, he had left physics for environmental atmospheric science with the hope of avoiding military work, and he would not turn his hand to it now.[75]

Ackerman's contribution to TTAPS was primarily in the areas of radiative transfer and climate modeling, while Toon and Turco worked mainly on the microphysics, particle properties, and atmospheric chemistry of the smoke and dust clouds resulting from nuclear explosions. All three of the younger TTAPS members cooperated closely on the theoretical and practical levels and were "heavily invested in computers," developing the codes for microphysical and climate models. As Ackerman was the most experienced in climate modeling, he was the one who ran the final simulations of climate impacts.[76]

Ackerman and Toon belonged to a generation of graduate students who grew up with large computers and learned computational methods. In his own field of radiative transfer, Ackerman regarded a 1966 paper by the British scientists Clive D. Rodgers and C. Desmond Walshaw on a computational approach to infrared radiative transfer as inaugurating a new age.[77] Making good use of their knowledge also pushed Toon and Ackerman ahead of Pollack, who had had mathematical, but not computational, training as a graduate student. While Pollack was the senior scientist, and originally regarded Toon and Ackerman as his subordinates and protégés, he ceded leadership and also ownership in the nuclear winter research to his younger colleagues, to assume in larger part the role of the careful adviser, guiding them and persistently asking critical and relevant questions.

TTAPS was far from being a coherent group, either politically or scientifically. It was rather an intergenerational assembly of scientists with quite different backgrounds, worldviews, qualifications, and strengths. The things that pulled the group together were a shared interest in certain politically pressing planetary (and mostly environmental) issues, on which they had previously worked together in various combinations, and the complementarity of their scientific knowledge and skills in disseminating their research.

TTAPS, NASA, AND POLITICS

In the early 1980s, two papers necessitated a revision of traditionally held scientific beliefs, one pointing to climate's distant past and the other to its possible future. The first was the 1980 paper by Luis Alvarez and colleagues that attributed the extinction of the dinosaurs to climate change caused by an asteroid impact and the subsequent dust in the atmosphere; the second was the 1982 article by Paul J. Crutzen and John W.

---

[74] Interview with Ackerman (cit. n. 70). In 2009 Ackerman did not define himself as a pacifist, seeing a necessity for military defense, and having selectively done some consulting for intelligence agencies.

[75] Ibid.

[76] Ibid.; Ackerman, 30 April e-mail (cit. n. 73).

[77] Rodgers and Walshaw, "The Computation of IR Cooling in Planetary Atmospheres," *Quarterly Journal of the Royal Meteorological Society* 92 (1966): 67–92.

Birks that looked at the possible long-term impact of fires and sooty smoke on Earth's climate in the aftermath of a nuclear war.[78] Both issues were widely discussed at all kinds of meetings before and after the publication of these papers. The TTAPS members were in the midst of all these developments and rose to the occasion, rushing into a new quantitative study of the climatic effects of nuclear war that also took soot into account and that ultimately resulted in the nuclear winter theory. There were three decisive steps in the dissemination of the TTAPS research results. The first was an attempt to present preliminary results at the 1982 American Geophysical Union (AGU) meeting, which was stopped by the NASA administration. The second was the so-called *Blue Book*, the 127-page provisional TTAPS study on nuclear winter, which was debated and reviewed by some hundred scientists at a closed meeting organized by Sagan in Cambridge, Massachusetts, in April 1983.[79] The third step was the first public presentation of results in late October and early November 1983, followed by an article on nuclear winter in *Science's* issue of December 23, 1983. As I am not discussing the public debate, I focus exclusively on the first two events.

The political context was more present than ever in the publication of the article that ultimately pushed nuclear winter research, the one by Crutzen and Birks, which had been commissioned for and published in a special double issue of the Swedish journal *Ambio*, a journal addressed to scientists and "to other interested readers" published by the Royal Swedish Academy of Sciences.[80] The American editor of *Ambio*, Jeannie Peterson, had been preparing this issue for two years. The editorial introduction to the special issue explicitly addressed its political motivation by referring to the 1980 UN report on nuclear weapons: "As there is no guarantee that the risk of war can be avoided, the need for nuclear disarmament is imperative."[81] Thus, from the outset, the issue of soot and its effect on climate was put forward as potentially serving the cause of a larger political movement.[82]

From the fall of 1982 on, the global consequences of nuclear war became part of the Washington political scenery. Representative Albert Gore, Jr., chair of the Subcommittee on Investigations and Oversight of the House Committee on Science and Technology, held a hearing titled "The Consequences of Nuclear War on the Global Environment," which the Department of Defense declined to attend, "probably because of expected criticism of the Reagan administration's statements and policies."[83]

Meanwhile, research on nuclear winter accelerated, and Turco intended later in the year to present first results at the meeting of the AGU in San Francisco.[84] However, Angelo Gustafero, deputy director of NASA Ames, called the AGU and prevented Turco's presentation at the last minute, arguing that the presentation had not been properly cleared and internally reviewed. The subject made NASA administrators worry, as they feared a political backlash and possible budget cuts from Washington.

---

[78] Alvarez et al., "Extraterrestrial Cause for the Cretaceous-Tertiary Extinction," *Science* 208 (1980): 1095–108; Crutzen and Birks, "The Atmosphere after a Nuclear War: Twilight at Noon," *Ambio* 11 (1982): 114–25.

[79] Richard P. Turco et al., "Global Atmospheric Consequences of Nuclear War" (draft, report no. U122878, R & D Associates, Marina Del Ray, Calif., March 1983).

[80] "Impressum," *Ambio* 11 (1982): 75.

[81] Jeannie Peterson et al., "About This Issue . . . ," *Ambio* 11 (1982): 75.

[82] Badash describes the history of the *Ambio* issue in detail, but misses this crucial point; *Nuclear Winter's Tale* (cit. n. 6), 50–3.

[83] Ibid., 94.

[84] Richard P. Turco et al., "Global Consequences of Nuclear 'Warfare,'" *Eos* 63 (1982): 1018.

Later, in 1983, they objected to the use of "nuclear war" in the title of the *Science* article, as "they wanted to minimize any suggestion that Ames was working outside its area of expertise," which might bring it into conflict with military laboratories and the Department of Defense.[85]

Turco complied with Gustafero's objection and withdrew the paper a day before the AGU conference. He was familiar with the political nature of atmospheric science not only through his policy-relevant work at RDA, but also through earlier work he had done at Ames on nuclear explosions. As part of his ozone research, he had engaged in an ongoing debate on the effects of the nitrogen oxide created during nuclear explosions on stratospheric ozone in the atmosphere.[86] His Ames research of 1975, published in the article coauthored with Whitten and Borucki, confirmed quantitatively John Hampson's qualitative conclusion that in a nuclear war, decreases in ozone "were large enough and lasted long enough . . . [to] have disastrous effects on the biosphere."[87] While the Ames article refrained from discussion of any political implications, Hampson, a former atmospheric researcher at the Canadian Armaments Research and Development Establishment and, in 1974, at the Centre de Recherches sur les Atomes et Molécules at Université Laval in Quebec, had made clear in his *Nature* article the political implications of the research by pointing to its relevance for future talks on arms limitation (thus anticipating one of the major issues at stake during the nuclear winter debate).[88]

Whereas the publication of the 1975 article had posed no problems, the political situation in 1982, with Ronald Reagan as president, was quite different. Toon remembers feeling that "the threat of being fired was real."[89] He also recalls that at other NASA laboratories researchers were strongly discouraged from taking up the issue—for example, James Hansen at the NASA Goddard Institute for Space Studies. Perhaps the member most affected by the political pressure on the TTAPS group was Ackerman, the youngest, who had joined Ames only in 1979. Ackerman remembers the enormous pressure during the nuclear winter years quite vividly: NASA discouraged TTAPS's efforts and cut the funding of the Ames laboratory in 1983. Later, when Ackerman was the technical coordinating editor of the 1985 SCOPE (Scientific Committee on Problems of the Environment) report, an international assessment of the nuclear winter theory, NASA did not allow him to work on the report in his laboratory during the day, forcing him to do it at home.[90]

Toon and Ackerman, as the youngest members of TTAPS, were (unsurprisingly) concerned much more with their careers and solving scientific puzzles than with politics. The Vietnam experience, as well as the drought of military money at the beginning of the 1970s, may have pushed them toward new and socially much more relevant research topics, but they had no explicit political beliefs that corresponded one-to-one with the science they did or drove them actively toward the nuclear winter research. Ackerman, in fact, at first did not think that there was much promise in it,

---

[85] Badash, *Nuclear Winter's Tale* (cit. n. 6), 99.

[86] Henry M. Foley and Malvin A. Ruderman, "Stratospheric NO Production from Past Nuclear Explosions," *J. Geoph. Res.* 78 (1973): 4441–50.

[87] Whitten, Borucki, and Turco, "Possible Ozone Depletions" (cit. n. 65), 38.

[88] John Hampson, "Photochemical War on the Atmosphere," *Nature* 250 (1974): 189–91.

[89] Interview with Toon (cit. n. 59).

[90] For the SCOPE report, see A. Barrie Pittock et al., *Environmental Consequences of Nuclear War*, 2 vols. (Chichester, 1985). SCOPE was part of the International Council of Scientific Unions.

and only changed his mind when he saw the first results coming from the calculations. While during their early careers they had shied away from political activity in order to be able to focus on research, Toon and Ackerman were now drawn into the middle of a highly political debate, a situation for which they were not prepared.[91] It is also an interesting twist that while Vietnam had made them steer clear of military research, now the issue of war had come back to them, not by way of a military contract, but through work done on the sidelines of their main research that had the potential to shake up Washington defense politics. In the end it seems that the anxious reaction to the nuclear winter research by Ames and its director, Clarence Syvertson, had the effect of bringing the TTAPS group closer together to defend itself.

Turco was better prepared to take on nuclear winter than his younger colleagues, not only because he had a secure footing outside Ames at RDA, but also because he knew from his own research on the atmosphere and nuclear weapons that these scientific issues were not innocent and unworldly, but rather had the potential to stir up political and public controversy. However, he had no experience in making political or public statements and avoided them in the beginning.

As the most media- and policy-savvy TTAPS member, Sagan stepped in to manage the political issues. He tried to make sure that despite all the pressure, Ames would remain the institutional home for TTAPS and would continue to provide computer time, something he negotiated with NASA administrator James M. Beggs.[92] Furthermore, he came up with the idea to counter NASA's criticism that there was a lack of internal review by organizing a review session for the nuclear winter research in April 1983.

This session brought together specialists from the various disciplines that had embarked on nuclear winter research; the intention was to leave the political dimension on the sidelines and focus exclusively on the science. The controversial potential of nuclear winter was obvious to Sagan and the steering committee, and so they decided "that political discussion, links to disarmament and arms control, and economic and social factors that might ordinarily be relevant to a conference on the impacts of nuclear war should not be a part of the proposed conference. . . . The Committee felt that the inclusion of other considerations such as nuclear strategy and economic, social, and political implications would detract from the central scientific message."[93] However, the composition of the scientific advisory board for the conference could be understood as a political declaration in itself. Among its sixty-one members were no representatives from the military or associated laboratories and think tanks, but Crutzen, Birks, and two special advisers for the *Ambio* issue, Joseph Rotblat and Henning Rohde, professor of chemical meteorology of the University of Stockholm, were all included, which tipped the balance strongly toward the disarmament camp. Rotblat, a British nuclear physicist who had briefly participated in and then distanced himself from the Manhattan Project, was the founder and former secretary general of the Pugwash Conferences on Science and World Affairs, an international organization created in 1957 to bring together scholars and public figures to explore ways to reduce nuclear arms and the threat of war.

---

[91] They clearly felt offended at the way they were described as peace-loving communists in the media; interview with Toon (cit. n. 59); interview with Ackerman (cit. n. 70).

[92] Badash, *Nuclear Winter's Tale* (cit. n. 6), 97.

[93] Paul R. Ehrlich et al., *The Cold and the Dark: The World after Nuclear War* (New York, 1984), xv.

Unsurprisingly, then, what appeared to the organizers to be an honest scientific enterprise looked to some of those who were not invited or did not participate like a conference with a foregone, politically motivated conclusion. By separating science and politics, the organizers had meant to choose the path of objective, disinterested science, adopting the rhetoric of their political opponents; the approach was intended to win over the largest possible number of doubters. However, whatever the organizers chose to do, atmospheric science *was* politics, and their opponents would make clear that the science of nuclear winter was too much imbued with uncertainty to turn around a skeptic. Furthermore, neither the April meeting nor the *Blue Book* were accessible to the larger public. Thus, the skeptics believed that the meeting itself had been a mere spectacle rather than following the standard procedures of scientific publication. Nor did it matter to them that the article later appeared in the peer-reviewed journal *Science*. The April meeting and Sagan's public campaign, which started before the scientific publication, had voided its content. Understandably, the younger TTAPS members, at the time not particularly preoccupied with the issue of disarmament, seem to have simply tried to ignore these issues as annoying distractions from the fascinating science. Ultimately, however, they had no choice but to spend endless hours during the following years trying to convince skeptics at professional meetings and in all kinds of publications.

SAGAN AND TELLER, SCIENTISM AND POLITICS

The profound transformations in the field of atmospheric science during the 1970s and 1980s shattered old certainties and alliances. Moreover, research done independently of direct military demand, and published not in laboratory reports but in easily accessible reviewed journals, threatened to undermine long-held positions of power within science as well as politics. One person who clearly had to worry was Edward Teller, a towering scientific and political presence over the previous four decades, very much occupied with promoting SDI research during the early 1980s. Teller's 1982 article in *Reader's Digest*, arguing against a nuclear weapons freeze and promoting defense measures and arms development, can be read as an indirect rebuttal of the Crutzen and Birks paper and of Congressman Gore's public hearing.[94]

Teller tried to stimulate research at LLNL that would invalidate or at least attenuate the nuclear winter results. Turco remembers:

> Edward always struck me as being very disturbed by the possibility that a small group of scientists outside the nuclear weapons community might ultimately be able to influence, or even determine, the direction of the US nuclear program through new discoveries long overlooked within his community. Hence, he mustered the considerable manpower and computer forces at Livermore to find alternative explanations and to propose mitigations where possible, and to downplay the new findings owing to uncertainty and the limitations of the original analysis. Yet, he seemed to me—at least in my encounters with him—to respect the fairly solid nature of the underlying physics which he was working hard to find flaws in.[95]

---

[94] Edward Teller, "Dangerous Myths about Nuclear Arms," *Reader's Digest*, November 1982, 139–44.

[95] Richard P. Turco, e-mail message to author, 25 February 2009.

Indeed, in 1983, LLNL researchers worked intensely on the problem of nuclear winter, particularly Michael MacCracken, Julius S. Chang, and Joyce Penner. Clearly, Teller tried hard to come up with scientific research that would refute the nuclear winter theory, as he seems to have admitted to Sagan that "Nuclear Winter was the only serious unanticipated consequence of nuclear war [he] was aware of."[96] Turco observes that the nuclear winter was "an embarrassment to these experts, and so the desire to disprove or at least minimize its importance. [There was a] natural reluctance to accept a new idea so far out of the mainstream of in-house thinking [and an] internal need to blunt the influence of these new and unproven ideas on policy, which was seen as a serious threat to the status quo in that field."[97]

An opportunity to play down the results came with the 1983 research by Starley L. Thompson and Stephen Schneider of the National Center for Atmospheric Research, who had developed a three-dimensional model that provided the most serious scientific challenge to the nuclear winter theory. They came up with results that attenuated the original estimates, later summarized in an article by the media-savvy Schneider as "nuclear 'fall'" or "nuclear autumn."[98] Nevertheless, despite skeptics who took it as an invalidation of the nuclear winter theory as a whole, this diminution of temperature effects did not mean an escape from possibly disastrous global consequences.

As the research at LLNL could not simply discard the nuclear winter theory, Teller made another attempt to take control of events, this time by inviting Turco personally to the annual International Seminar on Nuclear War in Erice, Sicily, of which he was a coorganizer.[99] He did so in the hope of presiding over the first presentation of the TTAPS results and thus preempting and containing the potential media reaction and promoting the LLNL research (which was presented at the meeting).[100] However, Turco, unwilling to have the research "scooped," declined the invitation and aligned himself with Sagan's strategy for publishing and publicizing their results. It is worth noting that—on the other side—neither Teller nor any other LLNL researcher had attended the April 1983 review meeting, nor had they been invited, probably, as they did not figure on the provisional invitation list. Political pressure had left its marks on cooperation in research and participation in scientific meetings.[101]

However, despite this entrenched political infighting and the personal antipathy between Sagan and Teller, they also had much in common. Thinking about planetary atmospheres and nuclear weapons meant thinking big. Both Teller and Sagan were fascinated by identifying difficult scientific problems that required up-to-date science, awaited bold technological solutions, and required broad public and political support for their successful implementation. Their uninhibited scientism, however, played out differently: Teller was much more a man of concrete action and direct ac-

[96] Carl Sagan to Edward Teller, 23 February 1984, quoted in Rubinson, "Containing Science" (cit. n. 6), 319.

[97] Turco, 25 February e-mail (cit. n. 95).

[98] See Starley L. Thompson and Stephen H. Schneider, "Nuclear Winter Reappraised," *Foreign Affairs* 64 (1986): 981–1005, on 993.

[99] Richard P. Turco, e-mail message to author, 3 August 2009.

[100] W. S. Newman and S. Stipcich, eds., *International Seminar on Nuclear War, 3rd Session* (Hackensack, N.J., 1992).

[101] I am grateful to Turco for providing me with a list of potential attendees and invitees to the conference. The list also provides information on who had received a copy of the *Blue Book* (at LLNL, only Chang and Penner). Teller does not figure on the list, but he seems to have obtained a copy directly from Sagan; see Rubinson, "Containing Science" (cit. n. 6), 319.

cess to political power, while Sagan loved thinking about science and the future of science for its own sake. With nuclear winter—as later on with global warming[102]— Teller chose a wait-and-see attitude, stressing the uncertainties, pushing study, and putting his hope in future technological solutions to be drawn from basic research. Turco remembers that "towards the end, [Teller] moved to embrace somewhat wild mitigation techniques, such as shooting rocket loads of fine metal needles into the stratosphere to trap global infrared radiation, as countermeasures that might render nuclear winter harmless."[103] Sagan, in contrast, put his confidence in the peer-reviewed TTAPS science and concurrently held to his political conviction that disarmament was the solution to a problem that science had successfully identified. Sagan and Teller both saw science as the sole guarantor of the future of Earth, but this shared conviction did not entail similar political positions.

Sagan, despite his association with the antinuclear movement and his political confrontation with Teller, seems never to have lost the dream of scientific intervention, of "planetary engineering," which provided the title of his last publication, written together with Pollack.[104] The article, expressing concern about global warming of Earth, returned to the theme of Sagan's 1961 *Science* article on Venus and discussed climate modification of the planets. Here we see these two scientists in the "can-fix-it" mode that characterized the attitude of the military think tanks in the 1950s and 1960s. Pollack and Sagan wrote, "After the discovery of the inclement high temperatures at the surface of Venus, it was natural to try to imagine a possible technological fix to make that planet more Earth-like."[105] The mixture of hubris and aesthetics in this article might have pleased Teller and put off many others: "In principle, anticipated advances in human technology might, within the next century, provide us with practical planetary thermostats. . . . We find . . . that planetary engineering generally entails severe ancillary environmental costs. This may be because we have not yet conceived of the really elegant methods."[106]

The two astronomers were much more consonant here with their nuclear winter critics, such as Teller, than with the younger members of TTAPS, who set aims for their careers that were no less ambitious, but certainly more down-to-earth, which they then pursued with perseverance. Contrary to what his biographer Keay Davidson claims, Sagan did not fit into the environmental movement that started with Rachel Carson.[107] As an astronomer, Sagan did think like the environmental movement, in the long term and on planetary scales. He hastened to point to long-term consequences, as observed in space, and then projected these on terrestrial developments, as in the case of nuclear winter. Unlike many environmentalists, however, he believed that science and the future of science would provide solutions to humanity's problems.

When Teller thought about environmental science at all, it was merely as an endeavor subservient to technology and nuclear armament, useful to show how to implement new technologies and weapons without permanently damaging Earth. Turco

[102] See Teller's article "The Planet Needs a Sunscreen," *Wall Street Journal*, October 17, 1997.

[103] Turco, 25 February e-mail (cit. n. 95).

[104] James B. Pollack and Carl Sagan, "Planetary Engineering," in *Resources of Near Space*, eds. John S. Lewis, Mildred Shapely Matthews, and Mary L. Guerrieri (Tucson, Ariz., 1993), 921–50.

[105] Ibid., 923.

[106] Ibid., 925–6. For Teller's enthusiasm for technological fixes, see Scott Kirsch, *Proving Grounds: Project Plowshare and the Unrealized Dream about Nuclear Earthmoving* (New Brunswick, N.J., 2005).

[107] Davidson, *Sagan* (cit. n. 51), 93.

describes his experience with the general attitude in military circles toward environmental problems and climate change thus:

> The nuclear weapons community of my time effectively ignored the potential for climate change. Some I knew would have chuckled over the idea that 0.5°C of temperature variation could actually be considered a significant side effect of nuclear warfare, in the face of massive direct destruction and widespread fallout (which many of them didn't see as a show-stopper anyway). In other words, climate was not really part of the discussion before nuclear winter.[108]

In the eyes of many military members and policy makers until the 1980s, severe environmental (and human) consequences of potential nuclear war were largely accepted as the unfortunate by-products of necessary military buildup, a worthwhile price for peace. The new environmental movement that rose in the 1970s, by contrast, attributed an intrinsic value to Earth and the environment. In his book, Wang underlines PSAC's "measured advocacy for the use of science and technology in solving environmental problems, which stood in clear contrast to the approaches of technological enthusiasts such as Edward Teller."[109] Research in atmospheric science from the 1960s on gave researchers a much better sense of the atmosphere's history, complexity, and fragility, which then—given the multiple uncertainties they were happy to acknowledge—made some in the field, though certainly not all, less inclined to propose grandiose engineering schemes or at least willing to delay them into a distant future.

CONCLUSION

In this light, the atmospheric sciences appear less interventionist and more modest than the nuclear science of someone like Teller, and they were ultimately embraced by environmentalists. Nevertheless, these sciences and the research topic of climate change both came out of cold war science. In fact, in the mid-1950s, cold war protagonists like John von Neumann were already "warning that 'climatological warfare' could become more potent than nuclear war."[110] Von Neumann clearly saw the future military and scientific potential of anthropogenic climate change, though it seemed still well in the future. Until the 1980s, nuclear war was about the destruction of cities and military installations all over the world, unprecedented in its global reach but ultimately local or regional in its effects. With nuclear winter, there was now the possibility of destruction of the whole Earth's current climate, lasting for several years, with disastrous environmental consequences for surviving human beings.

The dangers of nuclear war had been discussed before. The idea of "human self-destruction" had been addressed since the first atom bomb explosion and the horrors of Hiroshima and Nagasaki, reflected upon in rather general terms as a ghastly new possibility that had previously been beyond human reach. With nuclear winter, TTAPS claimed for the first time that in the case of a nuclear war between the super-

[108] Richard P. Turco, e-mail message to author, 15 December 2009.
[109] Wang, *Sputnik's Shadow* (cit. n. 32), 254.
[110] Weart, "Global Warming" (cit. n. 73), 335; Von Neumann, "Can We Survive Technology?" *Fortune Magazine*, June 1955, 106–8, 151–2, repr. in *The Fabulous Future: America in 1980*, ed. D. Sarnoff (New York, 1956), 33–48.

powers, climatological changes due to the resulting smoke would in all likelihood adversely affect life on a global scale. The TTAPS scientists made use of the research tools and topics that the cold war had set up. Like military researchers, they relied heavily on up-to-date computational technology and modeling. Like Teller with his weapons research, they profited from an environment "hospitable to concerns about global change" and framed their new object of research, nuclear winter and climate change, on a global scale, scientifically as well as politically.[111] But the cold war did not just set up a hospitable environment for high-tech global research, it also provided scientists with a welcome opportunity to assure their continuing authority over and indispensability to military and political issues.

With Teller, Sagan and Pollack shared a strong scientistic belief in the power of science to identify and solve the true problems of the future on Earth. In contrast, Turco, Toon, and Ackerman, less inclined to follow in these footsteps or to pursue interventionist adventures, focused on the diagnosis of the environmental consequences of human action. However, this shift from intervention to diagnosis did not weaken their status and authority in the long term. Rather, the atmospheric sciences and the debate over climate change would ultimately overtake Teller's nuclear science with the end of the cold war. For the latter three scientists, their turn to the atmospheric sciences entailed the major inconvenience of dragging them into political and scientific debates, for which they were little inclined and had no training, but it gave their research a new direction and meaning. Their atmospheric science emerged in the bright light of public scrutiny. Ultimately, nuclear winter research, and the coinage of the term "nuclear winter" itself, framed the consequences of nuclear war in terms of temperature, directing scientific and public attention toward this one parameter. In this sense, nuclear winter and the debate around it foreshadowed and structured the subsequent focus on anthropogenic climate change.

[111] Weart, "Global Warming," (cit. n. 73), 356.

Optimal Climate Change:
Economics and Climate Science Policy Histories (from Heuristic to Normative)

by Samuel Randalls[*]

ABSTRACT

Historical accounts of climate change science and policy have reflected rather infrequently upon the debates, discussions, and policy advice proffered by economists in the 1980s. While there are many forms of economic analysis, this article focuses upon cost-benefit analysis, especially as adopted in the work of William Nordhaus. The article addresses the way in which climate change economics subtly altered debates about climate policy from the late 1970s through the 1990s. These debates are often technical and complex, but the argument in this article is that the development of a philosophy of climate change as an issue for cost-benefit analysis has had consequences for how climate policy is made today.

INTRODUCTION

A growing historically informed literature marks the rise of climate change as an issue of public policy and concern.[1] With a few exceptions most accounts focus upon climate scientists and their relationships with policy makers, but it is also important to reflect upon the role that social scientists played in shaping climate change as a public issue from the 1980s.[2] These social scientists, generally economists, both supported and stifled calls for action to reduce carbon dioxide (CO_2) emissions, a primary cause of anthropogenic climate change, as highlighted by a 1979 National

---

[*] Department of Geography, University College London, Gower Street, London WC1E 6BT, U.K.; s.randalls@ucl.ac.uk.

[1] Some examples include James Rodger Fleming, *Historical Perspectives on Climate Change* (New York, 1998); David M. Hart and David G. Victor, "Scientific Elites and the Making of U.S. Policy for Climate Change Research," *Soc. Stud. Sci.* 23 (1993): 643–80; Myanna Lahsen, "Experiences of Modernity in the Greenhouse: A Cultural Analysis of a Physicist 'Trio' Supporting the Backlash against Global Warming," *Global Environmental Change* 18 (2008): 204–19; Diana M. Liverman, "Conventions of Climate Change: Constructions of Danger and the Dispossession of the Atmosphere," *J. Hist. Geogr.* 35 (2009): 279–96; Spencer Weart, *The Discovery of Global Warming* (Boston, 2004).

[2] Maxwell T. Boykoff, David J. Frame, and Samuel Randalls, "Discursive Stability Meets Climate Instability: A Critical Exploration of the Concept of 'Climate Stabilization' in Contemporary Climate Policy," *Global Environmental Change* 20 (2010): 53–64; Naomi Oreskes, Erik M. Conway, and Matthew Shindell, "From Chicken Little to Dr. Pangloss: William Nierenberg, Global Warming, and the Social Deconstruction of Scientific Knowledge," *Hist. Stud. Nat. Sci.* 38 (2008): 109–52; Steve Rayner and Elizabeth L. Malone, *Human Choice and Climate Change*, 4 vols. (Columbus, Ohio, 1998).

Academy of Sciences (NAS) report.[3] It is therefore instructive to revisit economic literatures since economic approaches are being actively debated and implemented today. This article provides a historical account of the use of cost-benefit analysis in climate change research from 1974 to 1996. It argues that economic research in this period developed from tentative claims about action on climate change to direct policy advice. In this process, heuristics became prescriptions and policy developed around ideals of optimal solutions. Rather than conceiving economic analyses of climate change as *merely* providing justification for (or against) economic responses to climate change, social scientists also implicitly reshaped the goals of climate change policy. Histories of climate change as a policy issue thus require examination of social science as well as scientific literatures, especially given the primacy of cost-benefit analyses in decision making.

Naomi Oreskes, Erik M. Conway, and Matthew Shindell suggest that economists provided a framework in the 1980s through which skeptics could claim support for not reducing carbon emissions immediately (it would be too costly). Thus, they suggest, "the carbon problem" (reducing emissions of CO_2) became "the climate question" (is climate change worth tackling?).[4] Maxwell T. Boykoff, David J. Frame, and I likewise support this hypothesis by suggesting that cost-benefit analysis helped lock in a policy that was framed around maximizing the use of climatic resources up to the point of dangerous anthropogenic interference rather than reducing CO_2 emissions to the lowest possible amount.[5] Economic analyses have played significant political roles to this day, and before turning to the historical material it is valuable to review the contemporary attention given to such analyses in the formation of climate change policy.

Nicholas Stern's *Review on the Economics of Climate Change*, commissioned by the U.K. government, represented an important example of climate change cost-benefit analysis when it was released in 2006, especially given its focus on being seen to be an ethically informed analysis rather than just an economically efficient one. Indeed, Cameron Hepburn's comments on the flap of the published version of the report attest to the belief that this extensive analysis represented an economic turning point in actions to limit climate change.[6] The *Stern Review* estimated that the cost of taking action now to prevent climate change would be much cheaper than the cost of future damages from climate change (1 percent of the world's GDP every year vs. 5–20 percent), a finding that has been subsequently used by the U.K. government under both Gordon Brown and David Cameron as a political anchor for the necessity of taking action. The review highlighted the political appeal of an economic analysis that would provide quantitative answers (even if sketched with some uncertainties) to the question of what to do about climate change. Cost-benefit analysis would prove the importance of stabilizing global climate. But Simon Dietz, Chris Hope, and Nicola Patmore were moved to ask why cost-benefit analysis should be used to make this political decision.[7] Terry Barker, on the other hand, saw the review as signaling

[3] Jule G. Charney, Akio Arakawa, D. James Baker, Bert Bolin, Robert E. Dickinson, Richard M. Goody, Cecil E. Leith, Henry M. Stommel, and Carl I. Wunsch, *Carbon Dioxide and Climate: A Scientific Assessment* (Washington, D.C., 1979).

[4] Oreskes, Conway, and Shindell, "From Chicken Little to Dr. Pangloss" (cit. n. 2).

[5] Boykoff, Frame, and Randalls, "Discursive Stability Meets Climate Instability" (cit. n. 2).

[6] Hepburn, comments on the flap of Nicholas Stern, *Stern Review on the Economics of Climate Change* (Cambridge, 2007).

[7] Dietz, Hope, and Patmore, "Some Economics of 'Dangerous' Climate Change: Reflections on the Stern Review," *Global Environmental Change* 17 (2007): 311–25.

a move away from narrow cost-benefit analysis to a more multidimensional form of economics, representing a change in values toward an ethical approach to climate change girded by a low discount rate. This perspective clashed with those economists, such as William D. Nordhaus, schooled in analyses that aimed to be more "neutrally scientific" or that valued merely "descriptive heuristics" (a term particularly favored in early climate change economics research).[8] Disagreement about the types of cost-benefit analysis that should be performed highlights the political nature of this economic work.

As historian Theodore Porter's work has demonstrated, cost-benefit analysis in the United States has been a part of decision making since the 1920s in projects ranging from dams to railroads, though it was the Flood Control Act of 1936 that concretized its role. Quantification and uniform analysis promised rational decision making within a bureaucracy that would enable seemingly transparent conclusions to be reached.[9] Porter suggests that it was during the 1950s that economists began to redefine cost-benefit analysis as a predominantly quantitative, economic exercise, and by the 1960s they were applying the concept to topics as diverse as urban renewal and public health:

> Although they [economists] routinely concede by way of preface that calculation can never replace political judgment, cost-benefit and risk analysts clearly want to rein it [political judgment] in as much as possible. . . . The ideal of mechanical objectivity has by now been internalized by many practitioners of the method.[10]

Feminist economist Julie Nelson associated the *Stern Review* with the entrenchment of cost-benefit analyses in contemporary political decision making. She suggested that too much attention in these analyses is paid to the questions of optimality, cost-effectiveness, the behavior of *homo economicus*, and mathematical precision (or attempts at it), rather than thinking through what she terms a "high-reliability economics."[11] Nelson's approach would place uncertainty, resilience, and complexity at the heart of climate economics as opposed to simple, intuitive estimates that were necessarily limited in their precision.[12]

Economist Stephen DeCanio likewise presents a critique of "traditional" economic models of climate change, in particular, their focus on quantitative cost-benefit ac-

---

[8] Terry Barker, "The Economics of Avoiding Dangerous Climate Change: An Editorial Essay on the *Stern Review*," *Climatic Change* 89 (2008): 173–94. An example of the heuristic claim can be found in Nordhaus, "A Review of the *Stern Review* on the Economics of Global Warming," *Journal of Economic Literature* 45 (2007): 686–702. Discount rates are used by economists to equate present and future costs; the lower the rate, the more expensive future costs appear over present costs and thus, for climate change analysis, the more likely that costly action should be taken now to prevent those future costs. Discount rates have been the source of much dispute in climate change economics, supporting Lester B. Lave's hypothesis that this was an ideal subject through which to examine the way economists treated discount rates more broadly; Lave, "Mitigating Strategies for Carbon Dioxide Problems," *American Economic Review* 72 (1982): 257–61.

[9] Theodore M. Porter, *Trust in Numbers: The Pursuit of Objectivity in Science and Public Life* (Princeton, N.J., 1995).

[10] Ibid., 189.

[11] Julie A. Nelson, "Economists, Value Judgments, and Climate Change: A View From Feminist Economics," *Ecological Economics* 65 (2008): 441–7, on 445.

[12] See also Christian Azar, "Are Optimal CO_2 Emissions Really Optimal?" *Environmental and Resource Economics* 11 (1998): 301–15.

counts that might result in misleading policy prescriptions.[13] He suggests that economic models lack appropriate accounts of time, consumer preferences, production, and forecasting performance. While the *Stern Review* reflects a greater economic focus on normative valuation rather than just optimal efficiency, DeCanio's account points the finger of blame for current climate policy problems at economists' resistance to examining the overprecise calculations and assumptions guiding their conclusions.[14] Jeroen van den Bergh has even questioned whether it is possible to perform a quantitative cost-benefit analysis of climate change.[15] There is thus some dispute as to whether the *Stern Review* represents an incremental improvement in economics (through, e.g., explicitly arguing for an ethical justification for discount rates) or rather perpetuates trust in quantitative decision making under conditions of uncertainty. What is clear, however, is that to understand the political and economic relevance of the *Stern Review*, it is important to reflect upon the historical emergence of a particular economic philosophical approach to climate change that embraced forms of cost-benefit analysis as a core part of climate policy.

Technically there are several types of cost-benefit analysis used in climate change studies, but two are important for this article. Optimization analyses compute the costs and benefits of climate change risks and mitigation to examine the most economically optimal stabilization policy for the climate. The focus is upon efficient economic outcomes rather than equity or least harm. A target-based analysis, on the other hand, examines the most economically efficient solutions to reach a predefined target that is specified external to the analysis (i.e., if scientists said the target should be a temperature increase of no more than 2°C, then that is what economists would use).[16] While these are clearly separable theoretically, in practice it may be harder to separate them if the target in the target-based cost-benefit analysis is at all derived from other forms of economic analyses (including optimization analyses).

Cost-benefit analysis provided a heuristic reasoning that initially offered descriptive accounts of the viability of potential options, but also became prescriptive about what the economic goal of climate policy should be. Understanding climate science and policy thus involves more than simply examining the role of scientists, politicians, policy makers, and special-interest groups; it requires a more detailed account of the work of social scientists, especially economists, from the 1970s. Clearly many topics of interest within climate change economics cannot be covered here, and most especially the article focuses upon neoclassical economists to the exclusion of economists from other theoretical persuasions (e.g., Kenneth Boulding and Herman Daly). This article is thus intended as a call for a more sustained historical engagement with

[13] DeCanio, *Economic Models of Climate Change: A Critique* (Basingstoke, 2003); DeCanio, "Descriptive or Conceptual Models? Contributions of Economics to the Climate Policy Debate," *International Environmental Agreements* 5 (2005): 415–27.

[14] Stern, *Stern Review on the Economics of Climate Change* (cit. n. 6); DeCanio, *Economic Models of Climate Change* (cit. n. 13).

[15] Van den Bergh, "Optimal Climate Policy Is a Utopia: From Quantitative to Qualitative Cost-Benefit Analysis," *Ecological Economics* 48 (2004): 385–93.

[16] M. Munasinghe, P. Meier, M. Hoel, S. W. Honig, and A. Aaheim, "Applicability of Techniques of Cost-Benefit Analysis," in *Climate Change, 1995: Economic and Social Dimensions of Climate Change*, eds. James P. Bruce, Hoesung Lee, and Erik F. Haites (Cambridge, 1996), 149–75.

social science research than has heretofore been found in many accounts of climate science and policy.[17]

CLIMATE AS RESOURCE

Attempts to establish the costs and benefits of living, working in, and visiting particular climates are not new to the late twentieth century. There is significant historical research examining and providing examples of the ways in which climates (or at least attitudes) were intentionally modified to provide more conducive environments, whether for skin complexion or agricultural production, since the early modern period; determining the optimal climate for production or health was an important imperial and commercial activity.[18] William Stanley Jevons in the nineteenth century examined the relationship between weather, crop production, and aggregate economic activity, arguing that there was a direct link (and cyclicity) between weather trends and national economic performance.[19] Other industry-weather links included research on worker health in the indoor industrial environment. Indeed, this was taken forward most especially from the 1920s with the rise of air pollution as an industrial hygiene concern and in the 1950s with climatologist Helmut E. Landsberg's promotion of the study of industrial and urban climates.[20] William Maunder's work from the 1960s further developed understandings of how weather and climate would affect economic productivity in a range of activities, from manufacturing and retail to transportation. Maunder saw his work as advancing a resource analysis of the atmosphere such that weather could be best managed, even if this was always likely to lead to some contradictory effects, because what was optimal for one industry would not be for another.[21] He favored a more complete economic analysis that would include weather information services, long-range forecasting, and weather modification. Maunder's work drew from a series of narrower studies of the effects of climate on particular industries. One example can be found in research from the 1930s through the 1960s exploring the optimal climate for sales in department stores using a set of quantitative and qualitative analyses from data provided by customers and staff.[22] An optimal climate for maximizing consumption could be hypothesized and then created in-store.

Outdoor climates could be unintentionally changed in ways that would have seri-

[17] Reviews of the literature include the chapter by Munasinghe et al., "Applicability of Techniques of Cost-Benefit Analysis" (cit. n. 16), in the 1995 International Panel on Climate Change's *Second Assessment Report*; David Demeritt and David Rothman, "Figuring the Costs of Climate Change: An Assessment and Critique," *Environment and Planning* A 31 (1999): 389–408; Samuel Fankhauser and Richard S. J. Tol, "The Social Costs of Climate Change: The IPCC Second Assessment Report and Beyond," *Mitigation and Adaptation Strategies for Global Change* 1 (1997): 385–403.

[18] Examples include Bernard Mergen, *Weather Matters: An American Cultural History since 1900* (Lawrence, Kans., 2008); Mark Carey in this volume; Fleming, *Historical Perspectives on Climate Change* (cit. n. 1); Richard H. Grove, *Green Imperialism: Colonial Scientists, Tropical Island Edens and the Development of Global Environmentalism, 1600–1860* (Cambridge, 1994).

[19] Other economists took up this work in the early twentieth century, including Henry Ludwell Moore; see Philip Mirowski, *The Effortless Economy of Science* (Durham, N.C., 2004).

[20] Christopher C. Sellers, "The Dearth of the Clinic: Lead, Air, and Agency in Twentieth-Century America," *J. Hist. Med. Allied Sci.* 58 (2003): 255–91; Landsberg, *Weather and Health: An Introduction to Biometeorology* (New York, 1969); Landsberg, *The Urban Climate* (New York, 1981).

[21] William J. Maunder, *The Value of the Weather* (London, 1970).

[22] E.g., J. W. Snellen, "The Optimal Climate in Department Stores," *British Journal of Industrial Medicine* 19 (1962): 165–70.

ous effects on productivity. Science administrator W. O. Roberts, the first head of the National Center for Atmospheric Research, suggested that a changing climate might create serious issues of responsibility for the United States and Canada if they were going to be able to continue their role as world food granaries.[23] Another example can be found in the work of natural resource economist Irving Hoch in the early to mid-1970s, which examined the differences in wage rates in different climates, arguing that these could be indicative of future welfare issues if the climate changed. As Hoch (with Judith Drake) stated, "A summer temperature around 74° is optimal, given average precipitation" (wetter weather would diminish the advantages of higher temperatures).[24] Climate could be an economic resource for laborers, although equally this advantage could be ceded under "threatening" (welfare-diminishing) climatic changes.

If climate could represent a resource, but also a threat, then how would one manage it most effectively? One answer is to balance the costs and benefits of changing climates, and this was taken up in research on air pollution. This approach to natural resources sought to maximize the efficiency of their use, because industry and jobs were in many cases dependent on the ability to pollute, while pollution levels needed to be restricted to preserve those atmospheric services required for survival.[25] Creating a resource map of the atmosphere was critical to determining those benefits that needed to be preserved.[26] Maunder stated that "we must *first* find out the value of atmospheric resources at the present time, polluted or unpolluted, for then and only then will we be able to decide whether they are worth controlling, and in what way they should be controlled."[27] Thus it would be possible to determine the level that pollution should not exceed so that it does not harm the economically beneficial services the atmosphere provides, while also balancing the economic needs for those services that polluting industry provides. Finding critical thresholds in acid rain policy (or health policy), or the upper limits of CO_2 concentrations in the atmosphere in climate change policy, could define maximum limits for pollution that could then provide a baseline for determining the most efficient use of that limited resource.[28] In other words, what this work suggested is that pollution levels should be optimized to the highest quantity permitted by the capacity of the

[23] Roberts, "Climate Change and the Quality of Life for the Earth's New Millions," *Proc. Amer. Phil. Soc.* 120 (1976): 230–2.

[24] Hoch and Drake, "Wages, Climate, and the Quality of Life," *Journal of Environmental Economics and Management* 1 (1974): 268–95, on 292. An extended version appears as Hoch, "Climate, Wages, and Urban Scale," in *The Urban Costs of Climate Modification*, ed. Terry A. Ferrar (New York, 1976), 171–215. This work was partially supported by the U.S. Department of Transportation and was published in the first volume of the *Journal of Environmental Economics and Management*, which was established by Ralph D'Arge and Allan Kneese (who in 1978 became the first president of the Association of Environmental and Resource Economists).

[25] See, e.g., R. G. Ridker, *Economic Costs of Air Pollution: Studies in Measurement* (New York, 1967); H. Wolozin, ed., *The Economics of Air Pollution* (New York, 1966).

[26] We see these resources discussed in comprehensive fashion in Jesse Ausubel and A. K. Biswas, eds., *Climatic Constraints and Human Action* (Oxford, 1980), and Maunder, *Value of the Weather* (cit. n. 21).

[27] Maunder, *Value of the Weather* (cit. n. 21). Reducing pollution was deemed a key control strategy.

[28] On acid rain, see Kristin Asdal, "Enacting Things through Numbers: Taking Nature into Account/ing," *Geoforum* 39 (2008): 123–32. On climate change, see Richard S. J. Tol, "Europe's Long-Term Climate Target: A Critical Evaluation," *Energy Policy* 35 (2007): 424–32. Examples of this in the work of Nordhaus and others are presented in the next section.

atmosphere rather than reduced completely (and at a high cost).[29] This is a rather
different form of cost-benefit analysis than that used to make decisions about the
provision of future weather information services, bearing more similarity to cost-
benefit analyses of weather modification in respect to the idea of control. Indeed, it is
interesting to note that climate stabilization meant forms of climate control for William W. Kellogg and Stephen H. Schneider in which one question would be how to
compensate the losers from any deliberate change in climate.[30] Climate stabilization
(i.e., stabilization of atmospheric CO_2 levels) was seen as a direct intervention in climate to produce conditions that would enhance well-being. Thus, how to establish an
equitable optimal climate would be a vital question if the successes of weather modification could be replicated on a broader scale. As the funding for frequently milita-
rized projects of weather modification dried up after 1979, the promise of controlling
weather through technology dissipated.[31] Unintentional climate change posed similar
challenges, assuming it was relatively controllable. Whether in the area of forecasting, pollution levels, or modification, weather and climate would be subject to an
economic analysis to decide how to most effectively use the resources provided by
the atmosphere.

<h3 style="text-align:center">CLIMATE CHANGE COST-BENEFIT ANALYSIS</h3>

Prior to the establishment of formal cost-benefit analysis procedures for the issue of
climate change and energy, two economists had formulated rather different conclusions on the likely outcome. Ralph D'Arge and K. C. Kogiku concluded in 1973 that
it would be prudent to restrict CO_2 emissions where practicable, while Nordhaus in
1974 discounted the greenhouse effect as not being significant enough to provide
major constraints on energy growth, at least not in the short term.[32] Partly this difference emerged from a focus on pollution in the former case and a focus on energy resources in the latter. Alvin Weinberg, director of the U.S. Office of Energy Research
and Development (from 1974) and founder and director of the Institute for Energy
Analysis at Oak Ridge National Laboratory (1975–84), took a different position on
energy, however, and in 1974 pointed out the implications of feared CO_2-induced
climate change for energy production, proposing that a global institute of climatology explore the problem.[33] Climate scientists Stephen H. Schneider and Warren M.
Washington, in their response in 1975, pointed out the urgency of the problem, while
Hermann Flohn in 1977 developed a set of climate "scenarios" that bore "alarming

[29] This is the same argument used by third alkali inspector William Nicholson from the early 1900s
for regulating industrial pollution in the United Kingdom. For more on the Alkali Inspectorate, see
John Sheail, "'Burning Bings': A Study of Pollution Management in Mid-Twentieth Century Britian,"
J. Hist. Geogr. 31 (2005): 134–48.
[30] Kellogg and Schneider, "Climate Stabilization: For Better or for Worse?" Science 186 (1974):
1163–72.
[31] James Rodger Fleming, "The Pathological History of Weather and Climate Modification: Three
Cycles of Promise and Hype," Hist. Stud. Phys. Biol. Sci. 37 (2006): 3–25; Fleming, Fixing the Sky
(New York, 2010).
[32] D'Arge and Kogiku, "Economic Growth and the Environment," Review of Economic Studies 40
(1973): 61–77; Nordhaus, "Resources as a Constraint on Growth," American Economic Review 64
(1974): 22–6. Neither D'Arge nor Nordhaus referenced the other's work extensively at any point during the following years.
[33] Weinberg, "Global Effects on Man's Production of Energy," Science 186 (1974): 205.

inferences . . . for the energy problem."[34] The problem with many of these studies, however, was a lack of precision in the estimates, given the uncertainties about the relationship between CO_2 and climate change and the role of industrial energy production within that. What was needed was a way of making a more "objective" decision about the problem that could be used for determining policy advice. Cost-benefit analysis provided one such tool.

D'Arge edited a U.S. Department of Transportation study in 1975 entitled *Economic and Social Measures of Biologic and Climatic Change*, a work that is cited as being the earliest cost-benefit analysis of greenhouse gas control, though it focused on ozone depletion.[35] From the mid- to late 1970s a number of other economists examined the issues presented by climate change. Jesse Ausubel and A.K. Biswas published a collection of articles on the atmosphere as resource and constraint in 1980.[36] Thomas Schelling, professor of economics at first Yale and then Harvard, chaired the first academy committee to examine the economic and social dimensions of anthropogenic warming in 1980.[37] Yet these authors did not provide any particular method of thinking through climate change as a policy issue.

Here, therefore, I focus mainly upon Nordhaus, because even though he was arguably much less influential than Schelling, he developed and published, in a transparent and accessible way, his heuristic for thinking about managing climate change.[38] He also has the virtue of having published the assumptions he made for his models. Nordhaus was trained as an economist at Yale (BA, 1963) and the Massachusetts Institute of Technology (PhD, 1967), before returning to Yale as an assistant professor and later, in 1973, becoming full professor of economics. A highly distinguished economist, Nordhaus was appointed a member of the NAS Carbon Dioxide Assessment Committee (1981–4) and has subsequently served on several National Academy committees on climate change, energy, and resources. He is particularly interesting for our topic because he published some of the earliest work on CO_2 economics for climate policy and because he was able to apply resource economics to heuristic, cost-benefit approaches to the problem of climate change.

In 1977 Nordhaus published a path-breaking paper in the *American Economic Review* that examined a heuristic for thinking about carbon stabilization. Assuming a future doubling of CO_2, his model calculated the most cost-efficient methods of meeting that target. He wrote, "As a first approximation . . . it seems reasonable to

[34] Schneider and Washington, "Letters: Energy Production and Climate," *Science* 187 (1975): 13; Flohn, "Climate and Energy: A Scenario to a 21st Century Problem," *Climatic Change* 1 (1977): 5–20, on 17.

[35] Ralph C. D'Arge, ed., *Economic and Social Measures of Biologic and Climatic Change: Climate Impact Assessment Program* (Washington, D.C., 1975). The claim for this being the earliest cost-benefit analysis on greenhouse gas control is made by Clive L. Spash, "Trying to Find the Right Approach to Greenhouse Economics: Some Reflections upon the Role of Cost-Benefit Analysis," *Analyse und Kritik* 16 (1994): 186–99. There were clearly preceding studies of air pollution (see n. 25 above), and Spash also neglects work in the 1980s, but ultimately the value of his claim comes down to the extent to which one includes partial or incomplete analyses and what counts in terms of climate change. Putting priority on this work is difficult given the different communities (pollution, energy, weather information) within which these economists were working during this period.

[36] Ausubel and Biswas, *Climatic Constraints and Human Action* (cit. n. 26).

[37] See Oreskes, Conway, and Shindell, "From Chicken Little to Dr. Pangloss" (cit. n. 2), for more on Schelling.

[38] Ibid. Note that there is no intention to single out Nordhaus here as some kind of scapegoat. Other economists made claims similar to his.

argue that the climatic effects of carbon dioxide should be kept well within the normal range of long-term climatic variation. A doubling of the atmospheric concentration of carbon dioxide is a reasonable upper limit to impose at the present state of knowledge."[39]

Nordhaus cited the latest research from climate modelers Syukuro Manabe and Richard Wetherald that suggested this doubling would lead to a 3°C temperature rise.[40] His heuristic was not novel in terms of other environmental and resource problems (carrying capacity and population limits to growth were founded on very similar modes of reasoning), but his approach had clear policy implications.[41] If an upper limit to CO_2 concentration levels could be defined, then economic models could be used to calculate optimal emissions pathways to ensure that this concentration level was not exceeded, without the cost of this action being prohibitive. This approach set in place a particular logic extrapolating backward from anticipated climate damages to CO_2 concentration levels to CO_2 emissions, with the realization that each relationship at the time was subject to uncertainties and scientific debate. The ability to do this is dependent upon the success of climate and carbon-cycle models, and economists did not have particularly sophisticated ones available to them.[42] There is a subtle but important framing here. The heuristic aims to maximize use of the atmospheric resource until the world reaches the climatic target dictated by scientists. It thus increases exposure to any uncertainties in the relationship between emissions, concentrations, and temperature should it be applied as a policy decision.[43] Nevertheless, it is critical to stress that Nordhaus explicitly suggested this should be a heuristic to assess the costs of climate action; it was not to be a direct, predictive guide to policy as such.

In 1979 Nordhaus produced his full cost-benefit analysis of greenhouse gas reductions especially in relation to energy, employing ideas he had developed on future

[39] Nordhaus, "Economic Growth and Climate: The Carbon Dioxide Problem," *American Economic Review* 67 (1977): 341–6, on 344. Nordhaus's farsightedness in choosing this limit at this time could be no more than a heuristic, but it resonates through many later climate change economic studies. See a critique of 1990s economic studies using doubling scenarios by Samuel Fankhauser and David W. Pearce, "The Social Costs of Greenhouse Gas Emissions," in *The Economics of Climate Change: Proceedings of an OECD/IEA Conference*, OECD (Paris, 2004), 71–86.

[40] Manabe and Wetherald, "The Effect of Doubling CO_2 Concentration on the Climate of a General Circulation Model," *Journal of Atmospheric Science* 32 (1975): 3–15. See James Rodger Fleming, "Climate Change and Anthropogenic Greenhouse Warming: A Selection of Key Articles, 1824–1995, with Interpretive Essays," Primary Articles Learning Environment, *NSDL Classic Articles in Context*, no. 1 (April 2008), National Science Digital Library wiki, http://wiki.nsdl.org/index.php /PALE:ClassicArticles/GlobalWarming (accessed 14 January 2011). It is beyond the scope of this article to trace the history of the varying temperature increases predicted from a doubling of CO_2 concentrations, but they have been remarkably stable between 1.5°C and 4.5°C, though many social scientists and policy makers since the 1970s have selected 2°C or 3°C (the mode) as the working figure in their analyses. A good analysis of climate sensitivity to a doubling of CO_2 concentrations is in Jeroen van der Sluijs, S. van Eijndhoven, Simon Shackley, and Brian Wynne, "Anchoring Devices in Science for Policy: The Case of Consensus around Climate Sensitivity," *Soc. Stud. Sci.* 28 (1998): 291–323.

[41] On carrying capacity, see Nathan Sayre, "The Genesis, History, and Limits of Carrying Capacity," *Ann. Assoc. Amer. Geogr.* 98 (2008): 120–34.

[42] Jae A. Edmonds and J. M. Reilly, "Future Global Energy and Carbon Dioxide Emissions," in *Atmospheric Carbon Dioxide and the Global Carbon Cycle*, U.S. Department of Energy report DOE/ ER-0239 (Washington, D.C., 1985), 215–45. The concentration-to-temperature relationship ("climate sensitivity") is also important, but given that concentration levels were used as proxies for climatic damages (as per the Nordhaus quotation discussed earlier), this relationship did not constitute a specific issue in economists' research, though it did add another layer of uncertainty.

[43] See Boykoff, Frame, and Randalls, "Discursive Stability Meets Climate Instability" (cit. n. 2).

constraints to energy growth.[44] In this more comprehensive study he effectively placed an external constraint on emissions levels and then asked what kinds of consequences different constraints would have for possible optimal carbon tax rates. This enabled the creation of a series of optimal emissions pathways. His work used a linear model of emissions that was updated in later work with Gary Yohe to include a more highly aggregated multivariable model.[45] This was based upon a probabilistic scenario approach, which was more weakly developed than models by the U.S. Department of Energy (DOE) or the International Institute for Applied Systems Analysis (IIASA).[46] It had the advantage, however, of being easy to understand and use, whereas nonspecialists could find the results of other models difficult to interpret. All the models were reliant on assumptions about future energy demand, and though the topic is not in the scope of this article, it is important to reflect that during this period energy projections were also being more comprehensively developed as complex tools. While the 1970s Oak Ridge scenarios predicted significant growth in demand, the new models from IIASA predicted lower demand on the basis of an assumption of greater energy efficiency.[47]

In 1980, the report of a two-year collaboration between the DOE and the American Association for the Advancement of Science included a contribution by Nordhaus, "Theoretical and Empirical Aspects of Optimal Control Strategies."[48] This provided an opportunity to further adjust his natural resource models to the case of CO_2; indeed, his success at this can be seen in the way later authors such as Jae Edmonds and J. M. Reilly discussed the results of the DOE's new models as improvements upon Nordhaus's earlier work.[49] Nevertheless, as the DOE's agenda report on current and required future research in 1980 goes on to state, determining the optimal resiliency (meeting the goals of climate "stability" and economic growth) is an important research issue, because the costs of achieving a complete reduction in CO_2 for maximum climate resiliency (least harm) would be too high.[50] In stating this priority, however, the analysis suggested which concentration levels were cost-effective and which were not, effectively suggesting which climatic damages should be allowed and which prevented. What this heuristic implicitly does is turn an analysis of optimal emissions taxes or pathways into one of optimal climate change. This gamesmanship was sketched out by A. M. Perry and colleagues in 1982: "Useful insights regarding available energy policy options can be gained by considering the combined implications of a range of plausible CO_2 limits, a range of plausible projections for

[44] Nordhaus, *The Efficient Use of Energy Resources* (New Haven, Conn., 1979); Nordhaus, "Resources as a Constraint on Growth" (cit. n. 32).

[45] The later work was published as Nordhaus and Yohe, "Future Carbon Dioxide Emissions from Fossil Fuels," in *Changing Climate: Report of the Carbon Dioxide Assessment Committee*, ed. William Nierenberg (Washington, D.C., 1983), 87–153.

[46] Edmonds and Reilly, "Future Global Energy and Carbon Dioxide Emissions" (cit. n. 42). For information on the background and strategies of IIASA, see Sonja Boehmer-Christiansen, "Global Climate Protection Policy: The Limits of Scientific Advice, Part 1," *Global Environmental Change* 4 (1994): 140–59; Brian Wynne, "The Institutional Context of Science, Models, and Policy: The IIASA Energy Study," *Policy Sciences* 17 (1984): 277–320.

[47] See Edmonds and Reilly, "Future Global Energy and Carbon Dioxide Emissions" (cit. n. 42).

[48] This work is reported in full detail in Nordhaus, "Thinking about Carbon Dioxide: Theoretical and Empirical Aspects of Optimal Control Strategies" (Department of Energy-Cowles Foundation discussion paper no. 565, Department of Economics, Yale University, New Haven, Conn., 1980).

[49] Edmonds and Reilly, "Future Global Energy and Carbon Dioxide Emissions" (cit. n. 42).

[50] DOE, *Environmental and Societal Consequences of a Possible CO_2-Induced Climate Change: A Research Agenda*, vol. 1, report no. DOE/EVV/10019-01 (Washington, D.C., 1980).

future world energy consumption and a range of possible energy-supply scenarios that could satisfy the CO_2 limits."[51]

In practice, economists reduced this range of limits into a single one that aided analysis but reduced options. As this heuristic subtly transformed into policy options, the fact that most studies took a doubling of CO_2 concentrations as their benchmark became important. Examples include the work of John A. Laurmann and of Perry and colleagues.[52] The economists followed the protocol of the climate modelers in adopting this doubling scenario as the heuristic.[53] The economists were to some extent dependent on the scientific advice for those upper limits to calculate the costs and benefits of action. Thus this doubling scenario rapidly became the (notional) target in economic analysis; instead of being descriptive heuristics, the analyses had become prescriptive in calculating how best to meet this target. In the 1990s authors such as Schelling and Samuel Fankhauser and David Pearce would reflect on the fact that doubling scenarios were really only arbitrary estimates or scenarios, not actual predictions.[54]

Thus far, economists had created simple heuristics rather than concrete prescriptions. Nordhaus in 1982 admitted that his studies had viewed "the CO_2 problem . . . as a pure exercise in optimal economic growth" and recognized that this ignored problems of externalities, uncertainties, and control techniques.[55] The NAS Carbon Dioxide Assessment Committee report, chaired by William Nierenberg, concluded in 1983 that most economists agreed that until further scientific research could be conducted into climate change, the most prudent strategy would be adaptation and further research, given the costs of reducing emissions and the unclear benefits of that action.[56] The report was in keeping with economic research at the time conducted by Nordhaus, Ausubel, and Yohe, but also with work by D'Arge and colleagues and by Klaus Meyer-Abich.[57] Some energy researchers concluded that there were more

[51] A. M. Perry, K. J. Araj, W. Fulkerson, D. J. Rose, M. M. Miller, and R. M. Rotty, "Energy Supply and Demand Implications of CO_2," *Energy* 7 (1982): 991–1004, on 991. Note that this is scenario-based analysis and does not provide an actual prescription of the optimal change.
[52] U.S. Department of Energy, *Carbon Dioxide Effects: Workshop on Environmental and Societal Consequences of a Possible CO_2-Induced Climate Change*, DOE/CONF/7904143 (Washington, D.C., 1980); Laurmann, "Optimizing Energy Transition Paths in CO_2 Emission Reduction Strategies," *Energy* 8 (1983): 845–58; Perry et al., "Energy Supply and Demand Implications of CO_2" (cit. n. 51).
[53] Commentaries are provided by Liverman, "Conventions of Climate Change" (cit. n. 1); Stephen H. Schneider, "'Natural Experiments' and CO_2-Induced Climate Change: The Controversy Drags On—an Editorial," *Climatic Change* 6 (1984): 317–21; van der Sluijs et al., "Anchoring Devices in Science for Policy" (cit. n. 40).
[54] Thomas Schelling, "Some Economics of Global Warming," *American Economic Review* 82 (1992): 1–14; Fankhauser and Pearce, "Social Costs of Greenhouse Gas Emissions" (cit. n. 39).
[55] Nordhaus, "How Fast Should We Graze the Global Commons?" *American Economic Review* 72 (1982): 242–6, on 244.
[56] Nierenberg, *Changing Climate* (cit. n. 45). See Oreskes, Conway, and Shindell, "From Chicken Little to Dr. Pangloss" (cit. n. 2), for an in-depth account of the Nierenberg report, but see also a paper-length response from Nicolas Nierenberg, Walter R. Tschinkel, and Victoria J. Tschinkel, "Early Climate Change Consensus at the National Academy: The Origins and Making of *Changing Climate*," *Hist. Stud. Nat. Sci.* 40 (2010): 318–49.
[57] Oreskes, Conway, and Shindell, "From Chicken Little to Dr. Pangloss" (cit. n. 2); Ralph C. D'Arge, William D. Schulze, and David S. Brookshire, "Carbon Dioxide and Intergenerational Choice," *American Economic Review* 72 (1982): 251–6; Meyer-Abich, "Socioeconomic Impacts of CO_2-Induced Climatic Changes and the Comparative Chances of Alternative Political Responses: Prevention, Compensation and Adaptation," *Climatic Change* 2 (1980): 373–85. It is interesting that Schneider added a note to the Meyer-Abich paper to express that its publication in draft form from the April 1979 Annapolis conference (the joint AAAS and DOE "CO_2 Effects Research Meeting") was

important future threats to energy that would result in reduced CO_2 emissions and hence eliminate the need to fear climate change.[58] Thus, for economists, by the early to mid-1980s climate change was a potential problem in need of further research, but not one that was imminently actionable through cost-effective emissions reductions. The cost-benefit analysis just did not add up in favor of action, though the economists recognized that there were significant uncertainties in all of these figures. Yet they had successfully reframed the debate by prioritizing economic analysis at the center of climate change decision making.

CLIMATE CHANGE POLICY DEVELOPMENT

The mid- to late 1980s concretized the significance of economic analysis, though this was a period in which many of the leading economists published less on climate change than they had in the early 1980s or than they would after 1990.[59] The policy significance of economic analysis, however, rose. There may still have been a schism between those who wanted to take early, precautionary action on climate change and those who adopted a wait-and-see approach, but it would be fair to state that many policy discussions were actively proposing some form of emissions-control policies.[60]

The Villach conference in 1985 is widely credited with being a key moment in raising the agenda for dealing with climate change for policy makers.[61] Sonja Boehmer-Christiansen noted that this conference was "more cautious about science, but more reckless about policy advocacy than the IPCC would be five years later."[62] Delegates argued for a maximum 0.1°C per decade temperature increase (on the basis that most ecosystems can adapt to such an increase without significant problems) and greater support for the analysis of energy policy options to reduce emissions.[63] Numerous scientific experts and advocacy groups gathered at the second conference in Villach in 1987 and discussed possible consequences for regions under climate change scenarios as well as a range of estimates of the requirements for emissions cuts to stabilize atmospheric concentrations of greenhouse gases.[64] The Bellagio meeting that followed in November 1987 reiterated the earlier conferences' conclusions

---

hastened "to generate discussion, debate and exchange on the very important, yet hardly explored, question of societal response to CO_2."

[58] Indeed, CO_2 emissions would decline as a result of market forces, because a high-energy, high-CO_2 future was only plausible if economically inefficient energy supply continued, according to Amory B. Lovins and L. Hunter Lovins, "Energy, Economics and Climate—an Editorial," *Climatic Change* 4 (1982): 217–20.

[59] E.g., Nordhaus published little on climate change between 1985 and 1989, a period that also saw him become provost of Yale University (1986–8).

[60] On the question of "activists" versus "muddlers-through," see William W. Kellogg, "Mankind's Impact on Climate: The Evolution of an Awareness," *Climatic Change* 10 (1987): 113–36.

[61] The conference was titled "Second Joint UNEP/ICSU/WMO International Assessment of CO_2 and Other Greenhouse Gases in Climate Variations and Associated Impacts."

[62] Boehmer-Christiansen, "Global Climate Protection Policy" (cit. n. 46), 156. The emergence of the IPCC from discussions within the World Meteorological Organization (WMO) and United Nations Environment Programme (UNEP) in 1988 established a scientific reporting process that would enable a consensus opinion on climate science to be formulated, but it did not include significant discussion of social science or policy agendas.

[63] *International Assessment of the Role of Carbon Dioxide and of Other Greenhouse Gases in Climate Variations and Associated Impacts*, conference statement (Villach, 1985).

[64] Michael Oppenheimer, "Developing Policies for Responding to Climatic Change: An Editorial," *Climatic Change* 15 (1989): 1–4.

but was more significant in terms of policy impact given the attendance of government representatives at that meeting. Both meetings were under the purview of the Advisory Group on Greenhouse Gases and the Stockholm Environment Institute.[65] Researchers including, most prominently, Rob Swart and Piers Vellinga produced a traffic light system for representing climate change, with the categories of serious disruptive damage (red), extensive damage (amber), and limited damage (green). Consideration was taken of the rate of change as well as sea-level rise, but it is interesting to note that 2°C was given as the threshold between amber and red and was promoted as the maximum allowable temperature increase.[66] Meanwhile participants at the Toronto conference in 1988, with an influential policy audience, called for the stabilization of CO_2 concentrations using an estimate from the Bellagio meeting that would require approximately a 50 percent cut in emissions, with an intermediate target of 20 percent of 1988 levels by 2005.[67]

Turning to economic analysis and policy advice, researchers both issued calls to remember that economics should be seen as a heuristic and offered deepening prescriptive policy advice too. In the former case, Steven T. Sonka and Peter J. Lamb, for example, suggested that economics should be restricted to providing a range of possibilities, especially given the uncertainties in areas such as agriculture. They called for more complex analysis that incorporated a better understanding of responses to change as change was underway, rather than comparing solely between the present and the final, changed state.[68] Laurmann (separately) agreed with the need for a better understanding of change, but returned to the original heuristic of finding the acceptable climate change level that would ensure not just climatic stability, but also economic and social stability.[69] Irving M. Mintzer, likewise, formulated a preventative approach to climate change that would help delay global warming and keep it within "prudent bounds," at the same time not sacrificing economic growth.[70] Mintzer's proposal was directly taken up in a Commonwealth report in 1989 that promised "to maintain rapid economic growth globally" while "slow[ing] down global warming to 'acceptable' levels."[71] Increased energy efficiency and coal taxes were primary policy options, and the researchers predicting low-carbon energy futures were now at the forefront of policy.[72] Within a different setting, the 1988 International Project

[65] Shardul Agrawala, "Early Science-Policy Interactions in Climate Change: Lessons from the Advisory Group on Greenhouse Gases," *Global Environmental Change* 9 (1999): 157–69.

[66] Piers Vellinga and Rob Swart, "The Greenhouse Marathon: Proposal for a Global Strategy," in *Climate Change: Science, Impacts and Policy*, eds. Jill Jäger and H. L. Ferguson (Cambridge, 1991), 129–34. The details of the 2°C temperature target are discussed more broadly in Samuel Randalls, "History of the 2°C Climate Target," *Wiley Interdisciplinary Reviews: Climate Change* 1 (2010): 598–605.

[67] *The Changing Atmosphere: Implications for Global Security*, conference statement (Toronto, 1988); Irving M. Mintzer, "Energy, Greenhouse Gases, and Climate Change," *Annual Review of Energy* 15 (1990): 513–50.

[68] Steven T. Sonka and Peter J. Lamb, "On Climate Change and Economic Analysis," *Climatic Change* 11 (1987): 291–311.

[69] J. A. Laurmann, "Scientific Uncertainty and Decision Making: The Case of Greenhouse Gases and Global Climate Change," *Science of the Total Environment* 55 (1986): 177–86.

[70] Mintzer, *A Matter of Degrees: The Potential for Controlling the Greenhouse Effect* (New York, 1987). This was a World Resources Institute research report.

[71] Commonwealth Secretariat, *Climate Change: Meeting the Challenge; Report by a Commonwealth Group of Experts* (London, 1989), 84.

[72] Stewart Boyle, "Global Warming—a Paradigm Shift for Energy Policy?" *Energy Policy* 17 (1989): 2–5.

for Sustainable Energy Paths (IPSEP) report, which drew from work funded by the
Dutch Ministry of Environment, found that concentrations should be maximized at
400 ppm (parts per million) as a policy target and suggested that this was supported
by the policy appeals made by scientific organizations (e.g., the German Meteorological Society), which estimated that temperatures should not be permitted to rise by
more than 1°C above the current level. A concentration of 380–430 ppm was thought
likely to produce a 1–3°C temperature increase. The IPSEP report suggested that
400 ppm would be feasible without significant economic impact, while any lower
limit would be politically indefensible in the absence of evidence of serious climatic
harm.[73] In the United Kingdom, the 50 percent Toronto target was considered a useful benchmark for assessing progress on reducing emissions, and, as the U.K. Energy
Committee argued in 1989, presaging the *Stern Review*'s arguments fifteen years
later, "the *risks* of not adopting targets such as these are so great, and the insurance
premia required to achieve them so modest, especially when expressed as a percentage of GDP, that it would be irresponsible to avoid the challenge."[74] Climate policy,
for these varying authors and reports, revolved around managing the changes for ecosystems *and* societies, and they argued, again anticipating the *Stern Review*, that this
was not prohibitively expensive but rather of modest cost or even a good deal (when
examining more broadly benefits from increasing energy efficiency). Thus in policy
discussions and decision making the economic justification was increasingly inter-
woven with the scientific justification. Stabilizing the climate for economic activity
and stabilizing economic activity for the climate became intertwined. There was an
optimal change (or target) that would balance climate and economy. Hence by 1991
it was not considered absurd for environmentalists Amory and Hunter Lovins, co-
founders of the Rocky Mountain Institute, which promoted efficient resource use, to
title their review paper on the latest thinking on climate change and resources "Least-
Cost Climatic Stabilization."[75]

A RISING TIDE OF CLIMATE CHANGE ECONOMICS

According to many commentators, economic analysis of climate change took shape
in the early 1990s, and while this is not strictly true, as I have shown, there was
clearly a vast increase in the number of researchers and publications on the topic
at this time.[76] Indeed, Nordhaus continued to develop his work on climate change,
but his initial papers in the early 1990s suffered from being rather simplistic in their
calculations of damages, with many damages not even quantified.[77] Nordhaus concluded in 1991 that his "hunch" was that the impact of climate change on human

[73] Florentin Krause, *Energy and Climate Change: What Can Western Europe Do? An Analysis of Climatic Imperatives, Climate Economics of Energy Options, and Implications for Energy Planning and Policy* (draft report, IPSEP, Richmond, Calif., 1988).
[74] U.K. Energy Committee, *Energy Implications of the Greenhouse Effect*, vol. 1 (London, 1989), lxii. Emphasis in the original.
[75] Lovins and Lovins, "Least-Cost Climatic Stabilization," *Annual Review of Energy and Environment* 16 (1991): 433–531.
[76] E.g., Robin Cantor and Gary Yohe, "Economic Analysis," in *Human Choice and Climate Change*, vol. 3: *The Tools for Policy Analysis*, eds. Steve Rayner and Elizabeth L. Malone (Columbus, Ohio, 1998), 1–103; Spash, "Trying to Find the Right Approach to Greenhouse Economics" (cit. n. 35); Van den Bergh, "Optimal Climate Policy Is a Utopia" (cit. n. 15).
[77] Robert U. Ayres and Jörg Walker, "The Greenhouse Effect: Damages, Costs and Abatement," *Environmental and Resource Economics* 1 (1991): 237–70.

activity would be at most 2 percent of total world output.[78] It was therefore potentially optimal for CO_2 emissions to triple over the next century, though Nordhaus stated in both 1990 and 1992 that his research suggested that "modest investments to slow the pace of climate change" were warranted (in 1990 he defined this as 20 percent reductions in CO_2 emissions in the long run).[79] Nordhaus was continually developing his model, however, and in 1992 he published his "dynamic integrated model of the climate and the economy" (DICE) in *Science*. It was a dynamic optimization model that could calculate emissions reduction pathways and enable decisions to be made about the most economically optimal ways of slowing climate change given assumptions about climatic damage functions and population and economic growth. This model became widely used by later economists (the paper has been cited over 150 times).[80] Climate scientists also appreciated its transparency as it allowed them to engage in vigorous discussion of economic factors and even imagine incorporating them into their own models.[81] Referenced frequently in draft form, the model was used as much by others to change the various assumptions or functions to produce different results as it was to guide policy directly in its original form.

In 1992 the Organisation for Economic Co-operation and Development (OECD) tried to bring some clarity to the various approaches and targets. It issued a report making clear that better separation was needed in two types of cost analyses in the literature: those arising from impacts (frequently represented in terms of percentage loss of GDP, with targets expressed as CO_2 concentrations) and those that focused on the costs of emissions reduction (generally expressed as emissions targets for the various pathways). The report also noted that there were at least three possible science-based targets for climate change mitigation: the stabilization of atmospheric concentrations of greenhouse gases, the limiting of temperature increase to 2°C above the present level (on the argument that this was the maximum level experienced during human existence on Earth), or the prevention of rates of temperature increase faster than 0.1°C per decade.[82] The OECD drew inspiration for these targets from the IPCC and, to a lesser extent, Pearce, though in considering the latter's argument it noted the difficulty of translating from temperature to concentration to emissions targets.[83] The OECD report also noted the difficulties of finding particularly

[78] Nordhaus, "To Slow or Not to Slow: The Economics of the Greenhouse Effect," *Economic Journal* 101 (1991): 920–37, on 933.

[79] Nordhaus, "An Optimal Transition Path for Controlling Greenhouse Gases," *Science* 258 (1992): 1315–9, on 1315; Nordhaus, "Count before You Leap," *Economist*, July 7, 1990, 21–4. In 1991–2 he also advocated low-cost geoengineering to offset warming; see Fleming, *Fixing the Sky* (cit. n. 31).

[80] Nordhaus, "Optimal Transition Path for Controlling Greenhouse Gases" (cit. n. 79). A more complete formulation was published in 1994; Nordhaus, *Managing the Global Commons: The Economics of Climate Change* (Cambridge, 1994).

[81] Some examples include William R. Cline, *The Economics of Global Warming* (Washington, D.C., 1992); Cline, "Costs and Benefits of Greenhouse Abatement: A Guide to Policy Analysis," in OECD, *Economics of Climate Change*, 87–106; Fankhauser and Pearce, "Social Costs of Greenhouse Gas Emissions" (Both cit. n. 39); Colin Price, "Emissions, Concentrations and Disappearing CO_2," *Resource and Energy Economics* 17 (1995): 87–97; Spash, "Trying to Find the Right Approach to Greenhouse Economics" (cit. n. 35); Mort Webster, Chris Forest, John Reilly, Mustafa Babiker, David Kicklighter, Monika Mayer, Ronald Prinn, Marcus Sarofim, Andrei Sokolov, Peter Stone, and Chien Wang, "Uncertainty Analysis of Climate Change and Policy Response," *Climatic Change* 61 (2003): 295–320.

[82] OECD, *Convention on Climate Change: Economic Aspects of Negotiation* (Paris, 1992).

[83] IPCC, *Scientific Assessment of Climate Change* (Cambridge, 1990); David Pearce, ed., *Blueprint 2: Greening the World Economy* (London, 1991). See also Boykoff, Frame, and Randalls, "Discursive Stability Meets Climate Instability" (cit. n. 2).

compelling reasons for any of these targets, though article 2 of the 1992 UN Framework Convention on Climate Change (UNFCCC) states that emissions should be stabilized at a level that would avoid dangerous anthropogenic interference with the climate system, while making clear that policies should be cost-effective. Its language was borrowed directly from the 1992 Rio de Janeiro UN Conference on Environment and Development's restatement of "the precautionary principle":

> Where there are threats of serious or irreversible damage, lack of full scientific certainty should not be used as a reason for postponing such measures [to minimize the effects of climate change], taking into account that policies and measures to deal with climate change should be cost-effective so as to ensure global benefits at the lowest possible cost.[84]

An economic analysis was critical to delivering this goal of cost-effectiveness. Nordhaus's DICE model included five different policy options, ranked by his preference: (1) geoengineering, (2) optimal economically efficient climate policies, (3) no control, (4) emissions stabilization, and (5) climate stabilization.[85] Emissions stabilization had the virtue of simplicity, but had "no particular scientific or economic merit," while climate stabilization (assuming a temperature increase of 0.1°C per decade) made no sense financially. Since geoengineering was so very speculative (and remains so), Nordhaus proposed a modest carbon tax as an efficient, economically optimal approach.

The DICE model highlighted the increased interaction between scientific and economic modeling of climate change. Researchers increasingly looked to integrate the ideal of cost-effectiveness into their discussions of climate model outputs to produce economically and environmentally coherent decision frameworks for mitigating carbon emissions, though they readily admitted the narrowness of these frameworks.[86] Science and economics on climate change had been joined together for policymaking. Indeed, as Fankhauser and Pearce noted, the estimation of costs of scenarios involving more or less than doubled concentrations of CO_2 was unclear. As they put it, once an optimal level is defined, even heuristically—in this case, at a doubling of CO_2 concentrations—"the social costs are then equivalent to the pollution tax required to keep emissions on the optimal path"; in other words, it is easy to lapse into economic analyses that make assumptions based on that scenario.[87] Their plea, against other models like that developed by Nordhaus, was for a better estimation of uncertainties in evaluating optimal emissions policies such that the expected increase in damages from rising CO_2 would be properly accounted for. Nevertheless, as energy economists Richard Richels and Jae Edmonds were only too aware, such

---

[84] This quotation is from the agreed text of UNFCCC, "Report of the Intergovernmental Negotiating Committee for a Framework Convention on Climate Change on the Work of the Second Part of Its Fifth Session, Held at New York from 30 April to 9 May 1992," document no. A/AC.237/18 (part 2)/Add.1 (Geneva, 1992), available at http://unfccc.int/resource/docs/a/18p2 (accessed 24 January 2011).

[85] Nordhaus, "Optimal Transition Path for Controlling Greenhouse Gases" (cit. n. 79), 1317.

[86] Tom M. L. Wigley, Richard Richels, and Jae Edmonds, "Economic and Environmental Choices in the Stabilization of Atmospheric CO_2 Concentrations," *Nature* 379 (1996): 240–3. See also Richels and Edmonds, "The Economics of Stabilizing Atmospheric CO_2 Concentrations," *Energy Policy* 23 (1995): 373–8.

[87] Fankhauser and Pearce, "Social Costs of Greenhouse Gas Emissions" (cit. n. 39), 76. Remember that Nordhaus, "Economic Growth and Climate" (cit. n. 39), used precisely this doubling scenario in 1977.

evaluation would need to take account of political imperatives too. It would be infeasible to expect international negotiators to accept proposals that did not demonstrate climatic benefits being achieved at lowest cost. The concentration target needed to be economically sensible or "the perceived price tag will undoubtedly influence the willingness of nations to abide by the treaty."[88] Reflecting the growth of literatures on an economics of climate change, in 1995 the IPCC included a discussion of predominant types of cost-benefit analysis in its *Second Assessment Report*.[89] This was a rather narrow analysis with limited economic figures compared to later IPCC reports, which paid more attention to uncertainties in the economics.[90] It did, however, clarify the difference between optimal and target-based cost-benefit analysis, even if the latter may have been to some extent influenced by the former.

By 1996 the EU commissioners crystallized not just a target for keeping 2000 emissions to 1990 levels, but also a plan to work toward a maximum allowable temperature increase of 2°C.[91] This target emerged partly from the IPCC *Second Assessment Report*, which was interpreted to have predicted a 2°C temperature increase by 2100 as its midrange scenario and as the point beyond which increasing levels of climatic "dangers" would be visible.[92] While economists such as Richard Tol did not accept the 2°C target as being supported by cost-benefit analysis, it is clear that the rationale for this target was generated in part from a conjoined economic-scientific philosophy. There is rather less scientific than economic merit in setting a temperature target to be used as the guiding policy for cost-effectively dealing with climate change.[93] Whatever the merits of this or any other target, the important conclusion is that the notion of climate change as an economic resource to be used in a cost-effective manner is pervasive. Even in (scientifically informed) target-based cost-benefit analysis, efficiency is still a key criterion. Tracing its roots through the 1970s and 1980s sheds light on some of the assumptions and actors involved in this particular construction of climate change. It is interesting to note how different this was from more idealistic proposals that policy should have as its ultimate objective "reducing carbon emissions to the level where natural processes return carbon dioxide concentrations to natural levels."[94] This formulation by Greenpeace came too late, as optimal resiliency had taken the place of least harm.[95]

[88] Richels and Edmonds, "Economics of Stabilizing Atmospheric CO_2 Concentrations" (cit. n. 86), 376.

[89] Munasinghe et al., "Applicability of Techniques of Cost-Benefit Analysis" (cit. n. 16).

[90] Barker, "Economics of Avoiding Dangerous Climate Change" (cit. n. 8).

[91] European Environment Agency, *Climate Change in the European Union* (Copenhagen, 1996).

[92] Liverman, "Conventions of Climate Change" (cit. n. 1).

[93] Tol, "Europe's Long-Term Climate Target" (cit. n. 28). To clarify, the point being made here is not that economists set the 2°C temperature target; that would be an absurd claim. Rather, a confluence of interests defined the problem as one in which scientists determined what would happen at different temperature levels and economists could then work out associated cost curves such that an optimal economic-climate policy could be forged. The figure of 2°C could then become a tolerable ecological and economic target that the economists could optimize emissions pathways to reach (but not go beyond). While this figure is politically and economically actionable, such a policy is fraught with pitfalls due to the uncertainties in the relationship between emissions, concentrations, and temperature. See Boykoff, Frame, and Randalls, "Discursive Stability Meets Climate Instability" (cit. n. 2); Randalls, "History of the 2°C Climate Target" (cit. n. 66).

[94] Greenpeace, "Memorandum 9," in *Energy Policy Implications of the Greenhouse Effect: Memoranda of Evidence*, U.K. Energy Committee (London, 1989), 54–68, on 61. It is unclear, however, exactly what concentration levels this refers to.

[95] This change is also noted in DOE, *Environmental and Societal Consequences* (cit. n. 50).

CONCLUSION

The policy framing from the early 1990s highlights just how far economics had progressed as a tool for public policy on climate change. Yet should an economics of climate change be considered misguided? Not at all. These studies have provided a wealth of material on the potential costs of climate change, the viability of emissions-control strategies, "what if" scenarios, and future valuations (discounting). Problems arise, as Christian Azar points out, when certain models of optimal climate change become truth machines rather than heuristics.[96] Arguably this is precisely the approach that many policy makers have taken to climate change, and it is manifested in the policies designed to achieve, but not exceed, maximum concentration or temperature targets (e.g., the 2°C target of the European Union, which was set despite the fact that some scientists and economists within and outside the region were uncomfortable with it). In a shift from the somewhat descriptive and scenario-based cost-benefit analyses of climate change undertaken since the 1970s, by the mid-1990s the economists' work had, to some policy makers at least, become prescriptive in formulating optimal climate change. In many respects Nordhaus and others merely offered tools for guiding policy makers, but their approach to the problem was particularly compelling to policy makers, politicians, and businesses that wanted quantitative answers.[97] Whereas in the early 1980s economists frequently concluded that action was too costly, by the mid-1990s many cost-benefit analyses supported moderate action to reduce emissions, though economic as much as climatic stability was the target. Indeed, reducing emissions was increasingly being marketed as good for business. So much for the mechanical objectivity, or justification, of cost-benefit analysis.

It is not surprising that climate change has been considered through the lens of cost-benefit analysis. In a world in which environmental financial markets are proposed as cost-effective solutions to environmental problems and science is increasingly organized through concerns for efficiency and profitability, these approaches clearly extend far beyond the case of climate change. Climate change may indeed have become a resource that is to be organized in the same way as any other resource, property, or good. Thus, that "the climate question" is posed as one to be optimally and efficiently solved should be expected. Nevertheless, there are problems with this approach. What happens if the economic solution starts to fall out of step with physical changes in the atmosphere? Would the effects of the cost-benefit analyses be different if climate science became more catastrophic or mild in its predictions, or if climate change were not occurring on human timescales? The critique presented in this article is not one that excludes economics, but rather one that highlights how a utilitarian, cost-efficient approach has led to policies that expose the world to greater uncertainties and a higher likelihood of experiencing damaging effects from climate change.[98]As stated earlier, this article has not considered a wide range of economists who are critical of the assumptions made by the more policy-influential neoclassical economists.

More research is needed to attend to the ways in which economic research was

[96] Azar, "Are Optimal CO$_2$ Emissions Really Optimal?" (cit. n. 12).
[97] While Mike Hulme argues that climate models dominate climate change research, I argue that economics has dominated climate change social science; see Hulme in this volume.
[98] Nelson, "Economists, Value Judgments, and Climate Change" (cit. n. 11).

taken up explicitly by policy makers (especially from the late 1980s onward), the geographical variation in the research produced (this article has focused more on U.S. research than European), and the variety of forms of economic analysis deployed as well as to investigate in more detail the relationship between specified climate change research and that being conducted by energy economists (especially at IIASA). Social science, in addition to science, should present a fruitful inquiry for historians interested in climate change. Not least, underrepresenting social scientists' agendas, debates, and influence risks ignoring an important part of the history of climate change as an issue for public policy. Thus, when economist Michael Grubb asks "Who's afraid of [big bad] atmospheric stabilization?" at least we can be clear about the historical logic of this question and the implications this has for conceiving of climate policy today.[99]

[99] Grubb, "Who's Afraid of Atmospheric Stabilisation? Making the Link between Energy Resources and Climate Change," *Energy Policy* 29 (2001): 837–45, on 837.

CLIMATE REDUX

Reducing the Future to Climate:
A Story of Climate Determinism and Reductionism

*by Mike Hulme**

ABSTRACT

This article traces how climate has moved from playing a deterministic to a re-ductionist role in discourses about environment, society, and the future. Climate determinism previously offered an explanation, and hence a justification, for the superiority of certain imperial races and cultures. The argument put forward here is that the new climate reductionism is driven by the hegemony exercised by the predictive natural sciences over contingent, imaginative, and humanistic accounts of social life and visions of the future. It is a hegemony that lends disproportion-ate power in political and social discourse to model-based descriptions of putative future climates. Some possible reasons for this climate reductionism, as well as some of the limitations and dangers of this position for human relationships with the future, are suggested.

> The general attitude of many critics, however, is in keeping with the reaction during the last few decades against the simple de-terminism [of Ellsworth Huntington] which led to what Latti-more . . . has described as "the romantic explanation of *hordes* of erratic nomads, ready to start for lost horizons at the joggle of a barometer in search of suddenly vanishing pastures."
>
> Gordon Manley, 1944[1]

> At plus-4 degrees, *hordes* of climate refugees would flee famine and extreme water scarcity. At plus-5 degrees, climate refugees would number in the tens of millions as massive uninhabitable zones spread.
>
> Joanne Ostrow, 2008[2]

* School of Environmental Sciences, University of East Anglia, Norwich NR4 7TJ, U.K.; m.hulme@uea.ac.uk.
Earlier forms of this article were presented at the workshop "Climate Matters" at the Uni-versity of Manchester in October 2008 and at the conference "Climate and Cultural Anxiety" at Colby College, Maine, in April 2009. The participants of those workshops, especially Jim Fleming and Vlad-imir Jankovic, are thanked for their helpful questions, comments, and suggestions that have improved this article. Jon Barnett and David Livingstone also carefully read a draft of the article and offered helpful criticism, as did three anonymous reviewers. The author alone, however, takes responsibility for the views contained here.

[1] Manley, "Some Recent Contributions to the Study of Climatic Change," *Quarterly Journal of the Royal Meteorological Society* 70 (1944): 197–220, on 220 (emphasis added).

[2] Ostrow, "'Six Degrees' Charts Climate Apocalypse in HD Television," Denverpost.com, Febru-ary 8, 2008, http://www.denverpost.com/ostrow/ci_8190284 (accessed 19 January 2011; emphasis added). Ostrow's article is a review of *Six Degrees Could Change the World*, a National Geographic TV documentary based on Mark Lynas's award-winning book *Six Degrees* (New York, 2007).

INTRODUCTION

Human beings are always trying to come to terms with the climates they live with. This is as true for the ways the relationship between society and climate is theorized as it is for the practical challenges of living fruitfully and safely with climatic resources and hazards. The story of how the idea of climate has traveled through the human imagination is well told in Lucian Boia's *The Weather in the Imagination*,[3] and an exemplary account of how a society seeks practically to live with its climate is William B. Meyer's *Americans and Their Weather*.[4] When people reflect on these relationships between society and climate, they frequently adopt two intuitive positions. On the one hand it is obvious that climates influence and shape human psychological, biological, and cultural attributes. This is true for individual behaviors, cultural practices, and environmental resources. Yet it is equally true that an enduring strand of human encounters with climate seeks both to tame these climatic influences and constraints and to live beyond them. Human beings change microclimates, insulate themselves against climatic extremes, and adapt technologies and practices for survival and prosperity.[5]

Attempts to understand and theorize the relationship between climate and society are therefore prey to two distinct fallacies. The first is that of "climate determinism," in which climate is elevated to become a—if not *the*—universal predictor (and cause) of individual physiology and psychology and of collective social organization and behavior. The second fallacy is that of "climate indeterminism," in which climate is relegated to a footnote in human affairs and stripped of any explanatory power. Geographers have at times been most guilty of the former fallacy, historians at times most guilty of the latter.[6] Yet not even historical geographers or environmental historians have been always able to hold these two opposing fallacies in adequate and creative tension.[7]

At the beginning of the twentieth century, the determinist fallacy achieved considerable salience and popularity in European and, especially, American thought, championed by the likes of the geographers Friedrich Ratzel, Ellen Semple, and Ellsworth Huntington.[8] Climate was viewed as the dominant determinant of racial character, intellectual vigor, moral virtue, and the ranking of civilizations, ideas that had earlier appealed to Greek philosophers and European rationalists alike.[9] However, the ideological wars of the mid-twentieth century reshaped the political and moral worlds that had nourished such thinking, and determinism became discredited and marginalized within mainstream academic geography.

---

[3] Boia, *The Weather in the Imagination* (London, 2005).

[4] Meyer, *Americans and Their Weather* (New York, 2000).

[5] William W. Kellogg and Stephen H. Schneider, "Climate Stabilization: For Better or for Worse?" *Science* 186 (1974): 1163–72.

[6] Gabriel Judkins, Marissa Smith, and Eric Keys, "Determinism with Human-Environment Research and the Rediscovery of Environmental Causation," *Geogr. J.* 174 (2008): 17–29.

[7] Oskar H. K. Spate, "Toynbee and Huntington: A Study in Determinism," *Geogr. J.* 118 (1952): 406–24.

[8] Richard Peet, "The Social Origins of Environmental Determinism," *Ann. Assoc. Amer. Geogr.* 75 (1985): 309–33; James Rodger Fleming, *Historical Perspectives on Climate Change* (New York, 2005); Innes Keighren, *Bringing Geography to Book: Ellen Semple and the Reception of Geographical Knowledge* (London, 2010); Georgina Endfield in this volume.

[9] David N. Livingstone, "Race, Space and Moral Climatology: Notes toward a Genealogy," *J. Hist. Geogr.* 28 (2002): 159–80.

Now, a hundred years later, and at the beginning of a new century, heightening anxieties about future anthropogenic climate change are fueling—and in turn being fueled by—a new variety of the determinist fallacy. Although this variety is distinct from the politically and ethically discredited climate determinism epitomized by Huntington and his followers, climate has regained some of its former power for "explaining" the performance of environments, peoples, and societies. In seeking to predict a climate-shaped future, proponents of this logic reduce the complexity of interactions between climates, environments, and societies, and a new variant of climate determinism emerges. I call this "climate reductionism," a form of analysis and prediction in which climate is first extracted from the matrix of interdependencies that shape human life within the physical world. Once isolated, climate is then elevated to the role of dominant predictor variable. I argue in this article that climate reductionism is a methodology that has become dominant in analyses of present and future environmental change—and that as a methodology it has deficiencies.

This way of thinking and analyzing finds expression in some of the balder (and bolder) claims made by scientists, analysts, and commentators about the future impacts of anthropogenic climate change. Here are some examples of claims that emerge from this climate reductionist form of analysis:

- "Every year climate change leaves over 300,000 people dead."
- "We predict, on the basis of mid-range climate-warming scenarios for 2050, that 15–37% of species . . . will be 'committed to extinction.'"
- "185 million people in sub-Saharan Africa alone could die of disease directly attributable to climate change by the end of the century."
- "The costs and risks of climate change will be equivalent to losing at least 5% of global GDP each year, now and forever . . . [rising to] 20% of GDP or more."
- "I think there will be substantial change [in climate] whatever we do. If we do nothing over the next 20 years it will be catastrophic. If we do nothing over the next 50 to 100 years it might even be terminal."[10]

Such reductionism is also contributing to the new discourse about climate change and conflict. For example, climate change is offered as an explanation of cycles of war and conflict in China over the last millennium: "It was the oscillations of agricultural production brought about by long-term climate change that drove China's historical war-peace cycles."[11] The civil war in Darfur is categorized in the media as a harbinger of future climate-driven disputes: "In decades to come, Darfur may be seen as one of the first true climate change wars."[12] The recent report *Climate Change as a Security Risk*, from the German Advisory Council on Global Change, was reported in similar neodeterminist tones in the media: "Climate Change to Cause Wars in North

[10] Respectively, these are quoted from Global Humanitarian Forum, *The Anatomy of a Silent Crisis* (Geneva, 2009), 1; Chris D. Thomas et al., "Extinction Risk from Climate Change," *Nature* 427 (2004): 145–8, on 145; Christian Aid, *The Climate of Poverty: Facts, Fears and Hope* (London, 2006), 3; Nicholas Stern, *The Economics of Climate Change: The Stern Review* (Cambridge, 2006), xv; Andrew J. Watson, "Gaia and Accelerating Climate Change," transcript of ABC Radio National program, broadcast January 20, 2007.
[11] David D. Zhang, Jane Zhang, Harry F. Lee, and Yuan-qing He, "Climate Change and War Frequency in Eastern China over the Last Millennium," *Human Ecology* 35 (2007): 403–14, on 413.
[12] J. Borcher, "Scorched," *Guardian*, April 28, 2007.

Africa" and "Climate Change 'Likely to Cause Wars.'"[13] And a team of agricultural and resource economists went even further in predicting the effect of temperature increases on future battle deaths in Africa: "[The] historical response to temperature suggests an additional 393,000 battle deaths [by 2030] if future wars are as deadly as recent wars."[14]

In a view related to this belief that climate plays an explanatory role in determining war, climate refugees are seemingly set to threaten global, regional, and national security in a rerun of the Mongol invasions of Europe alluded to by Owen Lattimore in his caricature of Huntington's climatic theory of world history, quoted in this article's first epigraph. The term "climate refugees" was invented by Norman Myers in a 1993 article,[15] and his estimate of between 150 and 250 million climate refugees by 2050 has been subsequently widely cited. It is a claim that easily translates into powerful rhetoric, as in this example from the Royal United Services Institute in the United Kingdom: "If we fail to stop polluting, we will be committed to catastrophic and irreversible change . . . which will directly displace hundreds of millions of people and critically undermine the livelihoods of billions."[16] And recent work has sought to quantify this climate change effect on migration more precisely: "By approximately the year 2080, climate change is estimated to induce 1.4 to 6.7 million adult Mexicans to emigrate [to the United States] as a result of declines in agricultural productivity alone."[17]

In this new mood of climate-driven destiny the human hand, as the cause of climate change, has replaced the divine hand of God as being responsible for the collapse of civilizations, for visitations of extreme weather, and for determining the new twenty-first-century wealth of nations.[18] And to emphasize the message and the mood, the New Economics Foundation and its partners have wound up a climate clock that is now ticking, second by second, until December 1, 2016, when human fate will be handed over to the winds, ocean currents, and drifting ice floes of a destabilized global climate: "We have 100 months to save the planet; when the clock stops ticking we could be beyond the climate's tipping point, the point of no return."[19] Such eschatological rhetoric offers a post-2016 world where human freedom and agency are extinguished by the iron grip of the forces of climate. Such a narrative offers scant

[13] See, respectively, German Advisory Council on Global Change, *Climate Change as a Security Risk* (London, 2008); "Climate Change to Cause Wars in North Africa," Jordan Environment Watch, January 19, 2008, http://www.arabenvironment.net/archive/2008/1/444843.html (accessed 6 July 2009; site now discontinued); "Climate Change 'Likely to Cause Wars,'" *Daily Telegraph*, December 10, 2007.

[14] Marshall B. Burke, Edward Miguel, Shanker Shatyanath, John A. Dykema, and David Lobell, "Warming Increases the Risk of Civil War in Africa," *Proceedings of the National Academy of Sciences* 106 (2009): 20670.

[15] Myers, "Environmental Refugees in a Globally Warmed World," *Bioscience* 43 (1993): 752–61; although see also Svante Arrhenius, *Worlds in the Making: The Evolution of the Universe* (New York, 1908), 53, where the concept, although not the language, is also mentioned.

[16] Nick Mabey, *Delivering Climate Security: International Security Responses to a Climate Changed World*, Whitehall Papers, no. 69 (London, 2007).

[17] Shuaizhang Feng, Alan B. Krueger, and Michael Oppenheimer, "Linkages among Climate Change, Crop Yields and Mexico-US Cross-Border Migration," *Proceedings of the National Academy of Sciences* 107 (2010): 14257.

[18] See, respectively, Jared Diamond, *Collapse: How Societies Choose to Fail or Succeed* (London, 2005); Vladimir Jankovic, "Change in the Weather," *Bookforum*, February/March 2006, 39–40; Stern, *Economics of Climate Change* (cit. n. 10).

[19] "One Hundred Months" Web site, New Economics Foundation, http://www.onehundredmonths .org (accessed 19 January 2011).

chance for humans to escape a climate-shaped destiny.[20] James Lovelock offers the most vivid melodrama of such a predetermined fate. We are traveling, he says, on "a rocky path to a Stone Age existence on an ailing planet, one where few of us survive among the wreckage of our once biodiverse Earth."[21]

My argument in this article is that these sentiments, and many others that invade contemporary public and political discourses of climate change, are enabled by the methodology of climate reductionism (i.e., a form of neoenvironmental determinism). Simulations of future climate from climate models are inappropriately elevated as universal predictors of future social performance and human destiny. I am not alone in making this argument, even if my focus here is exclusively on climate rather than on the role of the wider physical environment. For example, geographers Andrew Sluyter, Christopher Merrett, and Gabriel Judkins (lead author of a study with Marissa Smith and Eric Keys) have all detected evidence of a resurgence of the determinist fallacy, citing examples from the work of Jared Diamond in *Guns, Germs and Steel* and *Collapse* and Geoffrey Sachs in *The End of Poverty* and *Common Wealth*.[22]

After offering a brief account of how climate reductionism has come to prominence, I turn my attention to understanding why this should be. Why should an explanatory logic—if not an ideology—dating from earlier intellectual and imperial eras, a logic subsequently dismissed by many as seriously wanting, have reemerged in different form in a new century to find new and enthusiastic audiences? Rather than offering an explanation, and hence a justification, for the superiority of imperial societies, cultures, and races—as in past ideological variants of determinism—I will suggest here that climate reductionism holds a different attraction for contemporary audiences, and I will demonstrate how it has come to prominence.

I suggest that the hegemony exerted by the predictive natural sciences over human attempts to understand the unfolding future opens up the spaces for climate reductionism to emerge. It is a hegemony manifest in the pivotal role held by climate (and related) modeling in shaping climate change discourses. Because of the epistemological authority over the future claimed, either implicitly or explicitly, by such modeling activities, climate becomes the one "known" variable in an otherwise unknowable future. The openness, contingency, and multiple possibilities of the future are closed off as these predicted virtual climates assert their influence over everything from future ecology, economic activity, and social mobility to human behavior, cultural evolution, and geosecurity. It is climate reductionism exercised through what I call "epistemological slippage"—a transfer of predictive authority from one domain of knowledge to another without appropriate theoretical or analytical justification.

Before elaborating this proposition, I first offer a brief account of the decline in climate determinism through the twentieth century and illustrate the recent rise of reductionist thinking. I then defend my thesis—that climate reductionism results from the enterprise of climate prediction and the practice of epistemological

[20] Christina R. Foust and William O. Murphy, "Revealing and Reframing Apocalyptic Tragedy in Global Warming Discourse," *Environmental Communication* 3 (2009): 151–67; Stefan Skrimshire, ed., *Future Ethics: Climate Change and Apocalyptic Imagination* (London, 2010).

[21] Lovelock, *The Revenge of Gaia: Why the Earth Is Fighting Back—and How We Can Still Save Humanity* (London, 2006), 4.

[22] See, respectively, Sluyter, "Neo-environmental Determinism, Intellectual Damage Control and Nature/Society Science," *Antipode* 35 (2003): 813–7; Merrett, "Debating Destiny: Nihilism or Hope in *Guns, Germs and Steel*?" *Antipode* 35 (2003): 801–6; Judkins, Smith, and Keys, "Determinism" (cit. n. 6).

slippage—drawing upon key events, developments, and texts from the 1960s to 1990s. In particular I demonstrate the asymmetry between representations of future climate and social change that has persisted in the conduct of climate impact assessments. I conclude the article by placing this reductionist tendency within a wider cultural context of Western pessimism and loss of confidence about the future and by pointing toward some correctives that involve restructuring ideas about how the future can be imagined and made known.

THE DEMISE OF CLIMATE DETERMINISM

The story of environmental determinism, and especially the climatic variant on which I focus, is well known—at least it is well known to academic geographers who have had to wrestle with the difficult relationships between environmental conditions and human agency ever since the discipline took form in the later nineteenth century.[23] The argument of previous generations of determinists was that aspects of climate exerted a powerful shaping influence on the physiology and psychology of individuals and races, which in turn shaped decisively the culture, organization, and behavior of the society formed by those individuals and races. Tropical climates were said to cause laziness and promiscuity, while the frequent variability in the weather of the middle latitudes led to more vigorous and driven work ethics. Evidence of these discourses has been well reviewed for the period up to 1800 by Clarence Glacken and for the late nineteenth century in a series of papers by David Livingstone as well as by Mark Carey.[24]

This is a determinism that *makes* human and social character. There is also a form of climate determinism that *moves* people. Thus Lattimore's "hordes of erratic nomads" cited at the beginning of this article are driven, almost involuntarily, by climate variations in search of better pastures, while accounts of Viking arrivals and departures to and from Greenland have sometimes given the impression that they were driven solely by the oscillations of warmth and cold.[25] Both these manifestations of climate determinism—the making of character and the moving of people—emphasize the agency of climate over the agency of humans. In more extreme articulations of the idea—"strong determinism," according to Judkins[26]—the human will becomes hostage to the fortunes of climate, too passive and powerless to respond proactively, or even reactively, to changes in environmental fortune.

The apparent simplicities of climate determinism appealed to philosophers of the Grecian Empire (such as Herodotus and Hippocrates) and to rationalists of the European Enlightenment (such as Montesquieu and Hume).[27] They also appealed to late nineteenth- and early twentieth-century European and, especially, American geographers. The work of Yale geographer Ellsworth Huntington (1876–1947) best en-

[23] Harold MacKinder, "On the Scope and Methods of Geography," *Proceedings of the Royal Geographical Society* 9 (1887): 141–60.

[24] Glacken, *Traces on a Rhodian Shore: Nature and Culture in Western Thought from Ancient Times to the End of the Eighteenth Century* (Berkeley, Calif., 1967); Livingstone, "Tropical Climate and Moral Hygiene: The Anatomy of a Victorian Debate," *Brit. J. Hist. Sci.* 32 (1999): 93–110; Livingstone, "Race, Space and Moral Climatology" (cit. n. 9); Carey in this volume.

[25] Ian Whyte, *World without End? Environmental Disaster and the Collapse of Empires* (London, 2008).

[26] Judkins, Smith, and Keys, "Determinism" (cit. n. 6).

[27] Fleming, *Historical Perspectives* (cit. n. 8).

capsulates the theory's rise during the apogee of modern European and American imperialism. Huntington's major works—from his 1915 *Civilisation and Climate* to his 1945 *Mainsprings of Civilisation*—gained him a contemporary popularity, but among his academic colleagues he generated a range of contrary reactions. As one of his protégés, Stephen Visher, admiringly remarked in an obituary published in 1948, "His eagerness to arrive at the big truths, the ultimate principles that crown scientific work, was disturbing to cautious scholars."[28] For example, anthropologist Franz Boas was consistently irked by Huntington's simplistic statistical methods, which, Boas argued, offered merely a fig leaf of scientific credibility to Huntington's claims.

Based on his belief that there were optimal—and universally optimal—climates for physical and mental activity, Huntington drew upon a number of empirical studies of factory workers in America to suggest that 20°C and a humidity of 60 percent maximized productivity.[29] It was a short step from here to postulate that the energy and vigor needed to develop and sustain civilizations was also related to these climatic optima, giving rise to his "mainsprings of civilization" hypothesis. And for Huntington a further short step into the emerging field of genetic selection was to bring him in the 1920s under the influence of the American eugenics movement.

Huntington's determinism was centrally concerned with the tracing of patterns of climate in history, rather than with predicting the future fates of civilizations. The British politician and writer Sydney Markham, however, later developed and applied some of these determinist arguments in a different direction. In *Climate and the Energy of Nations*, published in 1942, Markham argued that climate variations could not only explain the rise and fall of past civilizations, but could also explain and predict the changing geopolitical balance of power in his mid-twentieth-century world. The dependence of contemporary social and economic factors such as trade, wealth creation, and human mortality rates on climate offered Markham a way of interpreting the tumultuous decade in which he wrote—the 1940s—and foreseeing the political prospects of nations such as Russia, China, and the United States as well as of Europe.[30]

As with its rise, there is no shortage of accounts of the demise of environmental determinism in geographic thought. Noel Castree claims that the "excesses of determinism" had been countered by the 1930s,[31] while others from different sectarian perspectives suggest earlier timelines for this demise. Kent McGregor suggests that environmental determinism was subjected to increasing skepticism from the 1920s onward and "by mid-century had run its course,"[32] and the climatologist Richard Skaggs also claims that "environmental determinism lost intellectual efficacy . . . during the 1920s."[33] The Marxist geographer Richard Peet puts it more bluntly: "Environmental determinism became increasingly socially dysfunctional in the 1920s after the main issues of imperialist domination of the world had been

[28] Visher, "Memoir to Ellsworth Huntington, 1876–1947," *Ann. Assoc. Amer. Geogr.* 38 (1948): 38–50, on 43.

[29] Ellsworth W. Huntington, *Civilization and Climate* (1915; repr., Honolulu, 2001).

[30] Sydney F. Markham, *Climate and the Energy of Nations* (London, 1942).

[31] Castree, *Nature* (London, 2005).

[32] McGregor, "Huntington and Lovelock: Climatic Determinism in the 20th Century," *Physical Geography* 25 (2004): 237–50, on 238.

[33] Skaggs, "Climatology in American Geography," *Ann. Assoc. Amer. Geogr.* 94 (2004): 446–57, on 447.

settled by World War I."[34] And from a cultural and political ecology standpoint, Judkins claims that the "historical moment" when determinism handed over to possibilism was around 1920.[35]

The strong form of climate determinism was therefore largely discredited and marginalized as the ideological wars of the twentieth century reshaped the political and moral worlds that had allowed it to flourish. Academic geography embraced more descriptive and reflexive conceptions of the relationships between nature and society. In the 1920s and 1930s the possibilism of Vidal de la Blache and Carl Sauer offered satisfying ways of keeping the role of climate and the environment at more comfortable distances from theories of social organization and cultural history. The consequence was, according to Sluyter, that "geographers abandoned any concerted attempt at nature-society explanations and most of them re-aligned with either the natural or the social sciences."[36]

Vestiges of Huntington's strong determinism nevertheless still lingered among those engaged in talking about and analyzing climatic data in the context of society and behavior. Meyer discusses the persistence of climate determinism in American thought and culture through the middle decades of the twentieth century in *Americans and Their Weather*.[37] For example, Huntington's final book—*Mainsprings of Civilization*—was published in America in 1945 and was criticized at the time by Oskar Spate for offering a "pattern to history too much determined by physical factors."[38]

Elsewhere, too, climatic determinism remained ingrained in the way some climatologists and other scholars wrote about climate and its role in the world. Sociologist Nico Stehr has deconstructed the 1938 essay "Kultur und Klima" by German social psychologist Willy Hellpach, thereby offering an insight into the relationship between Nazi ideology and climatic determinism.[39] Determinism in fact offered a softening of the strident Nazi racism, by claiming that people could be "improved" if they were put in the right environment; it was not all down to Aryan genetic ancestry.

In England, Austin Miller's classic textbook *Climatology*, which went through nine editions between 1931 and 1961, was still claiming in its ninth edition in 1961 that "the enervating monotonous climates of much of the tropical zone . . . produce a lazy and indolent people,"[40] while in the 1950s the prolific English climatologist Charles Brooks was taken to task for the determinist outlook pervading his bestselling book *Climate in Everyday Life*.[41] One reviewer of the book complained, "The author has apparently not realised that the fumbling, prejudice-ridden speculations on human climatology which marked the earlier years of this century must now be replaced by . . . more adequate enquiries and emancipated from the restrictions of a race-dominant culture."[42] And in 1958, also in England, Gordon Manley was writing

[34] Peet, "The Social Origins of Environmental Determinism," *Ann. Assoc. Amer. Geogr.* 75 (1985): 309–33, on 327.

[35] Judkins, Smith, and Keys, "Determinism" (cit. n. 6).

[36] Sluyter, "Neo-environmental Determinism" (cit. n. 22), 816.

[37] Meyer, *Americans and Their Weather* (cit. n. 4), 168–72, 206–14.

[38] Spate, "Toynbee and Huntington" (cit. n. 7), 414.

[39] See, respectively, Stehr, "The Ubiquity of Nature: Climate and Culture," *J. Hist. Behav. Sci.* 32 (1996): 151–9; Hellpach, "Kultur und Klima," in *Klima-Wetter-Mensch*, ed. Heinz Wolterek (Leipzig, 1938), 428–9.

[40] Miller, *Climatology*, 9th ed. (London, 1961), 2.

[41] Brooks, *Climate in Everyday Life* (New York, 1951).

[42] Douglas H. K. Lee, "Book Review," *Quart. Rev. Biol.* 27 (1952): 75–6, on 76.

about the revival of climate determinism,[43] even if a weaker variant, with a poorly disguised ambivalence about the adequacy of earlier deterministic theories: "It is an opportune moment to be reminded that man is still subject to a variety of constraints that may yet be imposed by Nature."[44]

THE RISE OF CLIMATE REDUCTIONISM

Notwithstanding these examples, the fortunes of strong determinism, both as an ideology and as an explanatory framework for climate-society relationships, waned over the twentieth century. Yet with the emergence over the last twenty-five years of anthropogenic climate change as a physical and social phenomenon of worldwide importance, the question of how the challenging relationship between climate and society is conceived has taken on fresh importance. Geographer William Riebsame has offered four ways of viewing physical climate in relation to human society: climate as setting, as determinant, as hazard, and as resource.[45] As Riebsame explains, seeing climate as a determinant requires the identification of "causal chains that link climate to specific elements or behaviours of biophysical and socioeconomic systems,"[46] whether these elements be crop yield, malaria risk, economic performance, or violent conflict. The burgeoning climate change impacts literature of the 1990s and 2000s has been dominated by research "identifying" such causal chains, as witnessed by some of the examples cited earlier in the article. Such claims have been driven by a methodological reductionism.

Reductionism is an approach to understanding the nature of complex entities or relationships by reducing them either to the interactions of their parts or else to simpler or more fundamental entities or relationships. In the case of climate change studies, this means isolating climate as the (primary) determinant of past, present, and future system behavior and response. If crop yield, economic performance, or violent conflict can be related to some combination of climate variables, then knowing the future behavior of these variables offers a way of knowing how future crop yield, economic performance, or violent conflict will unfold. Other factors that influence these future environmental, economic, or social variables—factors that may be more important than climate or perhaps just less predictable—are ignored or marginalized in the analysis. To illustrate such reductionism at work, I offer two instances selected from among many possible examples.

The way climate reductionism requires and seeks out simple chains of climatic cause-and-effect is perfectly illustrated in an empirical study of the relationship between climate change and economic growth published by the U.S. National Bureau of Economic Research.[47] The authors recognize that the question of whether climate change has a direct effect on economic development is contentious, but they claim nevertheless that their global analysis using data from over 180 nations reveals

[43] See Georgina Endfield in this volume.

[44] Manley, "The Revival of Climatic Determinism," *Geogr. Rev.* 48 (1958): 98–105, on 105.

[45] Riebsame, "Research in Climate-Society interaction," in *SCOPE 27—Climate Impact Assessment*, eds. Robert W. Kates, Jesse H. Ausubel, and Mimi Berberian (Chichester, 1985), 85–104.

[46] Ibid., 72.

[47] Melissa Dell, Benjamin F. Jones, and Benjamin A. Olken, "Climate Shocks and Economic Growth: Evidence from the Last Half Century" (working paper no. 14132, National Bureau of Economic Research, Cambridge, Mass., 2008).

a "substantial contemporary causal effect of temperature on aggregate [economic] output. . . . On average, a 1°C increase in average temperature predicts a fall in per-capita income by about 8 per cent."[48] Since they find this effect to be asymmetrical between richer and poorer countries, they are then able to extend their analysis to consider the impact of *future* climate change on economic performance. They conclude: "The negative impacts of climate change on poor countries may be larger than previously thought. Overall, the findings suggest that future climate change may substantially widen income gaps between rich and poor countries."[49] First the complex relationships that exist between climate and economic performance are reduced to a dependent relationship between temperature and GDP per capita, and then, using projections of future climate warming, future economic performance is predicted for the twenty-first century. The many subtleties and contingencies of national and regional economic performance are ignored or suppressed. Climate reductionism opens up the prospect of developing a narrative about future economic growth in which climate change becomes the primary driver of performance.

A second example of climate reductionism at work is Peter Halden's analysis of the geopolitics of climate change from an international relations perspective.[50] Halden, a social and political scientist working for the Department of Defense Analysis at the Swedish Defense Research Agency, takes as given the climate predictions from the Intergovernmental Panel on Climate Change (IPCC) for the year 2050. But he makes no attempt to envisage the possible social or political worlds of 2050, "a venture," he claims from his position as a political scientist, "that would be flawed at best and approaching hubris at worst."[51] He rejects the attempt to combine natural science forecasts with "speculative" social science in favor of a presumption of the social and political status quo. This reasoning hands the future over to Earth system models and their claims of revealing the impacts of climate change unfolding on a passive, unimaginative, and static humanity. Climate reductionism results in a narrative about future geopolitical movements in which, again, climate change becomes the primary driver.

Both these examples offer a one-eyed view of the future, yet it is one that pervades many recent academic analyses of climate change and social impact;[52] and consequently it is an account of the future that enters easily into public perception and discourse.[53] Inadvertently or not, such reductionist reasoning opens these analyses of climate impact to the charge of operating within neodeterminist explanatory frameworks. The two examples above are offered as archetypical illustrations of a widespread pattern of methodological climate reductionism as it is applied to many different dimensions of the imagined future: health, food production, biodiversity, tourism and recreation, human migration, violent conflict, and so on. The precise

[48] Ibid., 4, 6.

[49] Ibid., 27–8.

[50] Halden, *The Geopolitics of Climate Change: Challenges to the International System* (Stockholm, 2008).

[51] Ibid., 22–3.

[52] E.g., see the two recent studies cited earlier: Burke et al., "Civil War" (cit. n. 14), and Feng, Krueger, and Oppenheimer, "Cross-Border Migration" (cit. n. 17).

[53] Two examples among many references to the Burke et al. and Feng et al. studies are Tom Chivers, "Climate Change Will Lead to Civil Wars in Africa, Says Research," *Daily Telegraph*, November 25, 2009, and Nacha Cattan, "Climate Change Set to Boost Mexican Immigration to the US, Says Study," *Christian Science Monitor*, July 27, 2010.

numbers and fearful tones cited in the introduction to this article are the result of such reductionist reasoning and analysis. But given the demise of climate determinism described above, at least within large parts of the academy, how is it possible to have arrived back at an understanding of climate-society relationships that, I am suggesting, distorts and overemphasizes the causative role of climate in shaping the future prospects of society and the well-being of individuals?

THE HEGEMONY OF MODEL PREDICTIONS OF THE FUTURE

Sluyter offers one explanation for this resurgence of neoenvironmental determinism, or climate reductionism, in the cases I am looking at. He suggests that the Enlightenment dichotomy between nature and culture, so pervasive in Western thought and practice, began increasingly to be challenged in the 1980s and 1990s—for example, as described through Ulrich Beck's manufactured global risks and through Bruno Latour's entanglements of nature and culture.[54] In response to this move, Sluyter argues that environmental determinism offered one means for a "quick and dirty integration of the natural and social sciences."[55] As if the inadvisability of the dualistic thinking pervading Western thought were being belatedly realized, there was a rush to forge a new rapprochement between nature and culture. Determinist thinking was the simplest and most available ideology to hand. Sluyter is scathingly dismissive of such opportunism, however, and of the intellectual credulity exhibited by what he calls the "neodeterminists," authors such as Diamond and Sachs.

While I think there is some merit in his argument in the more general field of environment-society interactions, I wish to suggest a different line of reasoning that applies very specifically to the case of climate reductionism I have illustrated above. It is a line of reasoning that emerges from the way in which the understanding of climate change developed over the last decades of the twentieth century.

In summary, my argument concerns the hegemony held by the predictive natural and biological sciences over visions of the future. In the case of climate change, this hegemony is rooted in the knowledge claims of climate or Earth system models. In the absence of comparable epistemological reach emerging from the social sciences or humanities, these claims lend disproportionate discursive power to model-based descriptions of putative future climates. It thus becomes tempting to adopt a reductionist methodology when examining possible social futures: "Lots of things will change in the future, but since we have credible and quantitative knowledge about future climate let us examine, also quantitatively, what the consequences of these climates for society might be." The subsequent and derived climate impact modeling then boldly calculates, for example, the billions of people who because of climate change will become starving or thirsty, or the millions who because of climate change will be made destitute or homeless.[56] Climate reductionism is the means by

---

[54] Beck, *Risk Society: Towards a New Modernity* (London, 1992); Latour, *We Have Never Been Modern*, trans. C. Porter (New York, 1993).

[55] Sluyter, "Neo-environmental Determinism" (cit. n. 22), 817.

[56] Nigel W. Arnell, Melvin G. R. Cannell, Mike Hulme, R. Sari Kovats, John F. B. Mitchell, Robert J. Nicholls, Martin L. Parry, Matt T. J. Livermore, and Andrew White, "The Consequences of CO_2 Stabilisation for the Impacts of Climate Change," *Climatic Change* 53 (2002): 413–46; Norman Myers, "Environmental Refugees: Our Latest Understanding," *Phil. Trans. Royal Soc.* B 356 (2001): 16.1–16.5; Mabey, *Delivering Climate Security* (cit. n. 16).

which the knowledge claims of the climate modelers are transferred, by proximity as it were, to the putative knowledge claims of the social, economic, and political analysts.

This transfer of predictive authority, an almost accidental transfer, one might suggest, rather than one necessarily driven by any theoretical or ideological stance, is what I earlier defined as "epistemological slippage." If not quite the inexorable geometric calculus of Malthus, it nevertheless offers a future written in the unyielding language of mathematics and computer code. These models and calculations allow for little human agency, little recognition of evolving, adapting, and innovating societies, and little endeavor to consider the changing values, cultures, and practices of humanity. The contingencies of the future are whitewashed *out* of the future. Humans are depicted as "dumb farmers," passively awaiting their climate fate. The possibilities of human agency are relegated to footnotes, the changing cultural norms and practices made invisible, the creative potential of the human imagination ignored.

To give some substance to this argument I need to explore some of the historical contexts that have allowed climate models to claim such hegemony over the future and that have allowed climate reductionism to thrive. This requires an examination of the emergence of anthropogenic climate change as a matter of scientific concern in the 1970s and 1980s and as a matter of public policy debate in the 1980s and 1990s. There are three developments that are important for my argument: the retreat of the social sciences, and geography in particular, from working at the nature-culture interface; the emergence of a new epistemic community of global climate modelers; and the asymmetrical incorporation of climate change and social change into envisaged futures. Each of these three developments will be examined in turn.

The Absence of Theory about Climate-Society Interactions

The previous sections have shown how the academic discipline that had thought the longest and hardest about relationships between climate and society—geography— had by the 1960s become suspicious about grand theories of climate-society interaction, particularly those tinged with any trace of the old determinist ideology.[57] This reaction against the worst excesses of determinism "left geographers without a coherent conception of causality that would 'bridge' the social and natural sciences."[58] It also meant that the study of environment-society relationships "became a subject without an academic home, a stateless person in the world of sovereign disciplines."[59] It was in fact a small number of historians and atmospheric scientists, rather than geographers, who were the most willing to reengage substantively with questions about climate change and human society. Historians such as Emmanuel Le Roy Ladurie, atmospheric scientists such as Reid Bryson, and historical climatologists such as Hubert Lamb produced the most significant investigations during the 1970s into the nature of past interactions between climate change and social organization.[60] But

[57] John F. Hart, "The Highest Form of the Geographer's Art," *Ann. Assoc. Amer. Geogr.* 72 (1982): 1–29.

[58] Castree, *Nature* (cit. n. 31), 63.

[59] Meyer, *Americans and Their Weather* (cit. n. 4), 209.

[60] See, respectively, Ladurie, *Times of Feast, Times of Famine: A History of Climate since the Year 1000*, trans. Barbara Bray (New York, 1971); Bryson and Thomas J. Murray, *Climates of Hunger:*

they did so in the absence of any coherent theoretical framework to explain such interactions and certainly without any basis for prediction.

Against this background of disciplinary maneuvers and intellectual hesitancy, there were an increasing number of important questions emerging in the 1970s about how climate change and social change were related. Accelerated by the cultural background of a new environmental consciousness,[61] concerns were mounting over global food and energy security and the social impacts of drought in Africa and weather modification in America.[62] "Climate change and human affairs" were becoming entangled in new ways, as masterfully narrated by Crispin Tickell in his 1977 book of that title.[63] To understand these interactions required some grasp of both the dynamics of climate and the nature of human agency, whether individual, collective, or institutional. And as the story moved from the 1970s to the 1980s, it became increasingly clear that climates worldwide *were* changing, at least in part because of human activities. New questions were being asked by researchers, environmentalists, and policy makers about what these emerging and prospective changes in climate might mean for society.[64] Geographers and social scientists, however, remained rather poorly positioned to answer such questions, lacking agreed (or else acceptable) theories and tools for investigating the interactions between climate and society. Judkins, for example, describes how for geographers and social scientists the period from the 1960s to the 1980s was characterized by competing and contradictory theoretical accounts of environment-society interactions.[65]

The Epistemic Community of Global Climate Modeling

It was against this background of theoretical and methodological uncertainty about how society and climate were related that the methods and claims of a new community of climate modelers and global change scientists were emerging. The 1960s and 1970s had witnessed the development of the first computer-based simulation models of a universal and globally connected climate system.[66] Originally an extension of numerical weather prediction models, these new climate-oriented models allowed experiments with global climate to be performed in virtual reality that were not possible in physical reality. These models were constructed initially by meteorologists and atmospheric scientists in a small number of research centers in the United States, the United Kingdom, and Germany. They were later joined by oceanographers, atmospheric chemists, and biologists as the models extended their representation from simply the climate system (initially the atmosphere) to the deeply coupled components of the Earth system. This move was encapsulated in NASA's 1988 report *Earth*

Mankind and the World's Changing Weather (Madison, Wis., 1977); Lamb, *Climate: Past, Present and Future*, vol. 2: *Climatic History and the Future* (London, 1977).

[61] Sheila Jasanoff, "Image and Imagination: The Formation of Global Environmental Consciousness," in *Changing the Atmosphere: Expert Knowledge and Environmental Governance*, eds. Clark Miller and Paul N. Edwards (Cambridge, Mass., 2001), 309–37.

[62] Central Intelligence Agency, *Potential Implications of Trends in World Population, Food Production and Climate*, report no. OPR-401 (Washington, D.C., 1974).

[63] Tickell, *Climate Change and World Affairs* (Cambridge, Mass., 1977).

[64] E.g., Jill Williams, ed., *Carbon Dioxide, Climate and Society* (Oxford, 1978); Council of Environmental Quality, *Global Energy Futures and the Carbon Dioxide Problem* (Washington, D.C., 1981).

[65] Judkins, Smith, and Keys, "Determinism" (cit. n. 6).

[66] Paul N. Edwards, *A Vast Machine: Computer Models, Climate Data and the Politics of Global Warming* (Cambridge, Mass., 2010).

System Science: A Closer View, the so-called Bretherton report.[67] The report's lead author was Francis Bretherton, an applied mathematician and atmospheric scientist, and the goal of this new scientific mission was "to obtain a scientific understanding of the entire Earth system on a global scale"; predictions were to be secured by using "quantitative models of the Earth system to identify and simulate global trends."[68]

In barely twenty-five years—from the early 1960s to the late 1980s—scientific accounts of the causes and properties of climate had become progressively more complex. Climate was now viewed as the outcome of the functioning of an interconnected biogeophysical global system whose past, present, and future behavior could be modeled—and hence "predicted"—using mathematical equations and advanced computing technology. This marked a distinct break from the more varied conceptions of climate used by geographers, climatologists, and synoptic meteorologists earlier in the twentieth century. Clark Miller makes the interesting observation that the "First Annual Conference on Statistical Climatology" was held in 1979, and prior to this time there was no reason to refer to *statistical* climatology because there was no other form of climatology to distinguish it from.[69]

The more systemic concept of climate as Earth system science, together with the representation of this concept in simulation models, formed the twin bases around which a new epistemic community of global climate modelers coalesced. An epistemic community is a community of experts who share sets of beliefs about factual and causal understandings of particular phenomena.[70] Furthermore, these shared beliefs and values guide the community in drawing policy conclusions from its knowledge. By the 1990s "computer modelling had become *the* central practice for evaluating truth claims" for this community of global climate change scientists.[71] Yet as Miller has argued, epistemic communities and the knowledge they produce do not form in isolation from wider social, institutional, and political settings.[72] And the knowledge thus produced has a very distinctive geography of production. For example, the role of the cold war was crucial in the development of American climate science,[73] and by the 1970s and 1980s it was the growing political interest in human-induced climate change and the globalization of environmental politics that drove forward this new intellectual program.[74] The development of global climate and Earth system models and their application to examining questions about the future performance of a now "global climate," one being subjected to human-induced changes in atmospheric composition, occurred against the backdrop of the new environmental geopolitics of the post-Stockholm era.[75] June 1972 had witnessed the first

[67] NASA, *Earth System Science: A Closer View* (Washington, D.C., 1988).

[68] Ibid., 11.

[69] Miller, "Climate Science and the Making of a Global Political Order," in *States of Knowledge: The Co-production of Science and the Social Order*, ed. Sheila Jasanoff (London, 2004), 46–66.

[70] Peter M. Haas, "Epistemic Communities and International Policy Coordination," *International Organization* 46 (1992): 1–35.

[71] Paul N. Edwards, "Representing the Global Atmosphere: Computer Models, Data and Knowledge about Climate Change," in Miller and Edwards, *Changing the Atmosphere* (cit. n. 61), 31–65, on 53.

[72] Clark A. Miller, "Challenges in the Application of Science to Global Affairs: Contingency, Trust and Moral Order," in Miller and Edwards, *Changing the Atmosphere* (cit. n. 61), 247–86.

[73] David M. Hart and David G. Victor, "Scientific Elites and the Making of US Policy for Climate Change Research," *Soc. Stud. Sci.* 23 (1993): 643–80.

[74] Paul K. Wapner, *Environmental Activism and World Civic Politics* (New York, 1996).

[75] Clark A. Miller, "The Dynamics of Framing Environmental Values and Policy: Four Models of Societal Processes," *Environmental Values* 9 (2000): 211–33.

United Nations Conference on the Human Environment held in Stockholm, and this presaged a new era of international environmental diplomacy. The World Meteorological Organization's First World Climate Conference, held in 1979, and the 1983 report by the Carbon Dioxide Assessment Committee of the U.S. National Academy of Sciences were also evidence of the growing political saliency of climate change.[76]

The consequence of this coproduction of knowledge between what Miller has called "climate science and the global political order" was the foregrounding of model-based predictions of future climate change in academic and policy discourses.[77] Models came to be seen as "the only practical way to discern the effects of policy choices about climate change," and all important knowledge and choice about climate change seemed to revolve around such models.[78] The early battles about the credibility of anthropogenic climate change in the 1990s were therefore fought largely around the credibility of these models,[79] because both sides recognized the political significance of their knowledge claims about the future.

Yet to answer the demanding questions being asked about the significance of anthropogenic climate change for human society required more than mere knowledge of future climate. It demanded some translation of future changes in climate into future impacts for society. The *First Assessment Report* of the IPCC in 1990, for example, was organized into three separate volumes: one on climate science, one on climate impacts, and one on climate policy options. If climate modelers were by now offering credible predictions of future climate change, then before policies could be developed and evaluated, it was argued (implicitly, perhaps) that plausible accounts of the impacts of these changes on human society were needed. It was here that the asymmetry between the knowledge claims of the predictive natural scientists and those of geographers and other environmental social scientists emerged most acutely. Given the poorly developed and atheoretical understandings of climate-society relationships in the social sciences, how were these demanding questions going to be answered? How did the first IPCC assessments address these relationships?

The Asymmetrical Incorporation of Climate and Social Change into Envisaged Futures

The first studies assessing the consequences of future anthropogenic climate change for society were undertaken in the late 1970s and early 1980s; some of this work is summarized by Jill Williams and by William Kellogg and Robert Schware.[80] But to investigate the methodological challenges these new policy-driven questions were posing for academic environmental social science researchers in the 1980s and 1990s, I examine two seminal books published in this era. Both books were methodologically oriented, and, taken together, they illuminate how methodological space was created within which climate reductionism could emerge.

[76] National Academy of Science, *Changing Climate—Report of the Carbon Dioxide Assessment Committee*, ed. William A. Nierenberg (Washington, D.C., 1983).

[77] Miller, "Climate Science" (cit. n. 69).

[78] Edwards, "Representing the Global Atmosphere" (cit. n. 71), 63.

[79] Fred Pearce, "Greenhouse Wars," *New Scientist*, July 19, 1997, 38–43.

[80] See, respectively, Williams, *Carbon Dioxide, Climate and Society* (cit. n. 64); Kellogg and Schware, *Climate Change and Society: Consequences of Increasing Atmospheric Carbon Dioxide* (Boulder, Colo., 1981).

The first is the volume commissioned and published by the International Council of Scientific Unions' Scientific Committee on Problems of the Environment (SCOPE) on climate impact assessment.[81] This report, known as *SCOPE 27*, was a response to the new World Climate Impact Program (WCIP)—whose aim was "to advance our understanding of the relation between climate and human activities"—that had been agreed at the 1979 First World Climate Conference. This volume was one of the first outputs from WCIP and became a standard text in the field. I have already quoted from Riebsame's chapter in this volume, which offered four ways in which climate may be viewed. But the crucial methodological chapter was written by the respected geographer Robert Kates. The WCIP, *SCOPE 27*, and Kates's specific chapter are all therefore a direct response to the growing policy demand for credible and salient knowledge about what anthropogenic global climate change might mean for future society.

Kates laid out the methodological challenges of performing climate impact assessments, three of which are particularly relevant for the argument presented here. First, he acknowledged an explicit knowledge hierarchy between the "hard" sciences and the "soft" sciences. As one moves from understanding global heat balances to the impacts of climate change on nutrition, for example, there is "less predictability, more speculation and greater uncertainty."[82] Complexity increases, precision decreases, and uncertainties are compounded. The second challenge identified by Kates and of particular interest in the present context is that of linking very different methodologies: for example, modeling of global climate with analysis of energy trends or assessment of population dynamics. The poverty of theoretical and methodological development in this area was recognized by Kates: "As yet there has been no comprehensive study of the problems of integrating such scientific apples and oranges."[83] (This is the field that today is more commonly known as integrated assessment modeling [IAM] and that is still deficient in its ability to represent processes of societal adaptation.[84])

The third challenge therefore was how to develop even the most basic of analytical frameworks for performing such "linked studies" of climate impact assessment. Kates offered two schematic diagrams, one of which he called the "impact model" and the other the "interaction model" (both reproduced here in fig. 1). In the former, climate change determines the impact directly, while in the latter the impact is the joint product of the interaction between climate and social change. And it is the former model that Kates claimed was predominant in nearly all attempts at climate impact assessment, which went "directly from climate events to inferences of higher-order consequences."[85] Reflecting on the reasons for this paucity of studies that sought to embrace a more interactive framework of climate-society relationships, as opposed to the instinct to revert to a cruder deterministic or reductionist account, Kates remarked that it was due "partly to disciplinary isolation and partly to

[81] Kates, Ausubel, and Berberian, *SCOPE 27* (cit. n. 45).

[82] Robert Kates, "The Interaction of Climate and Society," in Kates, Ausubel, and Berberian, *SCOPE 27* (cit. n. 45), 3–36, on 4.

[83] Ibid., 5.

[84] See Hans-Martin Füssel, "Modelling Impacts and Adaptation in Global IAMs," *WIREs Climate Change* 1 (2010): 288–303.

[85] Kates, "Interaction of Climate and Society" (cit. n. 82), 31. Füssel states, "Adaptation has received only limited attention in global IAMs so far"; "Modelling Impacts" (cit. n. 84), 288.

A. Impact Model

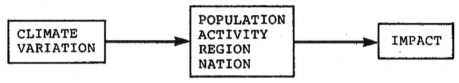

B. Interaction Model

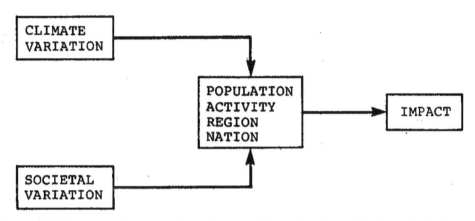

Figure 1. *Schematics of impact and interactive models are highly simplified graphic depictions of types of study methodologies. It was the more reductionist "impact model" that predominated in most impacts studies. Reproduced by permission from Kates, "Interaction of Climate and Society" (cit. n. 82), 31.*

the limited effort expended to date on the study of the interaction of climate and society as compared to the study of the dynamics of climate itself."[86]

SCOPE 27 therefore reveals, I suggest, how the idea of an explicit knowledge hierarchy, the lack of any theoretical frameworks for integrated analysis, and the preferred linear model of climate response contributed to a climate reductionism at work in impact assessments. At this crucial moment in the 1980s, when climate predictions were asserting their knowledge claims about the future and when policy was demanding knowledge about future consequences of climate change for society, it was easy for simple reductionist accounts of future climate change impacts to emerge.

Kates did not explicitly address the development of climate predictions or scenarios, which have become the pivotal component of so many climate impact studies. The second book I wish to examine, however, does so. Published in 1998 under the title *Climate Impact and Adaptation Assessment*,[87] this was a widely read guide to the IPCC approach to assessing climate change impacts and adaptations. This book offered "a readable guide" to the *Technical Guidelines for Assessing Climate Change*

[86] Kates, "Interaction of Climate and Society" (cit. n. 82), 31.
[87] Martin L. Parry and Timothy R. Carter, *Climate Impact and Adaptation Assessment: The IPCC Method* (London, 1998).

Impacts and Adaptations published a few years earlier by the IPCC,[88] guidelines that became widely cited and used internationally in the field. In these IPCC assessment guidelines, as interpreted by Martin Parry and Tim Carter in their guide, the default methodological assumptions and practices revealed by Kates in the 1980s were re-inforced. In this case it was done by privileging predictions of future climate over explorations of how the many other dimensions of cultural, social, and political life may change in the future. Climate reductionism through epistemological slippage was the result.

The IPCC method for impact and adaptation assessment had seven recommended steps (fig. 2), at the center of which—step 4—was the selection of future climate scenarios. With future climate(s) thus established, the method proceeded to estimate the consequences of climate change for both natural and social environments, before examining how such consequences might be adapted to. In their summary of step 4, Parry and Carter took care to emphasize the importance of recognizing social dynamics: "The environment, society and the economy are not static" in the absence of climate change.[89] But the subsequent practical guidance for how to incorporate such dynamism into scenarios of the future is limited. Out of twenty-three pages in this crucial scenario chapter, less than three are devoted to the representation of social change, whereas over fourteen pages offer guidance on how to develop future climate change scenarios. And most of this guidance refers to the use of data and results from global climate models.

The asymmetry evidenced in the Parry and Carter chapter between methods for depicting climate and social futures is merely representative of much wider practice in the field of climate impact assessment over the last twenty-five years. For the first twelve years of the IPCC process (1988–2000) there were no systematic attempts to develop methods or scenarios that represented future social, cultural, or political change, even though large amounts of effort were directed to advancing and distrib-uting model-derived representations of future climate. Only with the publication of the IPCC *Special Report on Emissions Scenarios* in 2000 was significant visibility given to the representation of different social futures in climate impact studies.[90] This deficiency contributed to the widespread adoption of what I call climate reductionist methods in climate impact assessment, the consequences of which have been earlier illustrated. The IPCC *Third Assessment Report* in one of its chapters lamented this practice: "Future socioeconomic . . . changes have not been represented satisfactorily in many recent impact studies," and "many impact studies fail to consider adequately uncertainties embedded in the scenarios they adopt."[91] And at a national scale, a re-view of the U.K. Climate Impacts Programme in 2005 noted that climate impact studies have seldom been able to incorporate alternative social futures, "preferring instead to concentrate on exploring 'climate-only' impacts," a direct illustration of climate reductionism at work.[92]

[88] Timothy R. Carter, Martin L. Parry, Hideo Harasawa, and Shuzo Nishioka, *IPCC Technical Guidelines for Assessing Climate Change Impacts and Adaptations* (London/Tsukuba, 1994).

[89] Parry and Carter, *Climate Impact* (cit. n. 87), 72.

[90] IPCC, *Special Report on Emissions Scenarios* (Cambridge, 2000).

[91] Timothy R. Carter and Emelio La Rovere, "Developing and Applying Scenarios," in *Climate Change, 2001: Impacts, Adaptation and Vulnerability*, eds. Jim McCarthy, Osvaldo Canziani, Neil A. Leary, Dave J. Dokken, and K. S. White (Cambridge, 2001), 145–90, on 181.

[92] Chris West and Megan Gawith, eds., *Measuring Progress: Preparing for Climate Change through the UK Climate Impacts Programme* (Oxford, 2005), 61.

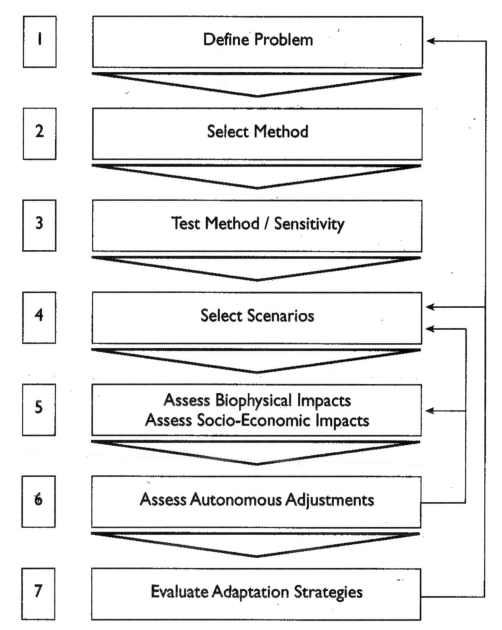

Figure 2. *Seven steps of climate impact assessment, adopted by the IPCC. Step 4 is the crucial stage of creating the future. Reproduced by permission from Parry and Carter,* Climate Impact and Adaptation Assessment *(cit. n. 87).*

Reactions against climate reductionism, notably the concepts of vulnerability and resilience, have emerged in the last decade or so from parts of the social science community.[93] The origins of these less reductionist conceptual and analytical paradigms are to be found in hazards research and ecology, respectively, and were introduced into climate change research in the late 1990s (vulnerability) and early 2000s (resilience). They offer ways of exploring sensitivities of socioecological systems to climate perturbations—and other environmental and social stresses—without being dependent upon the predictive claims of climate modeling. Although they have gained some visibility in recent climate change research, because vulnerability and resilience approaches to understanding climate-society relationships are less dependent on model-based climate projections, they have been slow to overturn the standard IPCC climate impact methodology.[94]

* * *

The combination of these historical developments—the rise of a powerful epistemic community of climate modelers, the asymmetrical incorporation of climate and social change into envisaged futures, and, confounding the whole enterprise, the lack of theory making around climate-society interactions—has allowed a form of climate reductionism to dominate contemporary analysis and thinking about the future. Although it is clear to many social scientists that "the impact of any climatic event depends on the local ecological setting and the organisational complexity, scale, ideology, technology and social values of the local population,"[95] current intellectual endeavors in this area unduly privilege climate as the chief determinant of humanity's putative social futures.

Quantitative climate predictions for the 2050s, 2080s, or even further ahead continue to be offered by a powerful community of climate modelers, most recently at very high spatial and temporal resolutions. For example, the latest climate projections from the U.K. government incorporate weather information at hourly intervals for regions as small as 25 square kilometers and for several decades into the future.[96] Yet the "complexity, scale, ideology, and social values" of future local populations and communities are for the most part ignored or assumed to be static. The study by Halden summarized earlier is a good example of this asymmetry in representations of the future. Quantified—and often unconditional—predictions of future climate change impacts therefore abound, such knowledge claims drawing power from the epistemic muscle of climate and Earth system models in a process of epistemological slippage.

And so the future is reduced to climate. By stripping the future of much of its so-

[93] Marco A. Janssen and Elinor Ostrom, "Resilience, Vulnerability and Adaptation: A Cross-Cutting Theme of the International Human Dimensions Programme on Global Environmental Change," *Global Environmental Change* 16 (2006): 237–9.

[94] For a discussion of some of the reasons why progress has been slow, see Karen L. O'Brien and Johanna Wolf, "A Values-Based Approach to Vulnerability and Adaptation to Climate Change," *WIREs Climate Change* 1 (2010): 232–42.

[95] Fakri A. Hassan, "Human Agency, Climate Change and Culture: An Archaeological Perspective," in *Anthropology and Climate Change: From Encounters to Action*, eds. Susan A. Crate and Mark Nuttall (Walnut Creek, Calif., 2008), 39–69, on 40.

[96] UK Climate Projections (UKCP09) Web site, http://ukclimateprojections.defra.gov.uk (accessed 20 January 2011).

cial, cultural, or political dynamism, climate reductionism renders the future free of visions, ideologies, and values. The future thus becomes overdetermined. Yet the future is of course very far from being an ideology-free zone. It is precisely the most important territory over which battles of beliefs, ideologies, and social values have to be fought. And it is these imagined and fought-over visions of the future that—in many indeterminate ways—will shape the impacts of anthropogenic climate change as much as will changes in physical climate alone.

PUTTING SOCIETY BACK INTO THE FUTURE

Climate reductionism—a form of neoenvironmental determinism—offers a methodology for providing simple answers to complex questions about the relationship between climate, society, and the future. In its crudest form it asserts that if social change is unpredictable and climate change predictable then the future can be made known by elevating climate as the primary driver of change. But such reductionism downgrades human agency and constrains the human imagination. So, looking back, Diamond claims that "history followed different courses for different peoples because of peoples' environments,"[97] while looking forward Lovelock fears that "despite all our efforts to retreat sustainably, we may be unable to prevent a global decline into a chaotic world ruled by brutal war lords on a devastated Earth."[98]

Although offering accounts of the past and the future that are more popular than academic, both Diamond and Lovelock adopt inadequate and impoverished reductionist frameworks for understanding the past and envisioning the future. Many of the statements concerning the impacts of future climate change emerging from the more analytical research community suffer from the same limitations. The consequence of such reductionism is expressed clearly in Karl Popper's attack from a generation ago on historicism and its deterministic roots: "Every vision of historicism expresses the feeling of being swept into the future by irresistible forces."[99] While Popper, writing in a different era, had historical materialism and the enemies of an open society in mind, his reasoning well applies to climate change today.

The allure of determinist thinking is that it offers the appearance of "naturalistic" explanations—even justifications—of cultural or economic dominance (as in past variants of determinism) or "naturalistic" accounts of the future that evacuate it of human agency (as I have contended is the case with climate change today). In contrast to earlier climate determinisms, which flowered in the ascendant and optimistic imperial cultures of classical Greece and of imperialist Europe and a youthful United States, I suggest that the climate reductionism I have described here is nurtured by elements of a Western cultural pessimism that promote the pathologies of vulnerability, fatalism, and fear.[100] It is these dimensions of the contemporary cultural mood that provide the milieu within which this particular form of environmental determinism has reemerged. By handing the future over to inexorable nonhuman powers,

---

[97] Jared Diamond, *Guns, Germs and Steel: The Fates of Human Societies* (New York, 1997), 25.

[98] Lovelock, *Revenge of Gaia* (cit. n. 21), 198.

[99] Popper, *The Poverty of Historicism* (London, 1957), 160.

[100] See Pat Devine, Andrew Pearman, and David Purdy, eds., *Feelbad Britain: How to Make It Better* (London, 2009); Rod Liddle, *Social Pessimism: The New Social Reality of Europe* (London, 2008).

climate reductionism offers a rationalization, even if a poor one, of the West's loss of confidence in the future.

These characteristics of Western culture have also been described by sociologist Frank Furedi in his book *Invitation to Terror*.[101] Furedi explains the confusion that has emerged in Western culture about the new international terrorism of this century and links it to a pessimism about the accomplishments of modernity and science and a fear of their legacy. Such pessimism evacuates the future of belief, vision, and promise. The knowledge claims of intelligence experts—or, in the case studied here, of climate modelers—are invited to fill the voids in the human imagination thus created. While Furedi's is a contested position—for many, the promises of new technologies remain as alluring as ever—Beck describes a similar phenomenon when he talks about the nonexistent and fictitious future replacing the legacies of the past as the basis for present-day action: "Expected risks are the whip to keep the present in line. The more threatening the shadows that fall on the present because a terrible future is impending, the more believed are the headlines provoked by the dramatisation of risk today."[102]

Climate reductionism is a limited and deficient methodology for accessing the future. In his poetic essay "The End of the World," environmental historian Stephen Pyne offers an insight into similar reductionist limitations with regard to the past:

> Reductionism is good for extracting resources and for creating instruments, medicines, gadgets; but it does not—cannot—tell us how to use them or when or why. It cannot convey meaning because meaning requires contrast, connections, context. . . . [Reductionism] cannot tell us what we need to know in order to write genuine history, even when that history involves nature.[103]

If reductionism is a limited form of reasoning for interpreting the past, then climate reductionism is even more inadequate with regard to telling the future. The epistemological pathways offered by climate models and their derived analyses are only one way of believing what the future may hold. They have validity, and they have relevance. But to compensate for the epistemological slippage I have described in this article it is necessary to balance these reductionist pathways to knowing the future with other ways of envisioning the future.

The "contrast, connections, and context" to which Pyne refers must be created by putting society back into the future. Since it is at least possible—if not indeed likely—that human creativity, imagination, and ingenuity will create radically different social, cultural, and political worlds in the future than exist today, greater effort should be made to represent these possibilities in any analysis about the significance of future climate change. Some of these futures may be better; some may be worse. But they will not be determined by climate, certainly not by climate alone, and these worlds will condition—perhaps remarkably, certainly unexpectedly—the consequences of climate change.

---

[101] Furedi, *Invitation to Terror: The Expanding Empire of the Unknown* (London, 2007).

[102] Ulrich Beck, "Global Risk Politics," in *Greening the Millennium: The New Politics of the Environment*, ed. Michael Jacobs (Oxford, 1997), 18–33, on 20.

[103] Pyne, "The End of the World," *Environmental History* 12 (2007): 649–53, on 650.

Notes on Contributors

Maria Bohn is a PhD student at the Division of History of Science and Technology at the Royal Institute of Technology in Stockholm, Sweden. Her graduate work is on climate science, monitoring, and modelling 1945-1970.

Mark Carey is Assistant Professor in the Clark Honors College at the University of Oregon, where he teaches history of science and environmental history. He won the Leopold-Hidy Prize for the best article published in the journal *Environmental History* during 2007, and in 2009 received Virginia's "Rising Star" Outstanding Faculty Award. His ongoing research on the history of climate science and natural disasters is funded by the National Science Foundation, grant #0822983. He is the author of *In the Shadow of Melting Glaciers: Climate Change and Andean Society* (Oxford University Press, 2010).

Deborah R. Coen is Assistant Professor of History at Barnard College, Columbia University. She is the author of *Vienna in the Age of Uncertainty: Science, Liberalism, and Private Life* (University of Chicago Press, 2007) and co-editor, with James Fleming and Vladimir Jankovic, of *Intimate Universality: Local and Global Themes in the History of Weather and Climate* (Science History Publications, 2006). Her current book project is *The Earthquake Observers: Scientists and Citizens in the Face of Disaster*.

Gregory T. Cushman is Assistant Professor of International Environmental History at the University of Kansas. He has published widely on environmental history and the history of environmental knowledge in Latin America, the Caribbean, and the Pacific Basin. This work has focused on the El Niño-Southern Oscillation and other climate phenomena and how geopolitical contexts shape the development of scientific knowledge. He is author of *The Guano Lords: Ecology & Empire in Peru and the Pacific* (Cambridge University Press, 2010).

Matthias Dörries is Professor of History of Science at the University of Strasbourg, France. He has worked extensively on the history of nineteenth-century French and European physical sciences. His more recent research and publications focus on the history of the geophysical sciences from the 19th to the 21st centuries, for example, the history of volcanology and climate change science ("In the Public Eye: Volcanology and Climate Change Studies in the 20th Century," *Historical Studies in the Physical and Biological Sciences, 37* (2006), 87-125). He is currently working on a book on the emergence of the Earth as an object of research.

Georgina Endfield is Reader in Environmental History in the School of Geography, University of Nottingham. Her research focuses mainly on the environmental and climate history of colonial Mexico and nineteenth century central, southern, and eastern Africa. She is the author of *Climate and Society in Colonial Mexico: A Study in Vulnerability* (Blackwell, 2008), and she is editor of the international journal, *Environment and History*.

James Rodger Fleming is Professor of Science, Technology, and Society at Colby College. He is a Fellow of the American Association for the Advancement of Science and the American Meteorological Society. His most recent book is *Fixing the Sky: The Checkered History of Weather and Climate Control* (Columbia University Press, 2010), and he is the co-editor of *Globalizing Polar Science* (Palgrave Studies in the History of Science and Technology, 2010). His current work involves connecting the history of science and technology with public policy.

Adrian Howkins is Assistant Professor of International Environmental History at Colorado State University. He received his doctorate from the University of Texas at Austin with a dissertation entitled "Frozen Empires: A History of the Antarctic Sovereignty Dispute between Britain, Argentina, and Chile, 1939-1959." He has published a number of articles and essays related to the history of Antarctica in scholarly journals such as *The Journal of Historical Geography*, *Environmental History*, and *The Polar Record*. He is currently working on a book manuscript on the environmental history of the Antarctic Peninsula.

Vladimir Jankovic is Lecturer at the University of Manchester, UK. He has just published a new book, *Confronting the Climate* (Palgrave Studies in the History of Science and Technology, 2010) on the history of British environmental medicine. He is the author of the critically acclaimed book, *Reading the Skies: A Cultural History of English Weather, 1650-1820* (University of Chicago Press, 2000) and the editor of *Weather, Local Knowledge, and Everyday Life* (Rio de Janeiro: MAST, 2009). He is now working on an international project on urban climatology and the resilience of cities to climate change.

Mike Hulme is Professor of Climate Change in the School of Environmental Sciences at the University of East Anglia and was founding director of the Tyndall Centre for Climate Change Research. His most recent books are: *Why We Disagree About Climate Change* (Cambridge University Press, 2009) and *Making Climate Change Work For Us* (Cambridge University Press, 2010) (edited). He is editor-in-chief of the new interdisciplinary review journal *WIREs: Climate Change*.

Ruth Morgan is a PhD student at The University of Western Australia. Her doctoral thesis is an environmental history of perceptions and understandings of rainfall decline in south-west Western Australia from 1945 to 2007. She teaches Australian history and is an editor of the online postgraduate journal of historical and cultural studies, *Limina*.

Samuel Randalls is Lecturer in Geography at University College London. His work specializes in history and sociology of late 20th century climate change debates and the commercialization of science and nature in the same period. Recently he has investigated weather futures markets and the historical development of approaches and policies for "climate stabilization."

Sverker Sörlin holds a PhD in the History of Science and is currently Professor of Environmental History at the Royal Institute of Technology, Stockholm. He has held visiting positions at Berkeley (1993), Cambridge (2004-05), and the University of Oslo (2006). His work includes co-edited books such as *Narrating the Arctic: A Cultural History of Nordic Scientific Practices* (with Michael T. Bravo; 2002), *Nature's End: History and the Environment* (with Paul Warde; 2009); and articles in journals such as *Osiris, European Review of History, Worldviews*, and *Journal of Historical Geography*. In Sweden he has published more than thirty books and in 2004 he won the prestigious August Prize (named for August Strindberg) for his two-volume history of European Ideas, 1492-1918.

Brant Vogel received his doctorate from the Graduate Institute of Liberal Arts at Emory University (America's oldest multi-disciplinary graduate program), for his dissertation "Weather Prediction in Early Modern England" (2002). His research interests are the history of the book, history of meteorology, early modern visual representation, and the history of scientific instruments. He is currently a Research Assistant on the *Selected Papers of John Jay* (Columbia University), working on cryptography and early aeronautics.

Index

269

SUGGESTIONS FOR CONTRIBUTORS TO OSIRIS

OSIRIS is devoted to thematic issues, conceived and compiled by guest editors who submit volume proposals for review by the OSIRIS Editorial Board in advance of the annual meeting of the History of Science Society in November. For information on proposal submission, please write to the Editor at osiris@etal.uri.edu.

1. Manuscripts should be submitted electronically in Rich Text Format using Times New Roman font, 12 point, and double-spaced throughout, including quotations and notes. Notes should be in the form of footnotes, also in 12 point and double-spaced. The manuscript style should follow *The Chicago Manual of Style*, 15th ed.

2. Bibliographic information should be given in the footnotes (not parenthetically in the text), numbered using Arabic numerals. The footnote number should appear as superscript. "Pp." and "p." are not used for page references.

   a. References to books should include the author's full name; complete title of book in *italics*; place of publication; date of publication, including the original date when a reprint is being cited; and, if required, number of the particular page cited (if a direct quote is used, the word "on" should precede the page number). *Example*:

   [1] Mary Lindemann, *Medicine and Society in Early Modern Europe* (Cambridge, 1999), 119.

   b. References to articles in periodicals or edited volumes should include the author's name; title of article in quotes; title of periodical or volume in *italics*; volume number in Arabic numerals; year in parentheses; page numbers of article; and, if required, number of the particular page cited. Journal titles are spelled out in full on the first citation and abbreviated subsequently according to the journal abbreviations listed in *Isis Current Bibliography*. *Example*:

   [2] Lynn K. Nyhart, "Civic and Economic Zoology in Nineteenth-Century Germany: The 'Living Communities' of Karl Möbius," *Isis* 89 (1999): 605–30, on 611.

   c. Journal articles are given in full in the first reference. For succeeding citations, use an abbreviated version of the title with the author's last name. *Example*:

   [3] Nyhart, "Civic and Economic Zoology" (cit. n. 2), 612.

3. Special characters and mathematical and scientific symbols should be entered electronically.

4. A small number of illustrations, including graphs and tables, may be used in each volume. Hard copies should accompany electronic images. Images must meet the specifications of The University of Chicago Press "Artwork General Guidelines" available from the Editor.

5. Manuscripts are submitted to OSIRIS with the understanding that upon publication copyright will be transferred to the History of Science Society. That understanding precludes consideration of material that has been previously published or submitted or accepted for publication elsewhere, in whole or in part. OSIRIS is a journal of first publication.

OSIRIS (ISSN 0369-7827) is published once a year.

Single copies are $33.00.

Address subscriptions, single issue orders, claims for missing issues, and advertising inquiries to *Osiris*, The University of Chicago Press, Journals Division, PO Box 37005, Chicago, IL 60637.

Postmaster: Send address changes to *Osiris*, The University of Chicago Press, Journals Division, PO Box 37005, Chicago, IL 60637.

OSIRIS is indexed in major scientific and historical indexing services, including *Biological Abstracts*, *Current Contexts*, *Historical Abstracts*, and *America: History and Life*.

Paperback edition, ISBN 978-0-226-02939-9

Osiris

A RESEARCH JOURNAL DEVOTED TO THE HISTORY OF SCIENCE AND ITS CULTURAL INFLUENCES

A PUBLICATION OF THE HISTORY OF SCIENCE SOCIETY

EDITORIAL OFFICE
DEPARTMENT OF HISTORY
80 UPPER COLLEGE ROAD, SUITE 3
UNIVERSITY OF RHODE ISLAND
KINGSTON, RI 02881 USA
osiris@etal.uri.edu